Dr Ron Nielsen was born and educated in Poland. He conducted research in the Department of Nuclear Physics at the Australian National University, and has worked in research institutes in Poland, England, Germany, and Switzerland. The author of numerous scientific papers in peer-reviewed professional international journals, Dr Nielsen is a fellow of the Australian Institute of Physics and an active member of the New York Academy of Sciences. He lives in Queensland, Australia.

THE LITTLE GREEN
HANDBOOK

seven trends shaping the future of our planet

Ron Nielsen, DSc

Picador

New York

To my wife, Lorna, and my children

www.picadorusa.com

Picador® is a U.S. registered trademark and is used by St. Martin's Press
under license from Pan Books Limited.

For information on Picador Reading Group Guides, as well as ordering,
please contact Picador.
Phone: 646-307-5629
Fax: 212-253-9627
E-mail: readinggroupguides@picadorusa.com

Edited by Ken Turnbull and Henry Rosenbloom

ISBN 0-312-42581-3
EAN 978-0-312-42581-4

First published in Australia by Scribe Publications

Printed on recycled paper

First Picador Edition: April 2006

10 9 8 7 6 5 4 3 2 1

Contents

Tables

Boxes

Preface

As a nuclear physicist, I am in some ways an unlikely environmentalist. However, throughout my life and career I have been fascinated by the physical environment, and have always been happy to share the results of my research into it with members of the public.

For many years, I have been speaking about environmental trends— to community groups, to college and university students, at public meetings, and on local radio programmes that I have been presenting for some time. The most common response I have experienced is that people want to know more. The radio programmes, in particular, have demonstrated a great deal of interest in the environment among listeners. As a result, many people have asked me for information on critical global issues and, specifically, for a copy of my 'book' on the subject. I originally self-published a short book for this purpose that I called *The Seven Thunders*, but it soon became clear to me that a more comprehensive volume was needed.

The Little Green Handbook is my attempt to provide a comprehensive summary of the essential facts and figures that we need to know in order to understand clearly global environmental changes, and to try to give a broader view of the implications for all of us if these trends continue. It surveys not only the deterioration of our physical environment but also economic, social, and political trends, including the increasing tensions and conflicts between nations.

I have tried to provide enough information for a general reader to gain an understanding of global and local developments without having to sift through numerous specialist publications. However, I also give a list of references for those who would like to pursue these issues in greater depth.

I have written *The Little Green Handbook* out of concern for the future of our children, because it is they who will be most affected by the consequences of what we do today. It is our duty to do what we can to understand the issues and then use whatever means at our disposal to ensure that their future is safe, nurturing, and sustainable.

As you will see, we have many complex and difficult problems to solve. The task is far from easy but it would be wrong to feel daunted or depressed. We need a mature response. We have to understand the problems, see them as they are, and concentrate all our energy on solving them.

Not all is yet lost. We have expertise and technology and we can create a better future for our children. We have many companies offering environmentally friendly goods and services. We have many organisations and scientists working on solving both environmental and social problems. We need to work with all these groups and individuals and give them our full support.

Never before was there a greater need to understand such critical issues. Never before have we faced dangers of such a magnitude, and never before was there a need for such a combined effort in steering away from danger.

I hope that this book will be of particular interest to people in their teens and early twenties, and to high school and university students. Young adults are endowed with energy and the potential for unconstrained creative ideas. The future belongs to them and, given a chance, they will be able to influence it in a positive way.

Finally, older generations—including policy-makers, decision-makers, and journalists—will also benefit from reading this book. I hope that it will encourage them to use their influence and lifelong experiences to help change what is clearly an undesirable course of events.

Introduction

'And I saw another mighty angel come down from heaven ...
and when he had cried, seven thunders uttered their voices.'

—Revelations 10:1–3

How much do we know about critical global trends? Alarming
events are taking place all around us, but in general we are not aware
of them. During the short span of a human life these events occur too
slowly to make a lasting impression on us. We do not notice them also
because we are too busy. We live in a rapidly changing, competitive,
and demanding world, without the time or the inclination to collect
information, analyse it, work out the right interpretation, and draw correct conclusions. The summary in this book should help to bypass such
obstacles.

These events generally do not occur at the same time and in the
same place. When we hear about them, they make only a fleeting
impression on us, and if something dramatic takes place where we live,
we see it only as an isolated event. We then concentrate our energy and
resources on solving the local problem. When it is solved we return to
our normal but busy lives.

We might even have become inured to them (see "False alarms"
below). We are lethargic and insensitive, or too sure of ourselves. We
know it all.

How can we identify critical global trends? The aim of this book is
to help readers understand critical global trends and events associated
with them. To this end, my first step was to arrange them in distinct
categories. I have followed three self-imposed criteria: all critical global trends and events should be included; they should be arranged in
categories that are as distinct as possible; and they should be arranged
in the smallest possible number of categories.

My aim was to make a complete, comprehensive, and concise list.
This led to the following classification:

1. The population explosion
2. Diminishing land resources

3. Diminishing water resources
4. The destruction of the atmosphere
5. The approaching energy crisis
6. Social decline
7. Conflicts and increasing killing power

These seven groups give a complete representation of the state of the world. Trends in the first group describe the unusual, undesirable, and critical increase of global and regional populations. The next four groups deal with changes in the physical environment, and the last two with changes in human interactions. Events categorised in one group are often echoed in another; they can be connected and overlapping, warning us of an approaching danger.[1]

Discussions of environmental issues are usually limited to the physical environment. This does not present a complete picture of critical global problems. To understand the developing global crisis it is necessary to include social aspects. Even though some global problems might already be out of our control we can identify a great number that could be resolved by rearranging priorities and changing our attitudes, aims, and styles of living.

For instance, it is now difficult—maybe even impossible—to slow down substantially the growth of global population. However, by changing our behaviour we could reduce the degree of environmental damage, improve the sharing of the planet's resources, and create a more secure future for our children.

The characteristic feature of critical global trends is that many of them are influenced strongly by only small groups of countries. A few nations are responsible for a large part of the global population increase, the consumption of resources is concentrated in a handful of them, and a small number contribute substantially to global warming. It is important to understand which countries play a central role in influencing global trends. Such identification can be used in strategies for avoiding global crisis by a suitable change of undesirable trends.

The convention I have adopted. I shall use the so-called American system of large numbers (see Appendix C) because it is widely accepted. Thus, a billion means one thousand million (1,000,000,000), and a trillion means one million million (1,000,000,000,000). Unless stated otherwise, the dollar sign refers to the United States currency. I have reserved square brackets for the abbreviations of units of measurements.

How reliable are the data I am using? There are at least two ways of writing: to tell or to show. It is relatively easy to tell about the state of the world, but it is much more difficult to show it, because one has to support the discussion with figures. It is easy to read words, but figures take a little more effort. However, figures are more rewarding, and they are more persuasive than words alone.

Yet there is a problem with such an approach because figures have to be from reliable sources or based on analysis of reliable data. This is not easy, because even when using authoritative sources one has to check whether the data make sense. It is also often difficult to consolidate data cited in various publications, and the differences can range from minor to significant.

In general, numerical data describing the state of the world are not based on precise measurements but on assessments, and the accuracy of the data depends on their type. It is impossible, for instance, to measure precisely our reserves of groundwater or crude oil. We do not know precisely how much energy various regions or countries are consuming, or how many people are dying of hunger or from water-borne diseases.

Statistical information is usually delayed by a few years and the data may depend on the way they are accumulated. We can use only what is available, and as long as we take data from respectable sources and check them carefully, we can have a fair degree of confidence in our conclusions. I present many sets of data in this book, in the text and in tables. Readers willing to spend the time studying them will benefit from much broader information. Each table contains a reference to sources and a guide on how to read the information.

False alarms. Critical trends and events of any kind can attract certain types of people who take pleasure in making dramatic and gloomy predictions. As long as we have had records there have been doomsday prophets. Such people cry 'wolf', and they keep on using their ridiculous, entertaining, misleading, irritating, or disturbing arguments in an attempt to draw attention to themselves. The trouble with doomsday prophets is that many people accept everything they say, or lump them in with those who point to real global problems—and ignore them all.

Are we being warned by respectable authorities? Warnings and expressions of concern have been emerging for more than 100 years.[2] They come from people who by the nature of their occupation are sceptics, who are in the habit of asking penetrating questions, who check information they receive, and who take great care in drawing their conclusions. Over the years, these warnings have been growing in intensity,[3] and now they are so numerous, so loud and clear, that it is hard not to heed them. It is difficult to list them all, but perhaps two should be mentioned: a joint statement by 58 scientific academies[4] and a statement signed by more than a thousand scientists, including many Nobel laureates in science.[5]

Can we make accurate predictions?[6] Discussing trends means talking about the future, but we can only make an educated guess based on careful analysis of the past. This is an acceptable procedure in the stock market, which uses complex methods to forecast events. We can also use various reliable procedures to predict global developments.[7] Forecasts of this kind give us only probable and conditional outcomes. However, even prophetic visions are often claimed to be provisional, so perhaps we are not so disadvantaged.

Our predictions depend on various options, conditions, and assumptions. For instance, the Global Scenario Group has projected four future developments: Market Forces, Policy Reform, Fortress World, and the Great Transition.[8] The United Nations Environmental Program (UNEP) calls them Market First, Policy First, Security First, and Sustainability First.[9] Market First is a market-driven future based on an assumption of an almost unlimited ecological capacity to support global development and industrialisation. Policy First requires a co-ordinated pro-environment effort by governments to create a better future for everyone. Security First is characterised by a self-centred attitude of protecting privileged groups, and of making their fortress safe and secure while ignoring the needs of the outside world. Sustainability First requires a radical change in our behaviour, reflected in selfless care for one another and dedication to the environment.

An interesting feature of all these projected developments is that they predict almost identical growth in the global economy. However, only Sustainability First (or the Great Transition) offers an improvement in a wide range of other areas, such as fewer conflicts, greater gender equity, lower carbon dioxide concentration in the atmosphere, increased forest areas, less water stress, greater international equity, and a dramatic reduction in the number of people suffering hunger. So it seems we still have a choice. With proper global management we can have global economic growth and a sustainable future.

We live in a unique period of the Earth's history described as the Anthropocene and characterised by the exceptionally strong human influence on the environment.[10] Our global future is determined not only by critical global trends and events we have already put in motion, but also by the way we respond to the problems we have created. The most likely response will be business as usual,[11] which means we will do little or nothing to change the progression of critical global events. This corresponds to the Market Forces or Market First development. This path is only one step away from an even more undesirable possibility: the Fortress World development.

The futures described in this book represent the most likely outcomes based on the assumption that we shall continue with business as usual. We still have a choice: to be actively involved and shape our future or to give up and allow for the critical global trends to shape it for us.

Bilbao can serve as a good example of what we can do. Destroyed by Industrial Revolution and abandoned, this city is now rising from the ashes like a Phoenix, with renewed strength and vitality. We can still repair much of the global damage, restore the environment, change the way we live, increase the ecological capacity of our planet and create a sustainable future.

We can criticise those who try to warn us about the dangers. We can look for faults in their arguments and can ignore a large body of persuasive evidence. We can try to win points, to prove who is right and who is wrong. However, we can also take constructive steps. We can accept what is convincing, correct what is not, and support—in whatever way we can—the efforts aimed at a sustainable future.

Chapter 1
Environmental Degradation

'Sooner or later, wittingly or unwittingly, we must pay for every
intrusion on the natural environment.'

—Barry Commoner (1917–) US biologist and environmentalist,
The Speaker's Electronic Reference Collection, AApex Software, 1994

The seven groups of critical global trends and events outlined in the
introduction have four common features:

1. They are associated with a hastened deterioration of the
    environment, both physical and social.
2. They show that, for the first time in human history, we are
    approaching and crossing global limits: the ecological limits of
    our planet, the limits of human suffering and deprivation, and the
    limits of human-induced damage to the planet.
3. They are happening in a relatively short time. They began about
    200 years ago, a small fraction of the time humans have been on
    the Earth.
4. They focus on an even shorter time: the second or third quarter
    of the 21st century. They show that we are likely to experience
    dramatic changes, with the possibility of a global collapse of life-
    supporting systems.

In general, our environmental awareness is increasing,[1] but we still
do not treat environmental and social issues with the seriousness they
deserve. We still think life will go on as usual, but we are facing an
unusual sequence of events. The Earth is the only planet in the solar
system that offers suitable living conditions for humans. Our foremost
priority should be to make this world safe, secure, and habitable.

What is the driving force behind critical global trends? The population explosion is the primary force behind the remaining six groups
of critical global events. Combined with it is another force, which is

less obvious, but adds to the aggravation of the global condition. It is our behaviour, a force that could be called human nature—or mind and culture.[2]

The important point is that by changing our behaviour we could change the progression of the critical global trends and events, and turn danger into opportunity. But to do this we would have to work against our natural tendencies and inclinations.

One aspect of our behaviour is relatively benign: we want to make our lives more comfortable and more pleasant. However, there is a more sinister aspect of human nature, reflected in greed, hatred, self-ishness, lack of cooperation, revenge, fighting, evil-doing, and pleasure in hurting others.

If we want a secure future for our children, we shall have to change the way we treat nature and one another. This will not come without effort.[3]

Are we getting richer? Prosperity is increasing, at least in a handful of countries. However, all our wealth comes from the planet, which is growing poorer through neglect and overexploitation. This will eventually reflect on our prosperity.

We can see the degradation of the physical environment everywhere: in the rapid depletion of fresh water, fertile soil, and biodiversity; in the pollution of land, water, and the atmosphere; and in ruthless deforestation and overfishing. Environmental degradation undermines global food security, increases health risks, and is responsible for current and future economic loss. Environmental risk factors affecting human health include unsafe water, the lack of sanitation, hunger, urban air pollution, exposure to toxic chemicals, and climate change.

Children under five years of age form only 10 per cent of the global population, but they share 40 per cent of the burden of disease caused by environmental factors. Hunger affects about 150 million children in this age group and kills about 3.4 million each year. About 5500 children die each day of diseases caused by polluted air, water, and food.[4]

Intensified agriculture. Global arable land resources are relatively small, and soil types are not equally productive. Unsustainable farming and mismanagement of land resources result in the degradation of land and the loss of soil productivity. To make more room for agriculture we are reducing the area of forests and pastures.

To grow food we need water, which is becoming scarce. To increase the productivity of arable land we use fertilisers. To protect crops we use pesticides, herbicides, and fungicides. In the process we kill organisms that make the soil rich, healthy, and productive. Into the bargain we are polluting the air and the water.

Toxic chemicals. It is estimated that there are more than 100,000 kinds of synthetic chemicals in use, and that between 200 and 1000 new chemicals are being added each year.[5] Many of them are toxic. Lyons (1999) lists 26 toxic synthetic compounds and two heavy metals (lead and mercury) that have been found in breast milk and in human body fat in the populations of Europe, the United States, and other regions. Worldwide, in the early 1990s, we were producing 400 million tonnes of hazardous wastes each year.[6]

In some areas, concentrations of toxic chemicals are high. The hotspots are the Great Lakes, British Columbia, Florida, Norway, Russia, Japan, Ethiopia, Pakistan, South Africa, and the Midway Islands.[7] The two groups of the most toxic and most dangerous chemicals, which are often discussed in the literature, are endocrine-disrupting chemicals (EDCs) and persistent organic pollutants (POPs).

Endocrine disrupting chemicals.[8] These chemicals act as endocrine disruptors. They are hormonally active—they mimic and interfere with the body's hormonal (endocrine) system.

Hormones are the messenger molecules that perform vital functions in the life of plants and animals. For instance, oestrogen and testosterone are sex hormones. Adrenaline is a hormone that mobilises the body for a quick action, known as "fight or flight". Thyroxine, produced in the thyroid gland, regulates many body functions such as the metabolic rate, reproduction, and resistance to infection. Growth hormone regulates the growth and development of children. Insulin regulates the level of sugar. Auxins regulate the growth of plants. Cytokinins regulate the growth of leaves and fruit.

Hormone-like chemicals are used as the means of communication between plants and animals. It is well known that pheromones are used by insects, mammals, and fish, but it is perhaps less known that plants emit volatile organic chemicals for communication, and that some insects have an ability to read these messages.[9]

EDCs block the receptors receiving hormonal messages, disturb communication, and thus influence the way organisms work and develop. For instance, they interfere with the sexual development of the human embryo, and with the sex-hormone assisted development of the brain, muscles, bones, immune system, organs, and tissues. EDCs are responsible for various observed congenital birth defects and for immune system dysfunction. They may be also influencing neurological, mental, and behavioural development.

EDCs may well be responsible for decreasing sperm counts and for the declining mobility and stamina of sperm cells. They could also be contributing to the development of testicular cancer and breast cancer. EDCs come in the form of pesticides (such as atrazine, DDT,[10] and endosulfan); industrial chemicals (such as dioxins, PCBs,[11] and phthalates); and heavy metals (such as lead and magnesium).[12] They can be present in food, drinking water, and certain types of plastics.

Persistent organic pollutants.[13] Persistent organic pollutants form a specific group of toxic chemicals characterised by persistency and mobility. They have various adverse effects on plants, animals, and humans. Some of them are also endocrine disruptors. POPs were identified in the 1960s when scientists observed unusual congenital deformities in wildlife.

POPs are present in food, water, air, soil, and household products. They are extremely toxic and can be active for decades, if not for centuries. They dissolve in fat and thus are stored in fatty tissue where they accumulate and are passed from mother to foetus. In this form they travel up in the food chain, from lower to more complex organisms, each time increasing their concentration. The process is known as bioaccumulation or biomagnification. The concentration of POPs in animal and human tissue can be hundreds or even millions of times higher than in the surrounding environment.

The sources of POPs are industry (the production and use of chemical compounds), agriculture (the use of pesticides and other chemicals), and low-temperature incineration. They also come from natural sources such as forest fires and volcanoes.

Many highly toxic and banned pesticides are used by developing countries because they are cheap. They evaporate in the warm climates and are deposited by rain or snow in colder parts of the world. POPs can travel thousands of kilometres and can be deposited in remote areas. They are therefore a global problem.

Intensified industry. Industry is an integral part of progress and civilisation. The Industrial Revolution began in the 18th century with a handful of countries and is continuing with a much larger group representing a greater percentage of the global population.

Industrialisation pollutes the air with sulphur dioxide, nitrogen oxides, carbon dioxide, hydrogen sulphide, hydrogen chloride, hydrogen fluorides, silica, mercaptans, lead, arsenic, vanadium, copper, nickel, selenium, zinc, mercury, magnesium, aluminium, chromium, hydrofluoric acid, soda ash, potash, polycyclic aromatic hydrocarbons (PAHs), polychlorinated biphenyls (PCBs), polychlorinated dibenzo-p-dioxins (PCDDs), polychlorinated dibenzofurans (PCDFs), benzene, and particulate matter (PM).

It pollutes water with suspended solids, oils, tars, benzene, acids, caustics, mercaptans, phenols, sulphides, sulphates, ammonia, cyanides, fluorides, cadmium, mercury, arsenic, lead, zinc, chromium, and PCBs.

It pollutes land with slag, sludges, oil, tars, grease residues, salts, sulphur compounds, hydrocarbons, and heavy metals.[14]

The industrialisation race.[15] Industrial production and the use of industrial products are still mainly concentrated in a small group of countries comprising about one-fifth of global population. However, poorer countries are also developing industries and one wonders how long the planet will be able to cope with the increasing demand for resources and the increasing level of pollution.

Table 1-1 illustrates the strongly unequal distribution of industrial activities among various countries. About a quarter of global industrial production is concentrated in just one country, the US, and about a half in just three countries (the US, Japan, and Germany). The largest country in the world, China, contributes only about 6 per cent to global industrial production.

In the group of developing countries, the top 10 producers account for about 80 per cent of production. China accounts for about 30 per cent, and three countries (China, Brazil, and South Korea) for about 50 per cent. However, the combined contribution of developing countries to global industrial production is still small: the top 10 industrial producers contribute only about 17 per cent to the global total. Developing countries are still far behind the industrialised countries with their industrial production, and we shall see later (in Chapter 7) that they always will be. Yet industrial production is increasing all over the world, putting enormous stress on the environment.

Between 1985 and 1995, the global value of industrial output per person increased by an annual average of 4.3 per cent.[16] The annual average increase in industrialised countries was 3.5 per cent and in developing countries 5.3 per cent. The greatest annual increase was in East Asia (7 per cent) and in particular in China (7.4 per cent). The smallest increase was in sub-Saharan Africa (0.8 per cent).[17]

Table 1-1 The top 10 industrial producers in 1998, globally, and in developing countries

All countries	World share		Developing countries	Developing countries' share		World share	
	(1)	(2)		(3)	(4)	(1)	(2)
USA	25.4	25.4	China	29.0	29.0	6.3	6.3
Japan	15.9	41.3	Brazil	12.3	41.3	2.7	9.0
Germany	8.5	49.8	South Korea	8.0	49.3	1.7	10.7
China	6.3	56.1	Mexico	6.7	56.0	1.5	12.2
France	5.0	61.1	Taiwan	6.0	62.0	1.3	13.5
UK	4.4	65.5	India	5.2	67.2	1.1	14.6
Italy	4.2	69.7	Argentina	4.3	71.5	0.9	15.5
Brazil	2.7	72.4	Turkey	3.6	75.1	0.8	16.3
Canada	1.9	74.3	Thailand	2.9	78.0	0.6	16.9
Spain	1.8	76.1	Indonesia	1.9	79.9	0.4	17.3

(1) – Percentage of global industrial production

(2) – Cumulative percentage of global industrial production

(3) – Percentage of industrial production by developing countries

(4) – Cumulative percentage of industrial production by developing countries

Example: The US share of global industrial production in 1998 was 25.4 per cent. China's share of industrial production by developing countries was 29 per cent, but only 6.3 in global production. About 41 per cent of global industrial production takes place in just two countries, the US and Japan.

Source: Based on UNIDO 2002

The industrial race can be also measured using the so-called competitive industrial performance (CIP) index. This index has four components: manufacturing value added per person, manufacturing exports per person, the share of medium and high-tech activities in manufacturing production, and the share of medium and high-tech products in manufacturing exports. The higher the CIP index, the more vigorous are the industrial activities.

The highest CIP index in 1998 was in industrialised countries (0.44) and was increasing. The CIP index in East Asia was 0.3 and was also increasing. In other regions of the world the CIP index was much lower.

Between 1985 and 1998, China improved its CIP by 24 points, the Philippines by 21 points, Indonesia by 19 points, Thailand by 12 points, Ireland by 12 points, and Egypt by 10 points. But there were also many countries that were lagging: Tanzania, Peru, Ghana, Venezuela, Zimbabwe, Saudi Arabia, Jamaica, Panama, Senegal, Oman, and Algeria. Their relative position in the industrial race worsened by between 10 points (Tanzania, Peru, and Ghana) to 20 points (Algeria and Oman).

Will the planet be able to cope with the increasing demands on it? Will it be able to cope with increasing pollution? How far can developing countries expect to progress? Do we have a secure future?

Mining.[18] Industrialisation is closely connected with mining, but mining creates enormous volumes of waste and pollution. We are turning the Earth inside out and destroying the environment. Mining operations are carried out all over the world, but the intensity varies between countries. Depending on the materials extracted, production is concentrated in small groups of countries (see Box 1-1).

For each tonne of useful metal, we have to dig out many tonnes of ore. The amount of waste depends on the ore type. For instance, for iron it is 60 per cent; for manganese, 70 per cent; for tungsten, 99.75 per cent; for zinc 99.95 per cent; and for gold, 99.99 per cent.

Mining creates mountains of tailings, which are either left lying around or disposed of somehow. In Latin America, mining activities have resulted in the scattering of about 5000 tonnes of mercury in forests and urban areas. For every tonne of extracted ore, a tonne of mercury is released into the Amazon River. Mining operations in South Africa are responsible for large emissions of sulphur. In China, gold mining is claimed to have resulted in large quantities of arsenic, mercury, and cyanide being scattered on the surface. In some places, tailings have been pushed into creeks, contaminating water used by villagers living downstream.

Most tailings contain sulphides. While sulphides are in the ground they cause no harm, but when they are brought to the surface and exposed to air they are converted gradually to sulphuric acid, which dissolves heavy metals in the tailings and releases them into the environment.

The oxidation of sulphides can go on for hundreds of years. Once the dirt is dug out, it will cause trouble for several generations. However, this dirt means money. It also means progress, of a sort, so we continue to dig everywhere. Mining contaminates the soil and waterways, kills fish and other forms of life, pollutes the air, and is harmful to people. Tailings from hundreds of mines all over the world are stored on land, and now many mining companies are dumping their waste at sea or are planning to do so.

Box 1-1
Top mining producers in 1999

Mining operations are carried out all over the globe. There does not seem to be any single specific place that can claim an outstanding distinction in this area of our activities.

Bauxite: Australia, Argentina, Brazil, Jamaica, China, India, Venezuela, Suriname, Russia, Guyana – 93 per cent of world production.

Copper: Chile, USA, Indonesia, Australia, Canada, Peru, Russia, China, Poland, Mexico – 83 per cent of world production.

Tin: China, Indonesia, Peru, Brazil, Bolivia, Australia, Malaysia, Russia, Portugal, Thailand – 97 per cent of world production.

Iron: China, Brazil, Australia, India, Russia, USA, Ukraine, Canada, South Africa, Sweden – 89 per cent of world production.

Cement: China, USA, India, Japan, S. Korea, Brazil, Germany, Turkey, Italy, Thailand – 65 per cent of world production.

Gold: South Africa, USA, Australia, China, Canada, Indonesia, Russia, Peru, Uzbekistan, Ghana – 79 per cent of world production.

Industrial diamonds: Australia, Democratic Republic of Congo, Russia, South Africa, Botswana – 95 per cent of world production.

Coal: China, USA, India, Australia, Russia, South Africa, Poland, Ukraine, Kazakhstan, Indonesia – 81 per cent of world production.

Oil: Saudi Arabia, USA, Russia, Iran, Mexico, Venezuela, China, Norway, UK, Iraq – 62 per cent of world production.

Source: Harrison and Pearce 2001

The idea behind disposal at sea is that oxygen is scarce in water, so oxidation of sulphides is slow. It is slow, but not absent. The disposal of tailings at sea creates impoverished marine ecosystems. The effect of extensive dumping can be devastating.

The dumping areas are mainly around the Philippines, Indonesia, and Papua New Guinea. There are about 20 sites in that small region of the globe, and the scale of environmental damage can be appreciated if we consider that in just one of them 22,000 tonnes of tailings are dumped each day, and that a single mine can produce about 270,000 tonnes of waste a day.

Intensified use of energy. Technology requires enormous inputs of energy. As we progress and develop, our use of energy increases. This creates environmental problems, because our main source of energy is in the form of fossil fuels, which generate large amounts of pollution.

In principle, cleaner sources of energy could be made available, but they would have to be developed. This is not a preferred option for profit-oriented operations, so we use what is easily available. We still have large deposits of fossil fuels, mainly in the form of coal. We prefer to use this source of energy, even though it harms the environment and in the long run can result in serious economic loss.

Four essential physical elements. The elements supporting our existence are land, water, atmosphere, and energy. Water is needed by humans and animals to support their physiological processes, but humans also need water to support agriculture and industry. The atmosphere is needed as a source of oxygen, but it also regulates climate, assists in the circulation of water, and protects against solar radiation. We need energy for our physiological processes, as well as for transport, industrial and agricultural production, communication, and many other activities.

Without any of these four elements, we could not live on this or any other planet. Now, for the first time in human history, we are rapidly destroying land, water, and the atmosphere. For the first time also we are facing an energy crisis. Thus, for the first time in human history we are facing the possibility of self-extinction.

Two social elements. Peace and cooperation are social elements essential for survival and prosperity. If we had peace and cooperation in the world, we could at least have time to reflect on global problems and could plan solutions. We could even do more than that. We could have the time and financial resources to improve living conditions on our planet.

However, we are trapped in a whirlwind of social and political conflict. This is where our energy and financial resources are wasted.

Even though we do not have a global military confrontation, the world is at war. Economically, the world is polarised, and the stronger group wins. We are experiencing a host of racial, ethnic, tribal, and ideological conflicts. Corruption, antagonism, violent confrontations, hate, and fear abound. This is not an environment for stable and sustainable development.

The future. For the first time in human history, we have a series of events that threaten not just the survival of one group of people or one country, but the survival of all the people in the world. Our positive and constructive response to critical global developments is disappointingly slow. We should be putting all our energy into solving global problems, but we are reluctant to make commitments. Some work has been done and some encouraging results have been achieved, showing that we could solve many of the existing problems, but we are not doing enough.

An increasing number of individuals and organisations are expressing concern about our common future, but making a profit and solving political problems seem to be more urgent and more important. The tendency is to postpone decisions on these issues or brush them aside.

In general, caught up in the strong current of life, oblivious to dangers ahead, we are enjoying the ride. Globally, there is a kind of peace. In general, we are experiencing progress nearly everywhere. We are proud of our technological achievements and have visions of endless progress. Meanwhile, global problems are increasing. It is not just one trend that points to the developing global crisis, but a whole complement of trends.

Chapter 2
The Population Explosion

'The hungry world cannot be fed until and unless the growth of its resources and the growth of its population come into balance.

'Each man and woman—and each nation—must make decisions of conscience and policy in the face of this great problem.'

—Lyndon Baines Johnson (1908–1973), 36th president of the United States.

The rapid growth in global population is not a problem on its own. However, it will be a problem if we cannot find enough room for everyone. Will the population keep increasing or will it reach a stabilisation point? What are the hidden problems of stabilisation? How can environmental limitations influence the growth of global population?

What is the carrying capacity of our planet? Various authors have tried to answer this question, but there is no consensus on how many people we can accommodate.[1] The long-term carrying capacity could be as low as two billion people, but for a limited period we might have as many as 12 billion.

The carrying capacity depends on the consumption levels of various countries and on the ecological capacity of our planet. However, it also depends on the definition of carrying capacity. For instance, how do we define the carrying capacity of a train? Should it be by the number of people who can sit comfortably inside, or by the number of people who can sit and stand there, or should it be by the number of people that can be packed as closely as possible inside the train, hanging on the sides, and crowded on the roof?

What consumption levels are appropriate in determining the number of people we can have in the world? Should we take consumption in the US, the average consumption in Europe, average global consumption, or consumption in Ethiopia?

This might be an interesting topic for academic discussion, but we would be missing the main point. The point is that for the first time in human history it makes sense to ask these kinds of questions, because for the first time we are close to the limit of the carrying capacity of our planet.

If you were driving on a narrow mountain road it would be foolish to try to find out how close you could come to the edge. You know by looking at the road that there is not much room for error. Likewise, with global population, we know that we cannot be complacent about its continuing growth. Each year we are inching closer to the edge.

There are already too many of us on this global "train". It is not so bad in the first-class carriages, but even here we can feel the pressure, with people knocking on the doors and windows and some even sneaking inside. It is hard to imagine what is happening in the second- and third-class carriages, and we prefer not to think about it. How long we can remain comfortable in our cosy little compartments is hard to tell. Those who claim that there is enough room for everybody on our planet should move to the overcrowded carriages, stay there, and share the discomfort.

Table 2-1 Explosive growth of global population (billions)

Year	Historical	Calculated explosive	Calculated exponential
1	0.3	0.1	0.1
1000	0.3	0.2	0.9
1500	0.5	0.5	2.3
1700	0.6	0.7	3.3
1800	0.9	1.0	4.1
1930	2.1	2.2	5.2
1958	3.0	3.0	5.5
1974	4.0	3.7	5.7
1987	5.0	4.6	5.9
1999	6.0	6.0	6.0

Historical – Historical global population in given years

Calculated explosive – Calculated global population using an explosive (hyperbolic) function

Calculated exponential – Calculated global population using an equivalent exponential function

Example: In 1958, the global population was 3 billion, which agrees well with the calculated value of 3 billion for an explosive increase, but does not agree with the value of 5.5 billion for an exponential increase. Global population has been increasing explosively and not exponentially, as is often claimed.

Source: Historical populations are based on Bongaarts and Bulatao 2000, Durand 1974, Haub 1995, and McEvedy and Jones 1978

How fast is global population increasing? Over countless generations, global population was below 100 million. The annual increase was negligible and there was always enough room for everybody. The first sign that this near-stable situation was about to change occurred during the second millennium AD. By 1500, global population had reached about 500 million, and 200 years later it was 600 million. Then, about 1800, it shot up to the 1000 million mark (see Table 2-1).

Even at these early stages of the growth of global population, this unusual increase was described as a population explosion. Little did we know then how accurate this description was. The explosion was confirmed about 1930 when we reached the second billion. From that time, global population increased at a breathtaking rate to reach the six billion mark at the end of 1999.

When presented in graphic form, global population shows a dramatic and rapid change from an almost horizontal line to an almost vertical one. Global population is shooting towards infinity, and as we cannot have an infinite number of people on the Earth, a breakdown in one form or another will have to occur—and soon.

Explosive v. exponential growth. You might have read or heard that global population increases exponentially. This is not true. The phrase "exponential growth" is often used to mean rapid growth, but this is misleading. Global population has been increasing explosively, not exponentially. Exponential increase is fast, but explosive increase is sudden and much faster. Even now, when the rate of increase is slowing, it cannot be described as exponential.

The time dependence of global population can be described by a simple mathematical function, but it is not an exponential function. Usually, to determine a function describing a set of data, we have to fit the data using a mathematical prescription. However, with global population this turns out to be unnecessary. All we have to do is take two sufficiently remote data points and calculate the function. It is a hyperbolic or reciprocal function—that is, a reciprocal of a linear function. This simple prescription reproduces the observed time dependence of global population remarkably well.

Table 2-1 lists the historical values of global populations for the years 1AD to 1999, and my calculated values for explosive (hyperbolic) and exponential growths. The two have been calculated using identical initial conditions. In both cases, the calculations were required to match 0.12 and 6 billion in 1AD and in 1999, respectively.

The calculations based on the assumption of explosive growth reproduce the historical values incomparably better than the calculations based on the assumption of exponential growth. For instance, about 1800 the global population was one billion, which is the same as the number of people calculated using the explosive (hyperbolic) function. However, the exponential function gives the population as 4.1 billion for that year. In the year 1500, global population was about 500 million, which is the same as that given by the explosive function. However, the exponential function gives the population as 2.3 billion for that year.

The problem with explosive growth is that there are no warning signs. With the calculated exponential growth here, global population would have reached the first billion about 1085. With explosive growth, this took place more than 700 years later. With exponential growth, we would have had enough time to analyse the situation and plan solutions. With explosive growth, the time is significantly shorter.

With the here-calculated exponential growth, we would have had 354 years between the first and the second billion. With explosive growth, we had only about 130 years. With exponential growth, we would have had about 200 years between the second and third billion. With explosive growth, we had only 28 years.

Before we know it, before we react to it, the explosion will be over, one way or another, and its effects will be devastating. This is what explosions usually do, and this is what the population explosion will do to us. It will leave us with damage that will be difficult or even impossible to repair.

An interesting feature of this mathematical exercise is that the time dependence of global population can be described so well by a mathematical function. How is it that people living in different eras and places, belonging to different social groups and customs, and having different religious beliefs have collectively matched such a simple mathematical rule in their procreation, and that their procreation pattern has not been affected by changing environmental factors?

We can hardly control global population growth now, however our behaviour can either intensify or soften its effect. We live in a time of pressing need to be less self-centred, to live with other humans and with nature, and to care for the environment. We have to broaden our outlook on life and learn a new way of caring for our children—not only by feeding them, dressing them, and sending them to school, but also by making sure we do not leave them a world unsuitable for human habitation.

Doubling time. The so-called doubling-time parameter for global population is inapplicable and makes no sense. We may say that global population has doubled over a certain period, but it is wrong to talk about the doubling time of global population because this implies a certain constant value. We can talk about a doubling time for an exponential function, but not for an explosive function.

For instance, the global population in 1930 was 2.1 billion, and by 1976 it had increased to 4.2 billion. The population therefore doubled in 46 years. However, the population in 1940 was 2.3 billion, and by 1982 it reached 4.6 billion—giving a doubling time of 42 years. In 1960 the figure was three billion, and by 1999 it was six billion— so it doubled in 39 years. The global population in 1500 was about 500 million, and rose to one billion about 1800—so it doubled in 300 years. Which of these doubling periods is *the* doubling time for global population?

Common misconceptions about the growth of global population. One misconception, which I have discussed, is that global population increases slowly and steadily. As we have seen earlier, the growth of global population is characterised by explosive behaviour, which means that we have no time to be complacent about it.

Another misconception is that dramatic events such as world wars or large-scale outbreaks of disease have had an influence on population growth. However, when we examine historical data for global population we can see that they follow closely a smooth mathematical curve. Any small, temporary, and local deviations, which may have been caused by such dramatic events, did not have the slightest effect on the overall trend of the growth in global population.

The third misconception is that better health care is responsible for a more rapid increase, but this hypothesis does not agree with recorded data.[2] The main thrust of the population explosion takes place in poor countries where health care is also poor. It does not take place in rich countries where health care is good.

The growth of the population is given by the *difference* between the birth rate and death rate. The larger the difference the faster is the growth of the population. So let us look at these two components.

The average birth rate is smaller in developed countries than in developing countries. The average birth rate is 11 per 1000 of the population in developed countries, where health care is good, and is 24 per 1000 in developing countries, where health care is bad. The birth rate in India is 26 per 1000, in Ethiopia 40, Eritrea 43, and Congo 44.

The birth rate is related to the fertility rate, which is defined as the number of children born per woman. Fertility rates are high in developing countries and low in developed countries. The average fertility rate in developed countries is 1.6 children born per woman, but the average in developing countries is 3.1—almost twice as high. The average fertility rate in the US is 2.1 and in India it is 3.2. The fertility rate in Canada is 1.5, in Australia it is 1.7, and in Egypt 3.5. China is an exception, but even there the average fertility rate is 1.8. The average fertility rate in Eastern Europe is 1.2, and in Africa it is 5.2. Ethiopia registers 6.3.

Now let us look at death rates. Here we are in for a big surprise. The death rates in developed countries are higher than in developing countries. This is the opposite of what we would have expected from different levels of health care.

The average annual death rate in developed countries is 10 per 1000 of the population, and in developing countries it is eight per 1000. Despite poorer health care in developing countries, a smaller fraction of the population die each year compared with the developed countries.

The average annual death rate in the US and western Europe is nine per 1000 of the population, the same as in India. The death rate in Canada and Australia is seven per 1000, the same as in Egypt and Northern Africa. The death rate in China is even lower, at six per 1000.

The average annual death rate in eastern Europe is 14 per 1000, the same as in Nigeria and Congo and for the whole of Africa, which is the poorest continent. The death rate in Eritrea is only 12 per 1000. The rates in Ethiopia and South Africa are the same, 15 per 1000, and yet these countries have different levels of health care.

There is no clear connection between the state of health care and birth or death rates. Birth rates in developing countries are higher than in developed countries, but death rates are lower. The difference between these two indicators is therefore higher in developing countries, which means that population growth rates are higher, and this has nothing to do with the state of health care. The growth rate is only 0.1 per cent in developed countries, but 1.6 per cent in developing countries. The average growth rate in Africa is 2.4 per cent, and in Eritrea and Congo it is 3 per cent.

If there is a correlation between health care and population increase it is an *inverse* correlation—the better the state of health care the lower the growth rate.

However, there is definitely a clear correlation between the state of health care and infant mortality,[3] which is higher in developing countries, where health care is poor. The number of deaths per 1000 live births is only seven in the more-developed countries, but 60 in the less developed countries. Infant mortality in the US is 6.6, in Europe it is five, but in India it is 68. In Canada it is 5.3, in Australia 5.2, but in Egypt it is 44, Northern Africa 55, and China 31. Infant mortality in eastern Europe is 13, but in Africa it is 86 and in Eritrea 97.

We should also understand the difference between child mortality and the death rate of the population. Child mortality is calculated for a narrow age range; the death rate of the population is calculated by averaging over all age groups. Child mortality rates are high in developing countries, but growth rates are also high because birth rates are high. Children who die are quickly replaced by an even larger number of newborn children.

Finally, we could question whether the average life span depends on the levels of prosperity and health care. Calculated average life spans include the number of children who die before the age of five. These numbers are high in developing countries, so average life spans are lower. However, those who survive are strong and live long, even though conditions are poor and health care is far from satisfactory.

How many people are we adding to global population each year?[4]
Every minute, 250 children are born, or about 130 million a year (see Table 2-2). About 90 per cent of the increase in global population is in developing countries.

Table 2-2 Population dynamics, 2002 (millions per year)

	World	MDC	LDC	Per cent
Births	133.14	13.28	119.86	90.0
Deaths	53.93	12.17	41.76	77.4
Increase	79.21	1.11	78.10	98.6
Infant deaths	7.25	0.09	7.16	98.8

MDC – More-developed countries

LDC – Less-developed countries

Per cent – Percentage of global total for the less-developed countries

Example: Worldwide about 90 per cent of children are born in the less-developed countries. About 99 per cent of infant deaths occur in developing countries, yet these countries contribute close to 99 per cent of the annual increase in global population.

Source: PRB 2002

Imagine a counter clicking at the rate of four clicks per second and try to count that fast. This is how rapidly children are being added to the global community. The counter goes on day and night, week after week, and year after year, relentlessly adding people to our already overcrowded planet.

Each minute, about 100 people die in the world, or about 50 million per year. The net result is that global population increases by about 80 million per year (almost exclusively in developing countries), which is the equivalent of adding the population of Germany to the global community each year. Ideally, we should also be adding a land area the size of Germany.

Of the 130 million children born each year, about seven million will die in the first year of life (almost exclusively in developing countries),[5] leaving about 123 million children, which is a little more than the combined population of the United Kingdom and France. This dramatically highlights the stress we are experiencing globally, the problems we have to cope with, and the challenges we have to face.

The demands of the young are far greater than the demands of the old. As the millions of children grow and develop physically and mentally, they need to be clothed, fed, guided, and educated. When they reach maturity, they expect employment, marriage, their own accommodation, and their own families. How much chance do they have for all this?

Many of the children who survive their first year will die later of hunger, malnutrition, and other forms of deprivation. Many of those who survive will be deprived of their basic rights because they have been born in the wrong place. Only a small fraction will be destined to enjoy a relatively good life. However, with the approaching global crisis, even this privileged minority living mainly in industrialised countries is facing an uncertain future. Its inheritance is likely to be a ruined planet.

How is global population divided according to levels of affluence? Officially, the six billion mark of global population was reached on 12 October 1999. We have not celebrated this event because we know that the population explosion is nothing to be happy about. By mid-2002, global population increased to more than 6.2 billion.

The mid-2002 populations of the most affluent areas—Canada, the US, northern and western Europe, Japan, Australia, and New Zealand—add up to a surprisingly small number of about 750 million, which is only 12 per cent of global population. If we include eastern and southern Europe and Russia, the total number in this group of richer countries is only 1197 million, which is about 19 per cent of global population.

Thus, about 1.2 billion people live in richer, industrialised countries and five billion in poorer, developing countries. Two regions account for about three-quarters of global population: Asia (61 per cent) and Africa (14 per cent).

It is useful to remember these figures: 750 million (0.75 billion), the approximate number of people in the affluent areas; and 1.2 billion, the approximate number in the industrialised areas. The figures can serve as reference points to help us later to appreciate levels of human deprivation.

Table 2-3 Projections of global population, 2000–2100

Year	Low			Medium			High		
	(1)	(2)	(3)	(1)	(2)	(3)	(1)	(2)	(3)
2000	1.4	82.8	6.05	1.4	80.6	6.05	1.2	74.2	6.05
2010	1.2	79.4	6.86	1.2	80.4	6.85	1.1	74.1	6.78
2020	1.0	72.0	7.62	1.0	77.5	7.65	1.0	72.3	7.44
2030	0.7	60.6	8.28	0.7	71.6	8.39	0.9	70.3	8.15
2040	0.5	45.6	8.81	0.7	62.8	9.06	0.8	67.2	8.91
2050	0.3	27.7	9.17	0.5	51.2	9.63	0.7	63.8	9.56
2060	0.1	8.0	9.33	0.4	37.3	10.06	0.6	60.1	10.18
2070	-0.1	-12.2	9.30	0.2	21.6	10.35	0.5	56.3	10.76
2080	-0.3	-31.7	9.07	0.05	4.8	10.48	0.5	52.3	11.30
2090	-0.6	-49.2	8.66	-0.1	-12.2	10.43	0.4	48.3	11.80
2100	-0.8	-63.7	8.08	-0.3	-28.7	10.22	0.4	44.4	12.26

Low, Medium, and High – the three levels of projection

(1) – Percentage annual growth rate of global population

(2) – Annual increase (if positive) or decrease (if negative) of global population in millions

(3) – Global population in billions

Example: The low-level projections show that in 2060 the annual growth of the global population will be only 0.1 per cent. Ten years later, in 2070, the annual growth rate will be negative, which means that global population will be decreasing. The increase in global population will be reduced from 82.8 million per year in 2000 to only 8 million per year in 2060. In 2070, global population will be decreasing at the rate of 12.2 million per year. Global population in 2060 will be 9.33 billion and in 2070 9.30 billion.

Source: Author's projections

How many people will live on our planet in the future? I have calculated future populations using annual growth rates.[6] We cannot use the hyperbolic function for this purpose because global population cannot grow to infinity. We are dealing here with an end of the population explosion, so we have to use a different method. Extracts of my projections are presented in Table 2-3.

I have carried out the calculations in two steps. First I analysed the annual growth rates using various mathematical functions, then I used the fitted functions to predict future populations. This resulted in three sets of projections: low-level, medium-level, and high-level. Two out of three projections show that global population is likely to reach a maximum during this century, but not earlier than in the second half. Results of these simple calculations are well within the range of predictions by other authors.[7]

For instance, my three predictions of global population for 2025 are about eight billion. Other sources give 7.1–8.8 billion. My predictions for 2050 are 9.2–9.6 billion. Other sources give 7.7–11.3 billion. My low-level predictions of 7964 million and 9165 million for 2025 and 2050, respectively, agree well with the values listed by PRB (2002) of 7859 million and 9104 million. My predictions are only 1 per cent higher.

The low-level projections. Global population will reach a maximum of about 9.35 billion in 2063 and will decrease to about eight billion by the turn of the next century.

The medium-level projections. Global population will cross the 10 billion line about 2059, and will reach a maximum of 10.48 billion about 2082. It will decrease to just over 10.3 billion in 2100.

The high-level projections. Global population will cross the 10 billion line about 2057 and will keep increasing. By 2095, it will pass the 12 billion mark and by 2100 it will reach 12.26 billion.

Annual growth rates and global population. The annual growth rates of global populations are decreasing. However, this does not necessarily mean that global population is also decreasing. Indeed, if we look again at the values in Table 2-3 we shall see that annual growth rates are decreasing for all three projections, but the corresponding populations are increasing. Global population growth is only slowing down.

Global population reaches a maximum when annual growth rates decrease to zero. For the low-level and medium-level projections, these events occur during the current century. Global population starts to decrease only when annual growth rates become negative.

We can see also (see Table 2-3) that, for the high-level projections, annual growth rates do not reach zero during the current century. These projections are interesting because they correspond to a case in which annual growth rates decrease forever. They become smaller, but they never reach zero.

For the high-level projections, global population reaches 13 billion in 2117, 14 billion in 2147, 15 billion in 2190, and 16 billion in 2278. It then increases exceedingly slowly to reach 16.5 billion in 2501.

We should also notice that the low-level and medium-level projections give almost identical results until 2030–2040. Likewise, the medium-level and high-level projections give almost identical results until 2050–2060. This means we shall probably not know for a long time whether global population is going to reach a maximum. There are some differences in the annual growth rates of global population for the three sets of projections, but these small differences do not seem to be helpful. This situation does not offer us much comfort.

How many people will live in various regions of the world? The numbers of people living in various regions of the world, now and in the future, are listed in Table A2-1 (see Appendix A). Asia accounts for close to two-thirds of global population and Africa about 14 per cent. Thus, nearly three-quarters of the world's people live in just two regions. The population in these regions will continue to increase. About one-third of the global population increase between 2002 and 2050 will be in Africa and about half in Asia.

At present, Asia adds about 49 million people each year to global population, Africa adds 20 million, Latin America–Caribbean nine million, and North America two million. Currently, Africa and Asia account for about 86 per cent of the increase. The population of Europe is decreasing at the rate of about 0.7 million per year.

Between 2002 and 2025, Africa is expected to add about 441 million to its population and Asia about 975 million. The population of North America will increase by only 63 million, and that of Latin America–Caribbean by 148 million.

Between 2002 and 2050, the population of Africa will increase by just over a billion, Asia by 1531 million, North America by 131 million and Latin America–Caribbean by 284 million.

The projected populations in developed and developing countries are shown in Table 2-4. My projections are only about 1 per cent higher than the relevant values, which can be derived from the PRB (2002) data (see Table A2-1, Appendix A) for the years 2025 and 2050.

Table 2-4 Populations in the less and more developed countries, 2000–2050

Year	Developing countries		Developed countries	
	Millions	Per cent	Millions	Per cent
2000	4 869	80.5	1 181	19.5
2010	5 621	81.9	1 241	18.1
2020	6 353	83.4	1 267	16.6
2030	7 003	84.6	1 277	15.4
2040	7 531	85.5	1 275	14.5
2050	7 926	86.5	1 239	13.5

Per cent – Percentage of global population

Example: In 2000, 4869 million people lived in developing countries, or 80.5 per cent of the global population. The number of people living in developed countries was only 1181 million, or 19.5 per cent of the global population.

Source: Author's projections

This table shows that although the population in developing countries will continue to increase rapidly, the population in developed countries will remain nearly constant. From about 2035, the population in developed countries will even start to decrease. The relative percentage of the population in developed countries will be decreasing throughout the current century, from 19.5 per cent of global population in 2000 to 13.5 per cent in 2050.

Currently, about 80.5 per cent of global population lives in less developed countries. By 2050, this will be 86.5 per cent. The worsening demographic imbalance between developed and developing countries will increase immigration pressures. By 2050, the combined population of Asia and Africa will be higher than the current global population.

The most populated countries, now and in the future. Only a handful of countries can be identified as having a distinctly large share in global population (see Table 2-5). These countries are mainly in the developing regions.

Table 2-5 The most populated countries in 2002, 2025, and 2050

Country	2002 (1)	(2)	(3)	Country	2025 (1)	(2)	(3)	Country	2050 (1)	(2)	(3)
China	1281	21	1281	China	1455	18	1455	India	1628	18	1628
India	1050	38	2331	India	1363	35	2818	China	1394	33	3022
US	287	42	2618	US	346	40	3164	US	414	37	3436
Indonesia	217	46	2835	Indonesia	282	43	3446	Pakistan	332	41	3768
Brazil	174	48	3009	Pakistan	242	46	3688	Indonesia	316	45	4084
Russia	144	51	3153	Brazil	219	49	3907	Nigeria	304	48	4388
Pakistan	144	53	3297	Nigeria	205	52	4112	Brazil	247	51	4635
Bangladesh	134	55	3431	Bangladesh	178	54	4290	Bangladesh	205	53	4840
Nigeria	130	57	3561	Mexico	132	56	4422	DRC	182	55	5022
Japan	127	59	3688	Russia	129	57	4551	Ethiopia	173	57	5195
Mexico	102	61	3790	Japan	121	59	4672	Mexico	151	58	5346
Germany	82	62	3872	Ethiopia	118	60	4790	Philippines	146	60	5492
Philippines	80	64	3952	Philippines	116	62	4906	Vietnam	117	61	5609
Vietnam	80	65	4032	DRC	106	63	5012	Egypt	115	62	5724
Egypt	71	66	4103	Vietnam	104	64	5116	Russia	102	64	5826
Ethiopia	68	67	4171	Egypt	96	65	5212	Japan	101	65	5927
Turkey	67	68	4238	Turkey	85	67	5297	Turkey	97	66	6024
Iran	66	69	4304	Iran	85	68	5382	Iran	97	67	6121
Thailand	63	70	4367	Germany	78	69	5460	Thailand	72	68	6193
UK	60	71	4427	Thailand	72	69	5532	Germany	68	68	6261
France	60	72	4487	UK	65	70	5597	UK	65	69	6326
Italy	58	73	4545	France	64	71	5661	France	65	70	6391
DRC	55	74	4600	Italy	58	72	5719	Italy	52	70	6443

Countries are arranged in descending order of population in the respective years

(1) – Population in millions

(2) – Cumulative percentage of global population

(3) – Cumulative population in millions

DRC – Democratic Republic of Congo

Example: The combined population of China and India in 2002 was 2331 million, or 38 per cent of global population.

Source: Based on PRB 2002

Currently, about half of the global population lives in just six countries (China, India, the US, Indonesia, Brazil, and Russia). By 2050, about the same number of people will live in only two countries (India and China). Currently, about six billion people are spread over 175 countries. By 2050, the same number will live in just 17 countries. By about 2034, China and India will have the same population—about 1.47 billion people each.

We now have 11 countries each with a population of at least 100 million. By 2025, the Philippines, Vietnam, Ethiopia, and the Democratic Republic of Congo will join this group, and Egypt will be added by 2050, with Turkey and Iran close behind.

The combined population of China and India is now about 2.3 billion, or about 38 per cent of global population. For every five people living in the world, one lives in China, and for every six people one lives in India

In South-East Asia, Australia's nearest region, the population will increase from 536 million in 2002 to 706 million in 2025. In Indonesia alone the figure will rise from 217 million to 282 million. Australia's population will increase from 20 million to only 23 million.

Further afield, the population of East Asia will increase from 1512 million in 2002 to 1690 million in 2025, and the population of South Central Asia from 1521 million to 2047 million. The population of China is expected to be 174 million greater by the year 2025, and the population of India about 313 million greater. The population of Pakistan will increase by 98 million, and Bangladesh by 44 million.

Which countries contribute most to the growth of the global population? Only a handful of countries can be identified as making large contributions to population growth (see Table 2-6). About half of the growth takes place in only six countries (India, China, Indonesia, Nigeria, Pakistan, and Bangladesh); about one-third occurs in just two countries, India and China. India adds about 18 million people each year, which is only a little less than the present population of Australia.

The largest annual growth rates in this group are in the Democratic Republic of Congo and Nigeria. The lowest—and almost the same—are in the US and China. However, China's population is so large that even with a small annual growth rate it still adds about nine million people each year. The US adds only about two million.

Stabilisation of global population. A stabilisation is desirable, but it will create a problem because in many regions a decreasing number of people who work will have to support an increasing number of non-workers—that is, children and the aged. The birth rate will be lower, but people will live longer.

Table 2-6 Countries contributing most to the growth of the global population, 2002

Country	Annual growth rate	Annual increase Millions (1)	(2)	Per cent (3)	(4)
India	1.7	17.9	17.9	22.9	22.9
China	0.7	9.0	26.9	11.5	34.4
Indonesia	1.6	3.5	30.4	4.5	38.9
Nigeria	2.7	3.5	33.9	4.5	43.4
Pakistan	2.1	3.0	36.9	3.8	47.2
Bangladesh	2.2	2.9	39.8	3.7	51.0
Brazil	1.3	2.3	42.1	2.9	53.9
Mexico	2.1	2.1	44.2	2.7	56.6
Philippines	2.2	1.8	46.0	2.3	58.9
US	0.6	1.7	47.7	2.2	61.1
Ethiopia	2.5	1.7	49.4	2.2	63.3
DRC	3.1	1.7	51.1	2.2	65.4

Countries are arranged in decreasing order of the number of people they have added to the global population in 2002.

Annual growth rate – Annual growth rates in percentage of local population

(1) – Annual increase in population, in millions

(2) – Cumulative annual increase in population, in millions

(3) – Percentage of the global increase

(4) – Cumulative percentage of the global increase

DRC – Democratic Republic of Congo

Example: The annual growth rate in China in 2002 was 0.7 per cent, or 9 million people. China and India were adding 26.9 million people per year to the global population. China's contribution to the annual growth of global population was 11.5 per cent. The combined contribution of China and India to the growth of global population was 34.4 per cent.

Sources: Based on the data of PRB 2002

Between 1500 and 1800, life expectancy in England and Wales was about 36 years. However, from about 1800, life expectancy has steadily and rapidly increased to reach about 80 years in 2000. (A nearly identical increase was recorded in Sweden.[8])

Between 1975 and 2000, the average life expectancy for low-development countries increased from 46 to 53 years, and for high-income OECD[9] countries from 72 to 78 years.[10] Life expectancy is projected to increase even further.[11] By 2050, in affluent countries it is projected to be more than 80 years; in Asia, close to 80; and in sub-Saharan Africa, close to 70. On average, people everywhere are expected to live longer.

Stabilisation of global population is expected to take place during the current century, because fertility rates are falling. At the same time, the percentage of people aged 65 years and over is increasing (see Table 2-7). In North America, the proportion of the population aged 65 years and over will increase from about 8 per cent in 1950 to about 22 per cent in 2050. In developing countries the increase will be from 4 per cent to 15 per cent, and in industrialised countries from 8 per cent to 26 per cent.

Table 2-7 Percentage of the population aged 65 years and over, 1950 and 2050

Region	1950	2050
North America	8.2	21.9
Latin America-Caribbean	3.7	16.8
Europe and Russia	8.2	27.6
Middle East and North Africa	3.9	13.5
Sub-Saharan Africa	2.1	4.1
South and Central Asia	3.7	14.2
East Asia and Pacific	4.4	21.1
Developing countries	3.9	15.0
Industrialised countries	7.9	25.9
World	**5.2**	**16.4**

Example: About 8.2 per cent of the population in North America in 1950 was aged 65 years and over. By 2050, this fraction is expected to increase to 21.9 per cent.

Source: UN 1999a

With the decreasing number of newborn children and with the increasing number of older generations, the number of the working group will be decreasing and the number of non-working groups will be increasing, thus putting an increasing burden on younger generations. This is already about to happen in many regions of the world.

The dependency ratio,[12] defined as the number of non-working people (children under 15 years of age and people over 65 years) relative to the number of working people, is now at its minimum in high-income OECD countries (see Table 2-8). A minimum in the dependency ratio is claimed to be good news, because with fewer children and older people to care for we have a better chance to grow richer.

However, it is also bad news, because eventually those who are now working will retire, and many will not be replaced by younger generations. We shall have an increasing number of people living longer and needing to be supported by a decreasing number of workers.

Table 2-8 Dependency ratios, 1970–2050 (per cent)

Region	1970	1990	2010	2030	2050
High-income OECD countries	57	50	49	68	77
East Europe and Central Asia	82	60	42	50	60
East Asia	59	54	42	51	66
South Asia	81	72	56	47	50
Middle East and North America	97	86	56	48	53
Latin America-Caribbean	87	68	50	48	56
Sub-Saharan Africa	91	93	80	59	47

Dependency Ratio – The number of non-working people (up to 15 years old and over 65 years) to the number of working people, in per cent

Example: On average, in high-income OECD countries, the dependency ratio is projected to reach a minimum of 49 per cent in 2010 (for every 100 working people there will be 49 non-working people). By 2050, the dependency ratio in these countries will be 77 per cent (100 working people will have to support 77 non-working people).

Source: WB 2003

The dependency ratio for high-income OECD countries will rise from 49 per cent in 2010 to 68 per cent in 2030, and 77 per cent in 2050. By 2050, for each 100 workers there will be 77 non-workers.

These countries will find it increasingly difficult to support non-workers. From about 2010, high-income OECD countries will have the growing problem of how to divide their national incomes between young children and the older non-workers.[13] It is possible that from about that year there will not be enough funds for retired people, and their pensions will have to be gradually reduced.

The dependency ratios in East Asia, eastern Europe, and Central Asia are about 50 per cent. These regions will reach a minimum of 42 per cent in 2010 and will climb back to about 60 per cent in 2050. Their burden of caring for non-workers will become a little lighter until about 2010. However, from about that year the burden will grow heavier.

The dependency ratios in South Asia, the Middle East and North Africa, Latin America–Caribbean, and sub-Saharan Africa are relatively high, but they are falling. The burden of caring for young and old in these regions is decreasing.

The pain of global population stabilisation will be felt first in high-income OECD countries, East Asia, eastern Europe, and Central Asia—beginning about 2010. Latin America will follow suit from about 2020, and other regions, with the exception of sub-Saharan Africa, from 2030 or 2040.

The average dependency ratio in sub-Saharan Africa will continue to fall in the first half of the current century. However, countries in this region are exceptionally poor, and the falling dependency ratios will not ease their burden.

Can environmental limitations reduce the growth of the global population? Projections of life expectancies and of future populations do not take into account the limitations imposed by environmental factors such as climate change, pollution, disease, hunger, and the decreasing availability of safe water. Environmental factors have a strong influence on human health and well-being.[14] Harmful environmental factors could influence mortality rates in certain regions, and thus population growth.[15] However, regional changes might not influence global population: only widespread and long-lasting harmful effects would have a substantial effect.

One example of reduced life expectancy due to disease emerges from the spread of HIV. In Zimbabwe, average life expectancy increased from about 40 years in 1950 to 57 in 1986, then went into decline. The average reduction of life expectancy as a result of HIV/AIDS is projected to be 14–26 years by 2005. The reduction in Botswana is projected to be 34 years, in South Africa 19 years, and in Kenya 17 years. By 2010–1015, the average reduction of life expectancy in Botswana, Kenya, Malawi, Mozambique, Namibia, Rwanda, South Africa, Zambia, and Zimbabwe is projected to be 17 years. People who would have been expected to live an average 64 years in the absence of HIV/AIDS are expected to live only 47 years.[16]

Africa and Asia are in the forefront of the demographic explosion, and they are likely to be the first to show the effects of environmental limitations. At present, their consumption is still relatively small, but it is increasing. We already have about three times more people in the world than we can comfortably support, and any notion of an acceptable future for everyone is unrealistic. It is possible that the population explosion will be terminated by a smooth transition to a state of equilibrium. However, it is also possible that it will be terminated by a huge increase in the number of people dying prematurely in the densely populated regions of the world.

The future. Annual global population growth is decreasing, and the population is likely to reach a maximum of 9–10 billion in the second half of the current century. There is also a strong possibility that in certain regions population growth will be influenced by a widespread and long-lasting rise in mortality rates caused by the shortage of resources and harmful environmental factors.

Problems we are already experiencing will intensify. They will be experienced mainly in developing countries, but industrialised countries will be also affected. Even now, industrialised countries are experiencing immigration pressures and terrorism. These and other related problems can be expected to increase in the future.

Chapter 3
Diminishing Land Resources

'As soils are depleted, human health, vitality and intelligence go
with them.'

— Louis Bromfield (1896–1956), US author, 1926 Pulitzer Prize winner

The land area of the planet supports almost all our consumption. Do we
have enough land to support current and future levels of consumption?
Millions of people are dying of hunger. We boost food production by
overuse of agricultural chemicals, but what effects do they have on the
land, on our health, and on the environment? How much forest area are
we losing each year? Human expansion is routinely carried out with no
regard for other species. How many are threatened with extinction and
how will that affect humans? Economic progress is measured by production, but at what environmental price? How can we improve profit
without damaging the environment?

How much land is needed to support our existence? To answer
this question we have to consider not only arable land for the production of food, but also the total biologically productive area to support
all forms of our consumption (see Box 3-1).

Global arable land area. The land area needed to feed one person
depends on such factors as diet (forced or adopted), soil productivity,
water availability, and the use of agricultural chemicals. The average
area of arable land needed to support the diet of people in industrialised countries has been estimated at 0.5 hectares per person [ha/p].[1]

The average arable land area required to feed one person on
a mainly vegetarian diet can be estimated by looking at the current
global distribution of arable land and at the degrees of hunger in various countries.[2] We also have to consider the financial capacity of a
country to supplement food requirements by imports. An approximate
area of arable land for this kind of diet seems to be 0.2 ha/p.

If we move to the extremely low level of arable land availability
we can use a definition from Canadian geographer Vaclav Smil, who

defines the borderline of arable land scarcity as 0.07 ha/p.[3] Countries with arable land scarcity are (in decreasing order of availability): Israel, Bangladesh, Republic of Congo, Switzerland, the Netherlands, Jordan, Egypt, Republic of Korea, United Arab Emirates, Qatar, Oman, Brunei, Bahrain, Kuwait, and Singapore. By 2025, Somalia, Liberia, Eritrea, Yemen, and Bhutan will join this group.[4]

Box 3-1
Required and available surface areas in 2003, 2025, and 2050

The surface area needed to support our consumption depends on adopted or forced lifestyles. We already do not have enough room for a high level of consumption, and the situation is growing worse. There is an urgent and increasing need to reduce consumption in affluent countries.

Arable land
Required:

To support the diet of industrialised countries	0.50 ha/p
To support a mainly vegetarian diet	0.20 ha/p
Absolute minimum	0.07 ha/p

Global average available in:	Year	Area
	2003	0.24 ha/p
	2025	0.16 ha/p
	2050	0.11 ha/p

Biologically productive surface area
Required to support:

The consumption of affluent countries	10.0 ha/p
The average consumption of Western Europe	6.0 ha/p
The consumption of very poor countries	1.0 ha/p

Global average available in:	Year	Area
	2003	1.8 ha/p
	2025	1.0 ha/p
	2050	0.7 ha/p

ha/p – hectare per person

I have included the loss of land areas through degradation, and I have also used my low-level projections of global population.

Source: Estimates of required arable land areas may be found in Engelman, et al 2000, Giampietro and Pimentel 1994, Lal 1989, and Smil 1993. Estimates of required biologically productive surface areas are based on ecological footprints (Wackernagel et al 2002).

With the present global availability of arable land we can support three billion people on the diet of industrialised countries, 7.5 billion people on a simple diet, and up to 20 billion starving people.

The global biologically productive surface area. The area of land and water required to support not only the production of food, but also all other forms of consumption depends on adopted or forced lifestyles. For average lavish consumption, one needs 10 hectares or more of the biologically productive surface area per person. For the average consumption of industrialised countries, one needs 5–6 hectares per person. For a very low level of consumption one would have to be satisfied with a little less than a hectare—however, this seems to be about the lowest limit.[5]

With the globally available biologically productive surface area, our planet can support only one billion people with the standard of living of the richest countries, two billion people with the standard of living of western Europe, or 11 billion people with the standard of living of Ethiopia or India.

The use of land resources. To understand the availability of land areas it is useful to look at the functional distribution of available land (see Box 3-2). The total land area used for agriculture is 1.5 billion hectares; for pastures, 3.4 billion hectares; and for forests, 3.9 billion hectares.[6] This makes a total biologically productive land area of only 8.8 billion hectares, which corresponds to 68 per cent of the estimated total land area of 13 billion hectares.[7] The land used for urbanisation is estimated at only 1–2 per cent of the total land area.[8] However, this is not a static situation: human-assisted degradation of land is slowly converting useful areas into wasteland.

Our consumption is supported not only by land, but also by a small fraction of water areas. Continental shelves form only two billion hectares, but they provide 95 per cent of the marine fish catch. Inland waters are 300 million hectares. Thus, the total global biologically productive water area is only 2.3 billion hectares.[9] If we add the biologically productive land area, we find that the earth's biologically productive surface area totals 11.1 billion hectares. This small area of the planet has to support the human race and many other species. It has been estimated that we need to reserve at least 12 per cent of the biologically productive surface area for the preservation of biodiversity.[10]

Box 3-2
Functional distribution of global surface areas

The total surface of our planet is 51.1 billion hectares [Gha], but only a small fraction of this surface can be used to support our existence.

Land areas

	Area [Gha]	Per cent (1)
Arable land	1.5	11.5
Pastures	3.4	26.2
Forest and woodland	3.9	30.0
Other land	4.2	32.3
Total land area	13.0	100.0

Global biologically available surface area

	Area [Gha]	Per cent (2)
Biologically productive land area	8.8	17.2
Biologically productive water area	2.3	4.5
Total biologically productive surface area	11.1	21.7

(1) – Per cent of global land area
(2) – Per cent of global surface area
Gha – gigahectare (billion hectares)

Source: Acosta et al 1999; Barbier 1999; Buringh 1989; Fischer and Heilig 1998; Wackernagel et al 1997, 2002; WRI et al 1998; UNEP 2002

The loss of land resources. The available land resources per person are decreasing not only because global population is increasing, but also because useful land areas are being degraded. Land degradation is a natural and continuing process, but we are contributing substantially to the loss.

Soil is a relatively thin and fragile layer on the land surface. A survey presented by the World Bank shows that most of the land areas of our planet are arid or ecologically fragile.[11] A good example is Australia, a large, but arid continent.

The causes of land degradation include deforestation by large-scale logging and clearing, overgrazing, fuel consumption (harvesting of wood for fuel), agricultural mismanagement, land conversion, industry, and urbanisation (including roads and highways). Land-destroying agents and processes include acidification (by pollutants from transport and industry), waterlogging (the over-saturation of soil caused by improper drainage and excessive irrigation), salinisation (a related problem often caused by irrigation), wind and water erosion, desertification (land degradation in arid or semi-arid areas caused by a combination of climate change and human activities), oil spills, grease residues, heavy metals, sulphur compounds, tars, and many more.[12]

How much land are we losing? Estimates vary, but it seems that we are losing about 45 million hectares of biologically productive land per year (see Box 3-3). At this rate, we shall lose all biologically productive land in 200 years.

Box 3-3
Estimated annual global loss of land

We are losing large quantities of productive land areas of our planet each year. We urgently need to reduce the loss.

Annual loss of:	[Mha/y]
Arable land	10
All productive land areas	45

Mha/y – megahectare (million hectares) per year

Source: Anderson et al 2001; Barbier 1999; Hinrichsen 1997; Oldeman et al 1990; Stocking and Murnaghan 2000; UNEP 2002; WRI 1992

It is claimed, for instance, that between 1945 and 1990 we lost 1964.4 million hectares of land through soil degradation, which corresponds to 44 million hectares per year. The loss has been mainly through overgrazing (35 per cent of the total), deforestation (30 per cent), and agricultural activities (28 per cent).[13]

Another report claims that we have lost 1967 million hectares of land through soil degradation over an unspecified period. The greatest land (soil) degradation is caused by overgrazing (680 million hectares), followed by deforestation (580 million hectares), agricultural mismanagement (550 million hectares), fuel consumption (137 million hectares), and urbanisation (19.5 million hectares). The types of degradation include soil and water erosion (56 per cent), wind erosion (28 per cent), chemical degradation (12 per cent), and physical degradation

(4 per cent). The worst land degradation is in Europe (23 per cent), Asia (18 per cent), and Africa (17 per cent).[14]

It has been also claimed that globally we have lost 23 per cent of the productive land area since 1945.[15] If we use the currently available biologically productive land area of nine billion hectares we can calculate that the loss has been 2690 million hectares. This corresponds to an average of 47 million hectares per year.

The rate of degradation of arable land has been estimated at about 10 million hectares per year.[16] This is practically a permanent loss because only about one centimetre of soil is restored in 200 years by natural processes.[17]

Dryland salinity in Australia.[18] A good example of human-induced land degradation is the dryland salinity problem in Australia (see Table 3-1). Dryland salinity is caused by the removal of the native vegetation that keeps the groundwater level down. When trees are removed, the groundwater level rises and releases large quantities of salt, which have been deposited over long periods by wind and rain. For instance, in Western Australia, salt has been deposited at the rate of 20–200 kilograms per hectare per year. The soil stores between 300 and 10,000 tonnes of salt per hectare.[19]

Table 3-1 Australian assets affected or threatened by dryland salinity in 2000, 2020, and 2050

Asset	2000	2020	2050
Arable land (hectares)	4 650 000	6 371 000	13 660 000
Perennial vegetation (hectares)	631 000	777 000	2 020 000
Total perimeter of lakes (kilometres)	11 800	20 000	41 300
Rail (kilometres)	1 600	2 060	5 100
Roads (kilometres)	19 900	26 600	67 400
Number of towns	68	125	219
Number of wetlands	80	81	130

Example: By 2000, close to 4.7 million hectares of arable land had been affected by dryland salinity. By 2050, 13.7 million hectares are expected to be affected.
Source: NLWRA 2001

By the end of 2000, in just three states—Western Australia, South Australia and Tasmania—dryland salinity had been responsible for the degradation of 5.7 million hectares, of which 4.7 million hectares was agricultural land. The projected loss by the year 2050 is 15 million hectares, of which 14 million hectares will be agricultural land. The worst affected areas are the south-west regions of Western Australia. This also happens to be one of the ecological hotspots hosting thousands of endemic plant species as well as many endemic species of birds, mammals, reptiles, and amphibians. Dryland salinity affects not only soil, but also roads, railways, and towns.

How much of the biologically productive surface area are we using to support our consumption? Consumption levels of various countries can be expressed as ecological footprints.[20] This idea is also useful in illustrating the problem of diminishing land resources.

The ecological footprint of a country is the total consumption of energy and materials (including food) expressed as the biologically productive surface area needed to support it. Ecological footprints can be expressed in hectares, but are usually expressed in hectares per person. The ecological footprint of a country depends on the standard of living. If it is high, consumption is also high and the ecological footprint is large.

The ecological capacity of a country is its biologically productive area. This is not the same as its geographical area. The ecological deficit or surplus is the difference between the ecological footprint and the ecological capacity.

If the ecological footprint of a country is larger than its ecological capacity, the country has an unsustainable style of living and lives on an ecological deficit—that is, beyond its means. However, this does not mean that the country is suffering deprivation. An ecological deficit can be offset by drawing on the resources of other countries or by overexploitation of domestic natural resources. The overexploitation of natural resources occurs everywhere, even in countries without an ecological deficit. It is an undesirable course of action that achieves a temporary solution to the problem of excessive consumption. In the long run it is economically devastating.

Many countries have an ecological deficit (see Table A3-1 in Appendix A). For instance, the ecological footprint of Italy is 3.8 hectares per person [ha/p], but its ecological capacity is only 1.2 ha/p. Thus, Italy's ecological deficit is 2.7 ha/p. However, trade knows no boundaries. Through trade, fragments of Italy's footprint are located in other countries and fragments of other countries' footprints are in Italy.

If the combined global footprint is smaller than global ecological capacity, we are still within the limits of sustainability. However, if our combined global footprint is larger, we are living beyond our means—that is, with a global ecological deficit. Here the situation is different than for individual countries, because we cannot distribute fragments of our global footprints elsewhere. We are entirely dependent on our planet for resources. To reduce the global deficit, the human race should be satisfied with less. If we continue increasing our global ecological deficit, we might eventually suffer global bankruptcy.

Ecological deficit can be also taken as a measure of overpopulation. The larger the ecological deficit of a country the greater the degree of overpopulation. Using this concept, we can find that many industrialised countries are more overpopulated than, for instance, India or China.

Overpopulation. The ecological capacity of the US is 5.3 ha/p, but its footprint is 9.7 ha/p. Thus, the ecological deficit is 4.4 ha/p. The US lives beyond its means, and its economy is not based on a solid, sound, and stable foundation. The US ecological deficit is nine times greater than China's. Even though China has the largest population in the world it is nine times less overpopulated than the US. People living in the US rely more heavily on the outside world and on the overexploitation of their own resources than do people living in China.

Ecological deficits in developing countries tend to be small. For instance, the ecological deficit in India is only 0.1 ha/p and in Bangladesh 0.2 ha/p. These densely populated countries have a simple style of living, their consumption per person is low, and their ecological footprints are small. In Indonesia, there is an ecological surplus of 0.7 ha/p. The small footprints of developing countries allow a handful of rich countries to make themselves more comfortable.

Using ecological deficits as a measure of overpopulation, we can see that developing countries are generally less overpopulated than many industrialised countries. However, this is changing because developing countries are increasing their consumption and population. And even though the population of industrialised countries is growing slowly, their consumption is increasing. Our footprints are therefore growing larger everywhere, and we have less and less room for all the people on our planet.

If China alone tried to adopt the US style of living, it would need 13 billion hectares of biologically productive surface area to support such an excessive level of consumption—that is, 17 per cent more than the total biologically productive area of our planet. With this high style of living, a substantial percentage of China's population would have to move to another planet—together with the rest of the global population.

The largest ecological footprint is in the United Arab Emirates, followed by the US. The highest ecological deficit is also in the United Arab Emirates, followed by Kuwait. However, we also have countries with an ecological surplus. The largest is in Gabon, followed by New Zealand.

Global ecological footprint. The current global footprint is 2.3 ha/p, but global ecological capacity is 1.8 ha/p, so we suffer a global ecological deficit. The global ecological deficit is 0.5 ha/p—that is, 28 per cent higher than global ecological capacity.

At the current average level of global consumption, our planet has enough room for only 4.8 billion people. It would have to be 1.3 times larger to support all the people that now live on it. We are already overcrowded, but we are adding more people, and the consumption per person is steadily increasing.

Calculations of ecological footprints are relatively simple, but they depend on the accuracy of the data for consumption and productivity. Differences in the calculated values for adjacent years are of little significance. However, the trend calculated over the past 40 years shows clearly that our global footprint is steadily increasing and that it is now 30 per cent larger than the global biologically productive surface area.[21]

About 1960, global ecological capacity was twice as large as the global ecological footprint. In that year, only 50 per cent of the planet was needed to support global population. By about 1985, the global ecological footprint reached 100 per cent of global ecological capacity, and by the end of the 20th century it was 125 per cent.

In about 1985, we should have put an end to our spending spree and should have controlled our withdrawals from the global ecology bank. Instead, it has been business as usual, an approach that will no doubt continue as long as we can escape the consequences.

Ecological capacity and immigration pressures. Rising population, dwindling land resources, and increasing consumption are some of the elements that add to immigration pressures. Resistance to migrants is also increasing, so it is important for receiving countries to estimate the maximum number of people they can accommodate. One way of doing this is by considering ecological footprints.

The maximum number of people a country can support depends on the level of its consumption (that is, on its ecological footprint), ecological capacity, unsustainable overexploitation of resources, and the support from other countries through trade or other means.

For example, Australia's consumption per person is high. Its ecological footprint is 7.6 ha/p, but its ecological capacity is 14.6 ha/p. Its ecological surplus is therefore 7 ha/p, which means that without changing its current level of consumption it can increase its population by 92 per cent—that is, by 18 million (see Box 3-4). This estimate is based on the assumption that consumption per person will not increase.

Box 3-4
How many people can live in Australia?

This summary presents an example of the calculations of the number of people a country can support within the limits of its ecological capacity. At its current level of consumption, Australia can safely support an estimated maximum of 38 million people.

Adopted level of consumption	(1)	(2)
Current (2003) level of consumption	38	18
Consumption level of Ethiopia	356	336

(1) – The maximum number of people (in millions) that can be supported within the limits of Australia's ecological capacity, depending on the adopted levels of consumption.

(2) – The maximum number of people that can be added to the current (2004) population (in millions), depending on the adopted levels of consumption.

Calculated using the footprint (the consumption level) estimated by Wackernagel et al 2002

Australia could support even more people, but only by increasing the overexploitation of its domestic resources, reducing its standard of living, exploiting the resources of other countries, or by a combination of all these factors.

Suppose that Australia decided to make a sacrifice and reduce its standard of living to the Ethiopian level. The maximum number of people it could then support would be 356 million—an addition of only 336 million. This would make no impression on human suffering in the overpopulated regions of the world. It is clear that such a sacrifice would have been pointless. Its citizens would be all equally poor, equally helpless, equally in need of support and assistance from the outside world, and equally longing for some land to which they could emigrate.

There is no easy solution to the problem of human suffering. However, a partial solution is still possible. We can help the poor to help themselves by supporting their development.[22]

We also have to understand that immigration pressures will increase, and we must be prepared to accept genuine refugees and asylum seekers. Closing the borders would be a step towards the development of a Fortress World, which represents the worst possible option for the future. This option may be forced on us by neglect and complacency, by our refusal to respond to the challenge of putting the future on a more secure basis.

The problem of immigration will not disappear and will not resolve itself. It will become more severe in the future, and we shall have to be prepared to keep a delicate and difficult balance between compassion and reason, between the dictates of the heart and the dictates of logic, in deciding who is admitted as a refugee or asylum seeker. There is also a possibility that we will not be able to control immigration pressures.

Human expansion as a threat to other species.[23] The range of plant and animal species in an ecosystem determines its biodiversity. In the quest to expand our ecological footprints we are sacrificing biological diversity. This is a deficit we shall never be able to reconcile. The human-induced loss of species is high. Once lost, they are lost forever, unless we assume that in a distant future we will be able to recover genetic information and reproduce some species.

Why is biodiversity important? The links between species is important in preserving a healthy environment. Biodiversity ensures a rich bank of genetic material, which we can use in various ways—for example, to have a wide variety of food, for medical applications, or to enjoy the beauty of nature.

The annual global market value of pharmaceutical products derived from the genetic bank of the Earth's biodiversity has been estimated at between $75 billion and $150 billion. The annual global market value of the direct use of botanical treatments such as herbal medicines is between $20 billion and $40 billion. Of the 150 top prescription drugs in the US in 1997, 86 were derived from biological sources.[24]

What we see as a lump of dirt is a genetic treasure trove. Soil is the home of millions of organisms that play an important role in the environment: amoebas, protozoa, nematodes, mites, termites, ants, earthworms, bacteria, and fungi. They preserve soil integrity, remove pollutants, and help in the purification of water. They also fight disease (by attacking or neutralising human and animal pests and pathogens), and regulate floods and droughts (through their influence on the amount of water absorbed by soil).[25]

UNEP is now sponsoring a study of the biological diversity of soil in seven tropical forests of Brazil, Mexico, Ivory Coast, Uganda, Kenya, Indonesia, and India. It is hoped that this rich genetic bank will provide new pharmaceutical products, including antibiotics, and extend the range of industrial products.[26]

How many species live on our planet? The total number of species is estimated between seven million and 15 million, but it could be higher.[27] The accepted estimate is 14 million species (see Table 3-2). Of this number, we have only 400,000 plant species,[28] and less than 50,000 vertebrate species. Together, these familiar groups make up only 3 per cent of all the species living on our planet.

Table 3-2 The number and types of species living on Earth

Group	Estimated	Identified	Per cent
Insects	8 000 000	1 000 000	12.5
Fungi	1 500 000	70 000	4.7
Bacteria	1 000 000	4 000	0.4
Arachnids	750 000	75 000	10.0
Viruses	400 000	4 000	1.0
Nematodes	400 000	25 000	6.3
Plants	400 000	40 000	10.0
Algae	350 000	300 000	85.7
Protozoa	200 000	40 000	20.0
Molluscs	200 000	70 000	35.0
Crustaceans	150 000	40 000	26.7
Chordates	50 000	45 000	90.0
Total	**13 400 000**	**1 713 000**	**12.8**

Estimated – The estimated numbers of species

Identified – The number of species identified and described

Per cent – Identified species as a percentage of the total estimated number in a given group of species. For instance, only about 12.5 per cent of the estimated number of insects have been identified and described.

Chordates – This group includes all the vertebrates

Source: Based on Stork 1999

Surprisingly, only 13 per cent of the estimated global number of species has been identified and studied. We still have much to learn about nature. The only consolation about the current accelerated loss of species is that we shall never know what treasures we have lost.

Insects form the richest group of species. The next largest group is made up of fungi. Two other large groups are bacteria and arachnids (air-breathing arthropods). These four groups account for 80 per cent of the estimated number of species. If we manage to destroy ourselves and most of the familiar species, the next organisms to rule the world will be insects, bacteria, and other lower forms of life.

Endangered species. Tropical forests cover only 7 per cent of the earth's surface, but they harbour about 50 per cent of all plant and animal species. However, we are now destroying systematically this important habitat. The extinction of species in tropical forests is three to eight times faster than the global average.[29] We may be planting new trees to replace those we have cut down, but we can hardly expect to restore complex ecosystems we have destroyed.

Table 3-3 The number of endangered vertebrate species

Region	Mammals	Birds	Reptiles	Amphibians	Fish	Total
Africa	294 (62)	217 (39)	47 (3)	17 (4)	148 (56)	723 (164)
Asia-Pacific	526 (68)	523 (60)	106 (13)	67 (15)	247 (48)	1,469 (204)
Europe	82 (7)	54 (6)	31 (8)	10 (2)	83 (13)	260 (36)
LAC	275 (33)	361 (59)	77 (22)	28 (7)	80 (24)	821 (145)
North America	56 (5)	65 (15)	29 (2)	25 (1)	134 (17)	309 (40)
West Asia	30 (3)	24 (2)	8 (1)	0 (0)	9 (0)	71 (6)
Polar	0 (0)	6 (0)	7 (0)	0 (0)	1 (0)	14 (0)
World	1 263 (178)	1 520 (181)	305 (49)	147 (29)	702 (158)	3 667 (595)

LAC – Latin America-Caribbean

The figures in parentheses show the number of critically endangered species. Considering that the areas of the 25 biodiversity hotspots are decreasing, the number of endangered species might be even higher.

Example: The estimated total number of endangered species in the five listed categories in Africa is 723, of which the total number of critically endangered species is 164.

Source: Compiled using data for the individual geographical regions listed in UNEP 2002. A summary table in this reference contains clearly incorrect values.

Human-induced extinction of species has been estimated at up to 140,000 per year.[30] At this rate, we shall wipe out half the existing species in 70 years. It has been estimated that close to four thousand mammal, bird, reptile, amphibia, and fish species are threatened with extinction (see Table 3-3). About 600 species are on a critically endangered list. Earlier estimates indicated that 13 per cent of global plant species are threatened with extinction, but the updated figure is much higher: 24–48 per cent.[31]

Biodiversity hotspots. The problem with the continuing extinction of species is that many of them are concentrated in a handful of niches called biodiversity hotspots, where humans can easily destroy them. A biodiversity hotspot is an area of land that contains at least 0.5 per cent of the estimated global number of higher plant species. One of the world's hotspots is in Western Australia, which contains more than 4300 endemic plant species.

A total of 25 biodiversity hotspots have been identified that require special protection.[32] However, it is possible that we shall soon destroy them, because many of them are in places where the population density and growth are high (see Box A3-1, Appendix A).

It is estimated that more than one billion people live in areas designated as biodiversity hotspots. Fourteen hotspots are in areas with a population density higher than the global average of 46 people per square kilometre [p/km^2]. In seven of them, the population density is between 100 p/km^2 and 340 p/km^2.

To appreciate these figures we should look at population densities in selected parts of the world. The average population density in developing countries is 60 p/km^2. The average in Indonesia is 114 p/km^2; in China, it is 134 p/km^2; in the Philippines, 266 p/km^2; in India, 319 p/km^2; and in Japan, 337 p/km^2.[33]

In nineteen of the world's 25 hotspots the population growth rate is higher than the world average of 1.3 per cent per year. In thirteen of them, the annual growth rate is between 2 and 4 per cent.

The average natural growth rate in Asia is only 1.3 per cent; in India, it is 1.7 per cent; in Africa, 2.4 per cent; in sub-Saharan Africa, 2.5 per cent, in Western Africa, 2.7 per cent; in Central Africa, 2.9 per cent; and in Niger, 3.5 per cent. Biological hotspots are attractive environments for human habitation.

It has been estimated that biodiversity hotspots originally amounted to 17.5 million square kilometres [Mkm^2]. Now the figure is two million. However, they still contain 44 per cent of all higher plant species and 35 per cent of all vertebrate species.

Ten of the 25 biodiversity hotspots require an exceptionally high level of protection. Each contains between 5000 and 20,000 endemic plant species. Unfortunately, all are in areas with a high population density or population growth rate, or both.

In addition to biodiversity hotspots there are many protected sites in the world. Here there is a glimmer of hope, because their number and combined area have been steadily increasing. The number of protected sites rose from 3,392 in 1970 to 11,496 in 2000. The combined area increased from 2.78 million square kilometres in 1970 to 12.18 million square kilometres in 2000. The increase was rapid in the 1970s, but it slowed towards the end of the 20th century. Between 1995 and 2000 there was hardly any increase in the number of sites and their combined area.[34]

How large are our remaining global forest resources?[35] A forest is an area of more than 0.5 hectares with more than 10 per cent canopy cover. A closed forest is an area with at least 40 per cent canopy cover. In both cases, the trees have to be at least five metres high.

As we have noted, forests cover 30 per cent of the global land area. Of this, natural forest makes up 95 per cent and plantations 5 per cent. Half the global forest area is in Europe and South America, and only 5 per cent in Oceania (see Table A3-2, Appendix A).[36] Only 12 per cent of the world's forests are protected. The areas are in North and Central America (20 per cent of the total), South America (19 per cent), Africa and Oceania (12 per cent), Asia (9 per cent), and Europe (5 per cent).

The quality of forests can be assessed by their area, the volume of timber they contain (expressed in total cubic metres), the volume density (cubic metres per hectare), the total biomass of timber (billion tonnes), or the biomass density (billion tonnes per cubic metre). The largest area of land covered by forests and the largest volume of timber is in Europe. However, the biomass density in Europe is the lowest in the world. Australia has the smallest forest cover and the smallest volume density in the world, but it has a higher biomass density than Europe. The largest volume density, total biomass, and biomass density is in South America (see Table A3-3, Appendix A). The largest above-ground woody biomass is in Brazil, followed by Russia, the Democratic Republic of Congo, the US, and Canada.

Close to 70 per cent of all forests are in 10 countries: Russia, Brazil, Canada, the US, China, Australia, the Democratic Republic of Congo, Indonesia, Angola, and Peru.

About 47 per cent of all global forests are tropical, 33 per cent are boreal, 11 per cent are temperate, and 9 per cent are subtropical.

Forests prevent erosion, control runoff and sediments going into rivers, preserve biosphere stability, and regulate the hydrologic cycle. Hundreds of millions of people in the world rely on forests for food and medicine. For instance, it has been estimated that in India, Indonesia, Nepal, the Philippines, and Thailand more than 600 million people depend on forests for their existence. Forests are also an important repository of genetic resources.

Forests regulate climate and absorb substantial quantities of carbon from the atmosphere. Soils store an estimated 2011 billion tonnes [Gt] of carbon, and all terrestrial vegetation stores 466 Gt, of which 77 per cent is in forests. Tropical forests store 212 Gt. By destroying forests we are releasing carbon into the atmosphere and contributing to global warming.

How extensive is deforestation? The loss takes place mainly in tropical forests and is estimated at 13 to 15 million hectares per year.[37] FAO estimates a loss of 14.2 million hectares per year (see Table 3-4).[38] At this rate, by the end of this century 1400 million hectares of natural tropical forest will be destroyed, along with the species that live in that environment.

Table 3-4 How much global forest area are we losing each year? (million hectares)

Type	Natural forests			Forest plantations	Total net change
	Loss	Natural gain	Net change	Change	
Tropical	-15.2	+1.0	-14.2	+1.9	-12.3
Non-tropical	-0.9	+2.6	+1.7	+1.2	+2.9
World	-16.1	+3.6	-12.5	+3.1	-9.4

The negative values indicate a loss; the positive values indicate a gain.

Example: The global loss of natural forests is 16.1 million hectares per year, but the natural gain is 3.6 million hectares per year, which makes a net loss of 12.5 million hectares per year.

Source: FAO 2000b

The greatest loss of forests is in Africa and Latin America, about 70 per cent of the global loss.[39] The largest destruction of closed forests is in Latin America—40 million hectares between 1990 and

2000. In Asia, the loss was 20 million hectares and in Africa 12 million hectares.[40]

It is claimed that we have already destroyed about 50 per cent of the original global forest cover.[41] Only about 50 per cent remains in Australia, and 25 per cent in New Zealand.[42] In Madagascar, 80 per cent has been wiped out, more than 50 per cent of the total loss having occurred in the past 40 years.[43] The river basins of the Ganges, Rhine, Danube, Yangtze, Parana, Indus, Huang, Jiang, Magdalena, Missouri and Mississippi, Brahmaputra, Po, and Rhone have lost more than 50 per cent of their original forest cover.[44]

Only eight countries still have more than 70 per cent of their original forest cover. Six of them (Brazil, Colombia, French Guyana, Guyana, Suriname, and Venezuela) are in South America. The other two are Russia and Canada.[45]

The causes of deforestation include the conversion of forest into cropland and pasture, commercial logging, firewood harvesting, the construction of dams, mining, oil exploration, urbanisation, and industrial development.[46]

In the last 40 years of the 20th century, the global trade in timber products—such as wood pulp, paper and paperboard, plywood, and wood panels—increased five-fold. Expressed in 1998 US dollars, the trade grew from $29 million in 1961 to $150 million in 2000.[47]

The clearing of forests for agriculture is often done by burning. We are therefore not only destroying the forests, but also polluting the atmosphere. In the long run, we are doing even more damage because the cleared land is soon converted into wasteland, and to compensate for the loss of land we have to cut more trees.

The Brazilian National Space Research Institute reports that an area of Brazilian Amazon forest larger than France has been cleared, mainly for agriculture, and one-third of this area has been abandoned as useless. The average rate of clearing is 18,000 square kilometres [km^2] per year; but between July 1994 and June 1995, 29,000 km^2 was cleared, the greatest area ever recorded. The clearing activity seems to be increasing again. Between July 2001 and June 2002, 25,500 km^2 was cleared.[48]

Another factor claimed to be contributing substantially to global deforestation is fragmentation. Many forests are in small isolated areas and are interrupted by roads and mining operations. The survival time for such forests is short. Earlier estimates indicated that 40 per cent of continuous stretches of forest would be lost in 10 to 20 years. New estimates by the same team of scientists indicate that the loss will occur much sooner.[49]

The cost of progress. Our progress is unsustainable because it is accompanied by excessive damage to the environment. We are interested in profit based on short-term solutions. The price of progress does not include the hidden cost of environmental damage. What is the cost of progress? It is hard to put a value on nature, but some attempts have been made to assess how much it costs us when we overuse nature's resources and how much better off we could be if we managed them more prudently.

In 1997, a team of scientists published an estimate of the value of services provided by nature.[50] They analysed 17 types of service such as soil formation, environment restoration, waste absorption, nutrient recycling by plants and micro-organisms, water regulation and supply, climate regulation, flood and storm protection, atmospheric gas balance, and pollination. They found that the value of nature's services is between $18 trillion and $61 trillion per year,[51] with an average value of $31 trillion per year. This is comparable with the gross world product—that is, with our global income, which we are also generating by using natural resources.

The value of the free services we receive from nature is enormous. By working with nature we can win, but by working against it we shall lose. Sooner or later we shall have to meet the cost of progress; and by the time we have to pay for all the damage, the price may be too high. Even now, we have begun to pay.

The calculations of Costanza et al (1997) have been criticised by a number of other authors who questioned the accuracy of the estimates.[52] Lost in details, we might again be missing more important points. The point is that the value of nature's *free* resources is substantial. The point is that by indiscriminate and thoughtless interference with nature and overexploitation of resources we are hurting ourselves. The point is that our survival and future depend on how we treat nature. The point is that by learning how to live with nature we can increase our prosperity. Indeed a new and follow-up study has shown that we can increase substantially our global profit by investing in the protection of nature and its resources.[53]

The new study was conducted by a team of 19 researchers: economists, biologists, and ecologists. The aim was to find out whether we gain or lose when we convert the natural environment into various human-engineered systems such as agriculture and fisheries—the so-called improvements to the natural environment. They also considered the cost of thoughtless exploitation of natural resources.

The researchers assessed and compared the total economic value (TEV) of goods and services we receive when we live with nature (that is, with minimal interference) and when we subdue and exploit nature (that is, when we convert the natural environment into human-designed projects). They found that interference results in serious economic losses.

For instance, in Malaysia the private gain from unsustainable, high-intensity logging is 14 per cent lower than the estimated social and global gain from sustainable management of forests. Converting the tropical forests of Cameroon into oil palm and rubber plantations involves a loss of 18 per cent when compared with sustainable forestry. Converting mangroves in Thailand into aquaculture means a loss of 70 per cent compared with the social benefits and global gains from the sequestration of carbon (removal of carbon from the atmosphere). Converting Canada's marshes into farmland results in a loss of 60 per cent compared with gains from sustainable hunting, fishing, and trapping. The gain from unsustainable and forcible fishing (often by using explosives) in Philippines reefs means a loss of 75 per cent compared with the social benefits of sustainable reef management and the estimated profit from tourism.

The researchers estimated that globally we are losing $250 billion per year through conversion of natural environments into human-designed projects and through thoughtless exploitation of nature's resources. They also estimated that an annual investment of only $45 billion in the conservation of natural environments would result in profit of $4000 billion to $5200 billion per year. That is a return of one hundred times the investment. This would not only boost global profit (gross world product) by about 10 per cent, but would also help to develop a sustainable global economy. With proper management of natural resources and proper care for nature, we could have economic growth *and* a sustainable future.

Changes in the global grain-harvested area. The global grain-harvested area increased from 687 million hectares in 1950 to 732 million hectares in 1981, then decreased to reach 674 million hectares in 1999 (see Table 3-5). When measured by land area per person, the global grain-harvested area has been decreasing steadily over the recorded period. It decreased from 0.23 hectares per person [ha/p] in 1950, to 0.11 ha/p in the year 1999.[54] A simple mathematical analysis of the data shows that by 2025 the global average will be only 0.07 ha/p. This will represent a three-fold decline since 1950.

Table 3-5 Changes in the global grain-harvested area, 1950–1999

Year	Area [Mha]	Area [ha/p]
1950	587	0.23
1960	639	0.21
1970	663	0.18
1980	722	0.16
1990	694	0.13
1999	674	0.11

Mha – megahectare (million hectares)
ha/p – hectares per person

Source: Gardner 2000a

Between 1950 and 1980, the global grain-harvested area was increasing at an average rate of 4.7 million hectares per year. It has since been decreasing at 3.3 million hectares per year. If the loss continues at this rate, by 2025 we shall have only 588 million hectares of grain-harvested land area in the world, or the estimated 0.07 ha/p.

Global distribution of the grain-production area is uneven. Using the estimated values for various regions,[55] we can calculate that developed countries have an average of 0.19 ha/p of grain-production area and developing countries have only 0.09 ha/p, which is about the average in India and China. Countries such as these survive because they buy food from other countries. In 1995, developing countries were buying 105 million tonnes of cereals per year from developed countries. China was buying 17 million tonnes per year. By 2025, developing countries will have to import 350 million tonnes of cereals per year. Asia will have to import 150 million tonnes per year, China 50 million tonnes, and India 10 million tonnes.[56] Will there be enough grain to feed the world in the future?

The future production of grain. Even though the global grain-harvested area is decreasing, it does not necessarily mean that less grain will be produced in the future. The combined grain area might be smaller, but if productivity is higher we should still be able to feed the world. Let us analyse this side of the story.

If we look at the yield of grain from one hectare of land, we shall find that productivity has been steadily increasing. The trend is very impressive. The global yield had increased from 1.2 tonnes per hectare in 1950 to three tonnes per hectare by the end of the last century.[57] If the trend continues, by 2025 we should be able to increase the yield to four tonnes per hectare. How to explain this steady increase?

Between 1960 and 2000, global food production more than doubled (see Box 3-5). This was accompanied by only a small increase in cultivated land area, but there was a substantial increase in irrigation and an enormous increase in fertiliser use.[58] The yield was also boosted by intensified use of agricultural chemicals to protect crops from insects and fungi.

Box 3-5
Changes in agriculture, 1960–2000

Between 1960 and 2000, global production of cereals, coarse grain, and root crops increased by 124 per cent. This was supported partly by an increase in the area of cultivated land, but more importantly by an increase in irrigated land areas and by a huge increase in the use of fertilisers.

	Increase Per cent
Global food production	124
Global cultivated land area	11
Global irrigated land area	74
Global use of phosphorous fertilisers	265
Global use of nitrogen fertilisers	730

Increase – the increase (per cent) between 1960 and 2000

Source: Based on the data of Tilman 1999

Irrigation helps to increase the productivity of land, but it also creates problems. Unskilful irrigation speeds up the process of salinisation and waterlogging. The data for the 1980s show that 24 per cent of irrigated land worldwide has been damaged by irrigation (see Table 3-6). The worst degradation is in India. Much of the damage is in the form of salt deposits (see Table 3-7). About half of the estimated cultivated land area damaged by salt in 1987 was in just two countries:

India and China. Irrigation is also associated with large-scale waste of water. Worldwide, 55 per cent of water used for irrigation is wasted.[59]

Table 3-6 Land damaged by irrigation in the 1980s

Country	Damage	Per cent
India	20.0	36
China	7.0	15
USA	5.2	27
Pakistan	3.2	20
FSU	2.5	12
World total	**60.2**	**24**

Damage – Land area damaged by irrigation (in million hectares)
Per cent – Percentage of irrigated land damaged by irrigation
FSU – Former Soviet Union

Source: Harrison and Pearce 2001

Table 3-7 Irrigated land damaged by salt, 1987

Country	Damage
India	7.0
China	6.7
USA	4.3
Pakistan	4.3
Uzbekistan	2.5
Iran	1.7
Turkmenistan	1.0
Egypt	0.9
Total	**28.3**

Damage – Irrigated land area damaged by salt (in million hectares)
Source: UNDP et al 2000b

Forcible farming methods have been increasing food production, but they have also been causing damage and losses. Our farming methods are unsustainable, and we may be reaching the limits of their applicability.

It is possible that the cereal yield in 2025 will be lower than four tonnes per hectare. Indeed, worldwide, the annual growth rate in cereal yields is coming down, and is expected to drop from 2.3 per cent in the 1980s to 1 per cent in the 2020s.[60] However, if we assume that the yield will rise to four tonnes per hectare, in 2025 we shall be able to produce

294 kilograms of grain per person per year [kg/p/y] from the available global grain-harvested area of 588 million hectares. How does that compare with the current global production of grain?

Global grain production per person reached a maximum of 380 kg/p/y in 1983. By the end of the last century it had dropped to 350 kg/p/y.[61] So, even if we assume that the yield per hectare will continue to rise, the best we can expect is that average production in 2025 will be 16 per cent lower than in 2000.

The increased yield will not help us to boost the availability of grain per person. On the contrary, the availability per person in 2025 will be lower than it is now. This does not augur well for feeding the world. If millions of people are dying of hunger now, we can hardly expect a smaller number to be dying in the future.

A simpler and more straightforward way of projecting future grain production is by using the global data.[62] Since 1983, global production of grain per person has been decreasing at the rate of 2 kg/p/y. A linear extrapolation gives 296 kg/p/y for 2025, which is in excellent agreement with the previously calculated value of 294 kg/p/y.

We need one tonne of grain to feed three people for a year—that is 333 kg/p/y,[63] which is close to the expected availability of grain in 2025. We might therefore have just enough grain in the world in 2025 to feed everybody. However, production is not distributed evenly: countries with grain sell only what they cannot use, and poor countries will not be able to buy as much as they need. So just enough grain is not good enough.

Food production. The production of grain is important, but we also have other sources of food. The global production of food, including grain, is still increasing. Between 1960 and 2000 the global production of cereals, coarse grain, and root crops more than doubled.[64] The production of meat nearly trebled (see Table 3-8). The total number of livestock rose from 7.3 billion in 1961 to 20.6 billion in 2000. Chicken constitutes a popular diet. The global number of ducks produced for consumption in 2000 was 886 million, and the number of geese was 235 million. However, the number of chickens was 14.3 billion.[65]

As for future food production, the situation is uncertain. For instance, it has been estimated that by 2020 land degradation will reduce food production by 15–30 per cent.[66] In South Asia, 50 per cent of land has been so degraded that it has lost its agricultural potential. In China, 27 per cent of land is affected by degradation. China loses 250,000 hectares each year through degradation.[67]

Table 3-8 Increase in meat production, 1961–2000

Livestock	Increase	Livestock	Increase
Buffalo	90	Goats	105
Cattle	41	Sheep	7
Chicken	267	Pigs	123
Ducks	357	Rabbits	370
Geese	553	Other	27
Turkeys	83	Total	180

Increase – Percentage increase between 1961 and 2000
Source: Halweil 2001

It has been also pointed out that, in 64 of 105 developing countries, the increase in the population has not been matched by the increase in food production.[68] It is possible that the signs of future decline in global food production can be seen in Africa, where food production has been decreasing for the past four years.[69]

Projections of food security for 66 countries in Asia, Africa, and Latin America, calculated by the US Department of Agriculture, show that the gap between the nutritional needs and food supply in these countries will widen. The worst region in the world is sub-Saharan Africa.[70]

Other methods of food production, such as hydroponics, may not be the answer to future food-related problems. It has been claimed that hydroponic production is 10 times more expensive than conventional methods.[71]

An estimated 829 million people are now forced to live with hunger, and 96 per cent of them live in developing countries. In the Asia-Pacific region, 526 million people live with hunger; in Africa, 209 million; and in Latin America–Caribbean, 57 million. In Africa, the number of undernourished people doubled in the last 40 years of the 20th century.[72]

Even if the richer countries did share their food we would not be able to eliminate hunger unless the food was free or sold at a low price, because the buying power of a large proportion of the global population is low.[73]

Food quality. To fight diseases and boost the production of food, farmers feed livestock with generous doses of hormones and antibiotics.[74] For instance, it is estimated that 70 per cent of antibiotics produced in the US are used for feeding farm animals.[75] The WHO claims that antibiotics are also used in fish farming and horticulture to promote growth and fight harmful organisms.[76]

It has been reported that in the US the consumption of antibiotics by poultry is three times greater than human consumption. Many of the antibiotics are similar to those used by humans. Poultry is affected by bacteria such as salmonella, enterococci, and campylobacter, and some strains have been found to be resistant to antibiotics used by humans.[77]

The global increase in food production comes at a price—environmental damage through the pollution of land and water, the overuse of water, loss of natural vegetation, loss of biodiversity, deforestation, and land degradation. Crop farming and animal facilities produce large volumes of harmful waste.

The overuse of pesticides. The trade in pesticides is increasing. Worldwide, exports increased nearly nine-fold between 1961 and 1998, from $1.3 billion in 1961 to $11.4 billion in 1998. The trade includes many banned, suspended, discontinued, and severely restricted pesticides. Sales rose from 31 thousand tonnes per year in 1992 to 39 thousand tonnes per year in 1994. Globally, the use of pesticides in agriculture increased from 0.4 kilograms per hectare in 1960 to two kilograms per hectare in 1999.[78]

The total use of pesticides rose from 50 million kilograms per year in 1945 to 2700 million kilograms in 1999. In addition, pesticide toxicity is now 10 times higher.[79] The number of insect and mite species that are resistant to pesticides increased from nearly zero in the early 1940s to 540 in 1999. The number of crop diseases resistant to pesticides increased from nearly zero in the early 1960s to 240 in 1999. The number of weed species resistant to pesticides has also increased.[80]

Pesticides in food, water, and air create health problems. Acute effects of pesticide poisoning include headaches, allergies, dizziness, flu-like symptoms, skin rashes, blurred vision, stillbirths, miscarriages, respiratory problems, nausea, inability to walk, and other neurological disorders.[81] Other effects include birth defects, lymphoma, leukaemia, and lung, pancreatic, and breast cancer.[82] Pesticides have also been shown to act as endocrine disruptors.[83]

We have sufficiently strong evidence that Parkinson's disease in people under 50 is caused by genetic factors. However, the origin of Parkinson's disease for people over 65 is less clear. A study carried out at Emory University in Atlanta, Georgia, suggests that pesticides are responsible for late-onset Parkinson's disease—pesticides used around the house or in agriculture. It also points out that even natural pesticides are not safe.[84]

The overuse of fertilisers. Worldwide, the use of fertilisers increased from 14 million tonnes per year in 1950 to 141 million tonnes per year in 2000. Much of this increase was in just two countries: China and India (from nearly zero in both countries in 1960 to 37 million tonnes per year in China and 18 million tonnes per year in India). Global use of fertilisers per person increased from 5.5 kilograms in 1950 to 23.2 kilograms in 2000.[85]

The use of fertilisers in developed countries declined to 80 kilograms per hectare [kg/ha] in 1998 from a maximum of 124 kg/ha that was reached in 1988. However, the use of fertilisers is increasing in developing countries. It rose from 8 kg/ha in 1963 to 76 kg/ha in 1988 and to 100 kg/ha in 1998.[86]

Until about 1991, the total use of fertilisers in developing countries was lower than in developed countries. At that time, the use of fertilisers in each of these groups had reached 70 million tonnes per year. By 1996, it had increased to 84 million tonnes in developing countries, but had decreased to 54 million tonnes in developed countries.[87]

The largest share in the use of fertilisers (measured in the percentage of global use) is in Asia, and the highest intensity (measured by kilograms per hectare) is in East Asia. However, if we consider the use of fertilisers in individual countries, the highest intensity in 1998 was in Switzerland, followed by Ireland, and the Netherlands (see Tables A3-4, A3-5, and A3-6 in the Appendix A).[88]

The overuse of fertilisers contributes to the nutrient pollution (eutrophication) of aquatic ecosystems. The resulting overproduction of organic matter, such as algae, reduces dissolved oxygen and causes the suffocation (hypoxia) of aquatic animal life. In addition, the growing algae produce toxins.

Eutrophication occurs in lakes, rivers, and dams near farmlands and in many coastal areas near the outlets of rivers. Marine eutrophication is a serious problem in many coastal regions in North America, Europe, and Japan. It also occurs in Australia and New Zealand.[89]

The future. Global land resources per person, including arable land, will continue to shrink, mainly because global population is increasing, but also because of continuing land degradation. The availability of land per person will continue to decrease much faster in developing countries than in industrialised countries because the population in developing countries is increasing much faster.

Deforestation and species loss are also expected to continue. Diminishing land resources are likely eventually to limit global food production.

About 75.5 per cent of global food production is through agriculture.[90] Between 1960 and 2000, the production of cereals, coarse grain, and root crops rose by 124 per cent. This increase was supported only slightly by an increase in the cultivation area, but more substantially by an increase in irrigation. However, the main support came from a huge increase in the use of agricultural chemicals.[91] Meat production has been supported by excessive use of growth hormones and antibiotics.[92] Can we continue increasing production by these methods? The answer seems to be in the negative—it has been pointed out that current farming methods are unsustainable.[93]

Climate change is also expected to have a big effect on the degradation of land and on the production of food in certain region. In Africa, grain yields are projected to decrease and desertification to increase. In Asia, agricultural productivity will decrease. In Australia and New Zealand, some changes might be initially beneficial in certain areas, but they are expected to be offset by negative changes later. In Europe, productivity is expected to decrease in southern and eastern areas. In northern Europe, productivity may be boosted by the increased concentration of greenhouse gases in the atmosphere. In Latin America, the yield of many important crops is projected to decrease, but it will probably increase in North America. On small islands, increased soil erosion will have strong negative effect on food production.[94]

Global production of food is still high, but a large percentage of the population suffers hunger because food is not available where it is needed or because people cannot afford to buy it. Attempts are being made to tackle the problem of hunger, but diminishing land resources are working against us.

It is possible that we could increase global food production. We have teams of scientists who could help us in this task, but they need generous financial support. This should come from public coffers and not from private corporations that would then control research and the distribution of food. Poor countries should be helped to grow their own food. Flooding them with products they do not need or want, and controlling what they can produce and sell, only increases their dependence on the outside world and intensifies their deprivation.

We could improve irrigation systems. We could reduce the wasting of water and salinisation, increase irrigated areas, and thus increase the production of food. We could reduce the use of agricultural chemicals and keep the soil healthy and fertile. We could reduce the disturbance of the soil in farming. We could look for and develop plants suitable for specific types of soil and climate—plants that would be more resistant to drought, disease, and pests.

In principle, our planet has the potential for a substantial increase of food production. It is claimed that we are using only 120 cultivated plant species to produce 90 per cent of global food.[95] Before we destroy the remaining reserves of plant species we could look for new sources of food and develop them. We still have extensive genetic resources for improving food production.

With international collaboration we could reduce the number of people living with hunger from the present 829 million to 200 million in 30 years,[96] and we could eradicate hunger in 130 years.[97] However, with the more likely business-as-usual approach, the number of people suffering chronic hunger will increase to a billion by 2030.[98]

We need land to support all forms of consumption. Global consumption is already larger than global ecological capacity, and it continues to increase. A global economy based on an increasing global ecological deficit will result in global crisis.

The drive to increase the consumption of energy and material resources in developing countries will continue. This will be reflected in their need to increase their ecological footprints. However, because of the physical limitations of the planet, and because the ecological footprints of industrialised countries are already large, the growing footprints in developing countries will not be balanced by the availability of the planet's biologically productive surface area. This is like making a growing child wear the same shoes all the time. Eventually something has to give—the foot becomes distorted beyond recognition or the shoe bursts. The increasing ecological deficit also increases the probability of global economic collapse.

Chapter 4
Diminishing Water Resources

'Water may seem to be everywhere, but for a rising portion of the world's population, there may soon be hardly a drop to drink—or to use for growing food, supporting industries and cities, and preserving life-giving ecosystems.'

—Sandra Postel, contemporary US scientist, *Dividing the Water*

Global withdrawals of fresh water are increasing. Do we have enough water to satisfy our growing demands? The planet is rich in water, but only a small percentage is suitable for human consumption. For the first time in human history we are facing a global water crisis. Many regions of the world are experiencing severe shortages, and the situation is getting worse.

How much water do we have on our planet?[1] The total volume of water on the Earth is estimated at 1386 million cubic kilometres [km^3], of which only 35.029 million km^3, or 2.53 per cent, is fresh water. An estimated 68.7 per cent (24.1 million km^3) of fresh water is in glaciers and permanent snow cover (see Table 4-1). Relatively large quantities of water are in underground aquifers. Water in rivers and lakes makes up only 0.266 per cent of global fresh water.

The circulation time of water in rivers is about 16 days; in lakes, 17 years. These reservoirs are therefore called renewable water resources. In contrast, the recharge time of groundwater reservoirs is about 1500 years.[2] They are therefore defined as non-renewable water resources. Water in glaciers and permanent snow cover is also non-renewable.

The annual global recharge rate of groundwater is unknown, but it has been estimated at only 2000 km^3/y.[3] Thus only a minute part of the total volume of groundwater is involved in the annual circulation of fresh water in these reservoirs. We can safely use this small amount of water, but we should not be using more than that. Groundwater aquifers are virtually static, permanent, fossil water reservoirs. Once emptied or polluted, they stay empty or polluted for a long time.

Table 4-1 Global freshwater resources

Location	Volume	Per cent
Glaciers and permanent snow cover	24 064	68.70
Groundwater	10 530	30.06
Ground ice and permafrost	300	0.86
Freshwater lakes	91	0.26
Soil moisture	16.5	0.05
Atmosphere	12.9	0.04
Marshes and wetlands	11.5	0.03
Rivers	2.12	0.006
Biota	1.12	0.003
Total	**35 029.14**	**100.00**

Volume – The volume of freshwater resources in thousands of cubic kilometres

Per cent – The percentage of global reserves of fresh water

Example: Close to 69 per cent (24,064 cubic kilometres) of global fresh water is stored in glaciers and permanent snow cover, but freshwater lakes contain only 0.26 per cent (91 cubic kilometres).

Source: Engelman and LeRoy 1993; UNEP 2002

Box 4-1
How much water is involved in annual global circulation?

Various authors give different estimates of annual global water circulation. Here is one of them, which seems to represent the most commonly accepted values. All the values are in cubic kilometres per year.

Global precipitation over oceans	458 000
Global precipitation over land	119 000
Total global precipitation	577 000
Global evaporation from oceans	502 800
Global evaporation from land	74 200
Total global evaporation	577 000
	Runoff water
Precipitation over land	119 000
Less evaporation from land	74 200
Less groundwater discharge	2 000
Total runoff water	42 800

Source: Shiklomanov and Rodda 2003. See also Gleick 1993

Runoff water.[4] The global volume of water involved in annual circulation is 577,000 km^3 (see Box 4-1). Total precipitation over land is 119,000 km^3, and total evaporation is 74,200 km^3. Runoff water (also called river runoff) is defined as the difference between precipitation over land and evaporation from land, less the groundwater discharge. The rates of global evaporation and precipitation vary from year to year. Consequently, global and local volumes of runoff water are not constant. Between 1920 and 1995, the volume of global runoff varied between 39,800 km^3/y and 44,800 km^3/y. The average value over this period was 42,800 km^3/y.

How much water is accessible?[5] We can use global runoff to measure global water availability or we can use local runoff to measure local water availability. However, we must understand that only a small part of runoff water is usable.

It is estimated that only 81 per cent of global runoff (34,700 km^3/y) flows in geographically accessible places. If we subtract the estimated global volume of water lost in floods and rivers, we find that only 9000 km^3/y of the so-called available runoff is usable. We can boost the availability of water by building dams. The estimated global volume of water recovered this way is 3500 km^3/y. Thus, the global volume of runoff that is usable can be put at 12,500 km^3/y. To distinguish it from total runoff we can call it the accessible runoff water.

The volume of global accessible runoff is claimed to be between 9000 km^3/y and 14,000 km^3/y;[6] however, the widely accepted value is 12,500 km^3/y. We can see therefore that accessible runoff is only 30 per cent of the global runoff. This percentage can be also applied to local runoffs.

Accessible runoff water should not be used exclusively for human consumption. Much of it should be reserved to support natural ecosystems.[7]

Various scales have been designed to measure water availability. They are described in Appendix B. In this book I use the simple and straightforward scale used by Shiklomanov (2000), which is based on dividing the volume of runoff by the number of people. In this scale, levels of water availability are defined as very high, high, adequate, low, very low, and catastrophically low.

Global and regional availability of water.[8] Currently, the available runoff water is about 6900 cubic metres per person per year [m^3/p/y], which in the scale used by Shiklomanov corresponds to an adequate level. If we could share fresh water evenly we would have just enough water for everybody. Unfortunately, water distribution does not match population distribution, and this deepens the water crisis (see Table 4-2). The uneven distribution of water contributes to its growing short-age in various parts of the world.

Table 4-2 Fresh water availability by region and globally

Region	Water availability			Population	Level
	Total [km³/y]	Per cent of global	Per person [m³/p/y]	Per cent of global	
Oceania	2 400	5.6	75 000	0.5	Very high
South America	12 030	28.1	33 983	5.7	Very high
North America	7 890	18.4	15 907	8.0	High
Africa	4 050	9.5	4 821	13.5	Low
Europe	2 900	6.8	3 984	11.7	Low
Asia	13 510	31.6	3 587	60.6	Low
World	**42 780**	**100.0**	**6 882**	**100.0**	**Adequate**

Regions are arranged in descending order of water availability per person per year; North America includes Central America and the Caribbean

Total – Runoff water in a given region in cubic kilometres per year [km³/y]
Per cent of global – Percentage of global runoff water
Per person – Availability of water per person in 2002, in cubic metres per person per year [m³/p/y]
Population – Percentage of global population in 2002 in a given region
Level – Level of water availability (see Appendix B, scale of water availability as used by Shiklomanov)

Example: The average volume of runoff water available in Europe is 2900 cubic kilometres per year, which in 2002 corresponded to 3984 cubic metres per person per year. In 2002, Europe had 11.7 per cent of the global population, but only 6.8 per cent of global fresh water. The average level of water availability in Europe in 2002 was low.

Source: The data on the availability of water in cubic kilometres per year (the first numerical column) are from Shiklomanov 2000. The availability of water per person was calculated using regional populations in 2002 as listed in PRB 2002.

The highest availability of runoff water is in Asia and South America. Each of these regions has roughly 30 per cent of global runoff. However, whereas the population of Asia is high, the population of South America is relatively low. Consequently, water availability per person in Asia is low, but in South America is very high.

Water availability is also low in Africa and Europe. However, the population in Europe is decreasing, which could help to improve availability—as long as we ignore water pollution. The situation is different in Asia and Africa where the populations are large and rapidly increasing. By 2025, many people in these densely populated regions will suffer a catastrophically low level of water availability. Currently, for 75 per cent of global population water availability is less than adequate, and for 35 per cent it is either very low or catastrophically low.

How much water are we using?[9] Global withdrawal of fresh water increased from 580 cubic kilometres per year [km^3/y] in 1900 to 3800 km^3/y in 1995. At the end of 2000, global withdrawal was 4000 km^3/y (see Table 4-3).

Table 4-3 Water withdrawals by region and globally, 1900–2050 (cubic kilometres per year)

Region	1900	1940	1960	1980	2000	2010	2025	2050
Europe	38	96	226	449	463	535	559	599
North America	70	221	410	676	705	744	786	856
Africa	41	49	89	166	235	275	337	440
Asia	414	682	1 163	1 742	2 357	2 628	3 254	4 297
South America	15	33	66	117	182	213	260	338
Oceania	2	7	15	24	33	36	40	47
World	**579**	**1 088**	**1 968**	**3 175**	**3 973**	**4 431**	**5 235**	**6 575**

North America – Includes Central America and Caribbean

Example: Water withdrawal in Europe in 1900 was only 38 cubic kilometres per year [km^3/y], but increased to 463 km^3/y in 2000.

Source: Withdrawals for the years 1900–2025 are from Shiklomanov 2000. Withdrawals for 2050 are based on my linear extrapolations.

It is estimated that by 2025 the global withdrawal of fresh water will reach 5235 km^3/y. Extending the trend further, we can make a tentative prediction that by 2050 the figure will be 6580 km^3/y. By that year the global withdrawal of fresh water will be about 11 times greater than in 1900.

Until about 1940, the withdrawal was increasing at the rate of 13 km^3/y, but in the following 60 years it was increasing at 50 km^3/y. The highest consumption of fresh water and the largest increase in consumption are in Asia.

Global withdrawal of fresh water per person was 380 cubic metres per year [$m^3/p/y$] in 1900, but it increased to 650 $m^3/p/y$ in 2000. In general, water withdrawal per person depends on the level of prosperity. On average, water withdrawal in OECD countries is 1000 $m^3/p/y$ and in developing countries 500 $m^3/p/y$.[10]

Table 4-4 Water withdrawals per person by region, 2000

Region	Withdrawal [$m^3/p/y$]
Europe	635
North America	1 453
Africa	294
Asia	642
South America	530
Oceania	1 052
World	**656**

North America – Includes Central America and Caribbean
$m^3/p/y$ – cubic metre per person per year
Source: Calculated using the data of Shiklomanov 2000

The highest water withdrawal per person is in North America and the lowest in Africa (see Table 4-4). In 2000, water withdrawal per person in North America was close to 1500 cubic metres per year, and in Africa 300 cubic metres per year. The global average was close to 660 cubic metres per person per year. The projected regional figures are not expected to be much different in 2025.

Table 4-5 Global water withdrawals by sector, 1900–2025

Sector	1900 Volume [km^3/y]	1900 Per cent	1950 Volume [km^3/y]	1950 Per cent	2000 Volume [km^3/y]	2000 Per cent	2025 Volume [km^3/y]	2025 Per cent
Agriculture	513	89	1 080	79	2 605	69	3 189	64
Industrial	44	8	204	15	776	21	1 170	24
Municipal	22	4	87	6	384	10	607	12
Total for sectors	**579**		**1 371**		**3 765**		**4 966**	
Reservoir	0		11		208		269	
Total	**579**		**1 382**		**3 973**		**5 235**	

km^3/y – cubic kilometre per year
Per cent – Percentage of water withdrawn for a given sector of economic activity
Example: Global water withdrawal for agriculture in 1900 was 513 km^3/y, or 89 per cent of total global withdrawals in that year.
Source: Based on data of Shiklomanov 2000

Water withdrawal by various sectors of economic activities.[11] In 2000, 69 per cent of global water withdrawal was for agriculture, 21 per cent for industry, and 10 per cent for municipal applications (see Table 4-5). We can therefore see that most water withdrawal is for the production of food.

' Water withdrawal by all three sectors is increasing. However, the share of water between the sectors does not remain constant. In 1900, 89 per cent of water withdrawal was for agriculture. The projected figure for 2025 is 64 per cent. Water withdrawal for industry was only 8 per cent, but is projected to be 24 per cent in 2025.

The percentage of water withdrawal for agriculture, industry, and municipal applications varies between countries, but on average there is a clear dependence on the level of prosperity (see Box 4-2).[12]

Box 4-2
Water withdrawal by sector as an indicator of economic status

The relative share of water withdrawals by the three sectors of economic activity depends on a country's economic status. The higher the income of a country, the lower the share of water taken by agriculture; conversely, the share of water by industry and municipalities tends to be greater in higher-income countries. However, there are also exceptions to these general trends.

Group	Agriculture (per cent)	Industry (per cent)	Municipal (per cent)
High-income countries	40	45	15
Upper-middle-income countries	65	19	15
Lower-middle-income countries	75	17	8
Low-income countries	93	4	3

Source: WB 1999a

Examples of exceptions to the general trends

Country	Agriculture (per cent)	Industry (per cent)	Municipal (per cent)
Australia	70	12	18
Japan	64	17	19
Gabon	6	22	72
Lithuania	3	15	82

Source: Calculated using the data from ABS 2000 and Harrison and Pearce 2001

Agriculture. In general, the percentage of water withdrawal for agriculture increases with a decreasing level of prosperity. The lower the income of a country the higher the percentage of water withdrawal for agriculture. On average, water withdrawal for agriculture varies between 40 per cent in high-income countries and 90 per cent in low-income countries. In order to save water, developing countries should be helped to improve their irrigation systems.

Industry. The lower the income of a country, the lower the percentage of water withdrawal for industry. In high-income countries, 45 per cent of water withdrawal is for industry; in low-income countries, it is only 4 per cent.

Municipal applications. The higher the level of prosperity, the larger the percentage of water withdrawal for municipal applications. In high-income and upper-middle-income countries, water withdrawal for municipal applications is 15 per cent; in low-income countries, it is 3 per cent.

However, there are exceptions to these general trends, and one of them is Australia, which belongs among the high-income countries. Its water withdrawal for agriculture is 70 per cent and for industry 12 per cent.[13]

How much water do we consume at home? The minimum amount of water we need to have to survive is estimated at 1.8 to 5 litres per person per day [l/p/d]. The basic requirement for all domestic needs (drinking, bathing, cooking, and sanitation) is only 50 l/p/d (see Table 4-6).[14] However, domestic water consumption is often much higher and it varies over a wide range of values for various countries. In many countries, domestic consumption of water is lower than 50 l/p/d. Gleick (1996) lists 55 such countries, with the lowest consumption in Gambia (4.5 l/p/d) and the highest in Zimbabwe (48.2 l/p/d).

Table 4-6 Basic water requirements (litres per person per day)

Drinking water	5
Bathing	15
Kitchen	10
Sanitation	20
Total	**50**

Source: Gleick 1996

THE LITTLE GREEN HANDBOOK

The average global domestic consumption of water is 83 l/p/d, with an average of 58 l/p/d in developing countries and 130 l/p/d in developed countries (see Box 4-3). The average domestic consumption in sub-Saharan Africa is 50 l/p/d.[15]

Box 4-3
Domestic water consumption

Average domestic water consumption in developing countries is close to the basic requirement. The average in developed countries is almost three times higher than the basic requirement.

Average in developing countries	58	l/p/d
Average in developed countries	130	l/p/d
World average	83	l/p/d
Basic requirement	50	l/p/d

l/p/d – litre per person per day

Source: The average values are from Rosegrant et al 2002. The basic requirement is from Gleick 1996.

Examples of domestic water consumption in OECD countries

People living in rich countries consume large quantities of water in bathrooms and toilets. The total domestic consumption in these countries is much higher than the basic required minimum of 50 litres per person per day.

	US [l/p/d]	(1)	Australia [l/p/d]	(1)	Sweden [l/p/d]	(1)	UK [l/p/d]	(1)
Bathroom	103	27	69	26	60	30	38	29
Toilet	91	24	53	20	40	20	40	31
Laundry	65	17	32	12	36	18	26	20
Kitchen	91	24	16	6	22	11	21	16
Other use	30	8	95	36	42	21	5	4
Total	**380**	**100**	**265**	**100**	**200**	**100**	**130**	**100**

l/p/d – Water consumption in litres per person per day
(1) – The percentage of total domestic consumption
Other use – Includes gardening, washing cars, and cleaning

Example: In the US, average water consumption in bathrooms is 103 litres per person per day, which corresponds to 27 per cent of the average domestic water consumption in that country.

Source: Calculated using data from ABS 2000, Clement et al 1997, Niemczynowicz 2000, POST 2000, and WSAA 2001

Domestic water consumption depends on the degree of access. People who have to carry water from distant places or from low-quality sources use 4–20 l/p/d of water for domestic purposes. People who carry water from easily accessible places or from plentiful sources use 10–40 l/p/d. People who have easy access to a single tap use 16–90 l/p/d, and those with multiple taps in their houses use 25–600 l/p/d.[16]

Examples of domestic water consumption in OECD countries are given in Box 4-3. Large quantities of drinking water in developed countries are not used for cooking or drinking, but for washing, showering, flushing toilets, washing cars, house cleaning, and for watering private gardens. Large volumes of water could be saved in these areas. Information on how to save water is widely available, but common sense is also a good guide. Rules and recommendations will not work unless people are willing to participate.

Virtual water. Virtual water is defined as that needed to produce goods or to support services imported from another countries.[17] For instance, the production of food requires the extensive use of water, therefore buying food from another country is like buying the water used to produce that food. Many countries depend on the supply of virtual water, including Kuwait, United Arab Emirates, Libya, Saudi Arabia, Jordan, Singapore, Yemen, and Israel.[18]

The dependence on virtual water in many countries is increasing. For instance, imports of wheat and cereals in the Middle East increased from 7.5 million tonnes per year in the 1960s to 40 million tonnes per year in the 1990s. Imports of food in Israel and Palestine increased nearly six-fold between 1960 and 1990.[19]

From about 1960, the Middle East and North Africa have had an increasing water deficit, hence an increasing dependence on virtual water. Between 1995 and 2050, the dependence on virtual water in these regions is projected to rise from 150 cubic kilometres per year [km^3/y] to 400 km^3/y.[20]

Boosting the availability of water by damming.[21] If large volumes of runoff water are lost in floods and overflows, why do we not build more dams to catch more water? We do, but this is a far from perfect solution. A much better one is to improve water management and thus reduce waste.

Many of the world's dams are so-called large dams, defined as holding at least three million cubic metres of water. Before 1900, there were fewer than a thousand large dams, but by the end of the 20th century there were 47,455 (see Table 4-7). Close to 50 per cent of all large dams in the world are in one country—China. The US has about 14 per cent. Three-quarters of the large dams are in only four countries: China, the US, India, and Japan.

Table 4-7 Countries with the greatest number of large dams, 2000

Country	Number of dams	Per cent of global	Cumulative per cent
China	22 000	46.4	46.4
USA	6 575	13.9	60.2
India	4 291	9.0	69.3
Japan	2 675	5.6	74.9
Spain	1 196	2.5	77.4
Canada	783	1.7	79.1
South Korea	765	1.6	80.7
Turkey	625	1.3	82.0
Brazil	594	1.3	83.3
France	569	1.2	84.5
Others	7 372	15.5	100.0
World	**47 455**	**100.0**	

Countries are arranged in descending order of the number of large dams

Example: The US has 6575 large dams, or 13.9 per cent of the global number of large dams. About 60 per cent of global large dams are in China and the US.

Source: Based on data of WCD 2000

About 40,000 large dams were constructed in the second half of the last century. On average, we were adding 800 large dams per year during that period. Construction peaked between 1970 and 1975, when we were adding an average of 1000 large dams each year.

Worldwide, between 160 and 320 large dams are being added each year. They are being constructed mainly in India, China, Turkey, South Korea, Japan, and Iran.

It has been estimated that at the end of the last century we had 800,000 dams of all sizes around the world. China has 60,000 medium and small dams, and the US 70,000.[22]

Dams play an important part in the economic activities of more than 140 countries. They are used for irrigation, flood control, water supply, and electricity generation (see Table 4-8). In Africa and Asia, dams are used mainly for irrigation. In Australasia, they are used mainly for water supply. In South Africa, South Korea, and Malaysia, close to 60 per cent of irrigated land is served by water from dams, in Uzbekistan close to 80 per cent, and in Egypt over 90 per cent. Dams serve as important reservoirs for municipal and industrial water consumption. In Europe, a large percentage of dams are used for electricity generation. Worldwide, about 20 per cent of electricity is generated using water from dams.

Table 4-8 Regional applications of dams (per cent)

Purpose	Africa	North America	South America	Asia	Austral-asia	Europe
Flood control	1	13	18	2	2	3
Irrigation	50	11	15	63	13	19
Water supply	20	10	13	2	49	17
Hydropower	6	11	24	7	20	33
Multipurpose	21	40	26	26	14	25
Other	2	15	4	0	2	3

South America – Includes Central America and Caribbean

Australasia – Australia, Fiji, Indonesia, New Zealand, and Papua New Guinea

Example: About 50 per cent of dams in Africa are used for irrigation, but in North America only 11 per cent.

Source: WCD 2000

Problems caused by dams.[23] Important and useful as they may be, dams are ecologically, socially, and politically problematic. We may be gaining in some areas, such as partially solving water and energy problems, but we are losing in others.

Large dams rob us of much-needed land. They result in a loss of forests, pastures, and fertile land. They degrade aquatic biodiversity and bring about the loss or displacement of terrestrial species. Dams block the movement of fish and other aquatic species, and cause the degradation of downstream and upstream fisheries. It is estimated that 20 per cent of freshwater fish species have been wiped out or seriously endangered by dams.

It is estimated that 40–50 million people have been displaced by the building of large dams. Upstream, they are forced out of their land to make room for catchment areas. Downstream, they suffer hardship because the rivers do not have enough fish. The number of ecological refugees in the world is rising. This increase is causing not only problems within countries, but also between countries. Large dams account for more than 60 per cent of all ecological refugees.

A total of 261 rivers cross political boundaries. These international rivers account for 80 per cent of global river flow, serve 45 per cent of the earth's surface, and affect people in 145 countries.[24] Dam building aggravates the problem of the international sharing of water.

Dams of all sizes contribute to global warming by emitting greenhouse gases from the rotting of flooded vegetation and from carbon flowing from catchment areas. The intensity of emission varies between dams. It may be as low as the emission from natural lakes or as high as that from thermal energy plants.

Table 4-9 Withdrawals of groundwater in the 1980s

Country	Groundwater withdrawals [km³/y] National	Cumulative
India	150	150
USA	101	251
China	75	326
FSU	45	371
Pakistan	45	416
Iran	29	445
Mexico	28	473
Japan	13	486
Italy	12	498
Bangladesh	7.7	506
Indonesia	3.7	509
Philippines	3.0	512
Australia	2.8	515
Thailand	0.9	516

Groundwater withdrawals [km³/y] – Withdrawals of groundwater in cubic kilometres per year

National – Withdrawals of groundwater per year in a given country

Cumulative – Cumulative (combined) withdrawals of groundwater per year

km³/y – cubic kilometre per year

Example: The withdrawal of groundwater in India in the 1980s was 150 cubic kilometres per year. The combined withdrawal of groundwater in India and the US was 251 cubic kilometres per year.

Source: Shah et al 2000

Groundwater. To satisfy the growing demand for fresh water we are helping ourselves generously to groundwater, which is used for supplementing drinking water, for irrigation, and for industrial processes. It is estimated that global withdrawal of groundwater is between 600 and 700 cubic kilometres per year [km³/y].[25] The highest level of groundwater withdrawal is in India, the US, and China. These three countries accounted for 330 km³/y of withdrawal in the 1980s (see Table 4-9).[26]

Groundwater is an important source for irrigation, and in some countries it serves a large percentage of irrigated land—for instance, more than 50 per cent of arable land in India. Other countries that depend strongly on groundwater for irrigation include Saudi Arabia, Bangladesh, Tunisia, Syria, Iran, and Pakistan (see Table 4-10).[27]

Table 4-10 Groundwater as a source for irrigation

Country	Per cent of share		Irrigated area [kha]
	Ground water (Per cent)	Surface water (Per cent)	
Saudi Arabia	96	4	1 610
Bangladesh	69	31	3 750
Tunisia	61	39	310
Syria	60	40	640
India	53	47	50 100
Iran	50	50	7 260
Pakistan	34	66	14 330
Morocco	31	69	1 090
Mexico	27	63	5 370
Argentine	25	75	1 550
China	22	78	48 000
South Africa	18	82	1 270

Groundwater – Percentage of irrigation served by groundwater

Surface water – Percentage of irrigation served by surface water

kha – Thousand hectares

Example: About 96 per cent of irrigated land in Saudi Arabia is served by groundwater. The total irrigated area in this country is 1610 kha.

Source: UNEP-DEWA 2003

It has been estimated that in the near future about a quarter of India's primary production will be in crisis because of the depletion of groundwater reservoirs.[28] Similar situations can be expected in other countries that depend strongly on groundwater for irrigation.

Groundwater serves as an important source of drinking water in many parts of the world (see Table 4-11). In Europe, 75 per cent of drinking water is from this source and in the US the figure is 51 per cent.[29] About a half of the urban water supply in India comes from groundwater aquifers.[30]

Table 4-11 Groundwater as a source of drinking water

Region	Per cent	Population
Asia-Pacific	32	1 000–2 000
Europe	75	200–500
CSA	29	150
USA	51	135
Australia	15	3
Africa	NA	NA
World		**1 500–2750**

Per cent – Percentage of drinking water served by groundwater in a given region or country

Population – Number of people (in millions) that rely on groundwater for their drinking water requirements

CSA – Central and South America

NA – Data not available

Example: About 51 per cent of drinking water in the USA comes in the form of groundwater. About 135 million people in that country use groundwater as the source of drinking water.

Source: UNEP-DEWA 2003

Problems with the withdrawal of groundwater. If we compare the estimated annual global recharge of groundwater of 2000 km³/y with annual global withdrawals of 600–700 km³/y we can see that, on average, the withdrawals are not yet excessive. However, even though the average is low, local withdrawals are high in some places, and they are causing problems

We have seen that India, the US, and China account for about half of the global withdrawals of groundwater, and the excessive withdrawal rate is creating serious problems in these countries. For instance, in cities the recharge areas are small, but the quantities of water drawn from underground aquifers are large.

Rural areas may also differ significantly in their use of groundwater. About 68 per cent of China's groundwater and 36 per cent of its farmland are in the south, so this region does not have a problem with water supply. However, 64 per cent of the farmland and only 32 per cent of groundwater are in the north, and this imbalance creates severe groundwater problems.[31]

Different recharge rates of underground aquifers are also causing problems. In some places recharge rates are so low that even large aquifers are depleted too fast. For instance, the Ogallala aquifer in the US is claimed to be the largest in the world. It stretches over 453,000 square kilometres and contains 3700 cubic kilometres of water. Yet its recharge rate is so slow that it cannot cope with the rapid rate of withdrawal.[32]

In Yemen, withdrawal is about four times greater than recharge.[33] In India and Pakistan, millions of wells are in operation and many more are added each year.[34] The number of wells in India rose from four million in 1951 to 17 million in 1997, and in Pakistan from 25,000 in 1964 to 360,000 in 1993.[35]

Several countries suffer large annual deficits in groundwater. The largest is in India (104 km^3/y), followed by China (30 km^3/y), the US (14 km^3/y), North Africa (10 km^3/y), and Saudi Arabia (6 km^3/y).[36]

In North China, groundwater tables are falling at an average rate of one metre per year. In the Fuyang River basin, tables fell by up to 50 metres between 1967 and 2000. In some areas of India, groundwater tables are falling 2–3 metres per year.[37]

Between 1940 and 1980, the water level of the Ogallala aquifer fell by an average of three metres, but in some parts of Texas it fell 30 metres.[38] This aquifer has been depleted by 325 cubic kilometres (that is, by 9 per cent of its original volume.)

Between 1978 and 1984, the groundwater level of the Umm Er Radhuma aquifer in Saudi Arabia dropped more than 70 metres, and salinity increased substantially because of the intrusion of seawater. A similar situation exists in the United Arab Emirates, Syria, Jordan, and Lebanon.[39]

The pollution of groundwater.[40] The pollution of groundwater is widespread, but the highly polluted regions are the US, Europe, India, and China. The pollutants are fluorides, heavy metals (such as arsenic and lead), nitrates, pesticides, herbicides, bacteria, viruses, hydrocarbon (including petrol and oil), radioactive wastes, salts, solvents, and many other toxic chemicals.[41]

Groundwater pollution occurs by surface contamination or by direct drilling. Pollutants can also reach groundwater via abandoned wells. Surface contamination comes from local or non-local sources. Local sources include accidental spillage or deliberate dumping. Non-local sources include farming, industrial, municipal, and domestic activities that cause a gradual diffusion of pollutants.

The leading source of pollution is agriculture, through the overuse of fertilisers, fungicides, and pesticides. In Australia, about 2.2 million tonnes of fertilisers are used each year in addition to $500 million of fungicides and herbicides. They seep through the ground and pollute groundwater.[42] Groundwater pollution is also caused by livestock facilities.

We pollute groundwater through industrial and mining activities, either by discharging pollutants on the surface of the Earth or by injecting them deep into the ground. In the US, thousands of injection wells each year put 34 billion litres of a toxic cocktail of solvents, heavy metals, and radioactive material directly into the ground.[43]

Other sources of groundwater pollution include sewage treatment plants, leaking petrol tanks in petrol stations all over the world, workshops, restaurants, dry-cleaning facilities, rubbish dumps—and even domestic gardening through the overuse of fertilisers, weedkillers, and pesticides.

A threat to large cities by excessive withdrawal of groundwater. Many large cities such as Beijing, Shanghai, Manila, Bangkok, Jakarta, Dhaka, Calcutta, Tehran, Cairo, Lima, Buenos Aires, and Mexico City depend strongly on groundwater, but dwindling supply threatens their survival. In addition, the overdrawing of groundwater causes the ground to subside in many cities and inflicts structural damage.[44]

In Bangkok, the ground sinks at 5–10 centimetres per year. Groundwater tables around Beijing, Xian, and Tiankjin sink at one metre per year.[45] Water tables around Seoul, Pusan, and Daegu in South Korea dropped by up to 50 metres over 30 years.[46]

An intrusion of seawater, caused by the overdrawing of groundwater, is a serious problem in many cities. The Philippines and the coastal regions of China, India, and of the Gulf of Mexico are badly affected.[47] In China, 36 cities were reported to have serious problems with the intrusion of seawater and with the sinking of land.[48]

In Manila, the overdrawing of groundwater caused a drop in the water table of 80 metres, which in turn increased the salinity of groundwater to a prohibitive level. Many wells in Bangkok have been abandoned because of the intrusion of seawater. It is claimed[49] that as little as 2 per cent of seawater added to groundwater makes the groundwater unsuitable for consumption.[50]

The extent of the developing water crisis.[51] The rapid decline in the availability of fresh water began about 1950. In the 1950s, only some countries in the northern parts of Africa suffered from a very low level of water availability, but no region of the world had yet reached the catastrophically low level (see Shiklomanov's scale of water availability, Appendix B).

By 1995, water availability in the northern part of Africa had fallen to the catastrophically low level. The Middle East and most parts of Central and South Asia fell to the very low level, and parts of Asia, Europe, and some countries of East Africa slid to the low level.

By 2025, the whole Middle East, as well as parts of Central and South Africa, will join North Africa and experience the catastrophically low level of water availability. The rest of Asia, almost all Africa, and Central America will become even dryer than in 1995. Water availability in various countries of these regions will be between low and very low.

The overall picture shows a gradual deepening and widening of water shortage from the northern parts of Africa to other parts of the world. In terms of water availability per person per year, the world is getting dryer over larger areas. Most of the global population is in Asia and Africa. These regions are already experiencing serious water problems, and their situation will grow worse.

It is helpful to imagine how we would feel if our water availability happened to be low. The very low level of availability corresponds to about 13 times less water than in Australia or about seven times less than in the US. The catastrophically low level corresponds to about 22 times less water than in Australia or 11 times less than in the US.

In general, industrialised countries are not expected to be affected as strongly as developing countries by the future shortage of water. By the end of the 20th century, water availability in industrialised countries has dropped to 60 per cent of the 1950 level, and it is not expected to fall much lower because the population in these countries is nearly stable.

In contrast, water availability in developing countries has dropped to 35 per cent of the 1950 level, and by 2030 it will be 20 per cent.

People living in arid climates are experiencing an even faster decline. For them, water availability has fallen to 30 per cent of the 1950 level, and it will fall to 10 per cent by about 2030.

This dramatic decline in developing countries means that by 2030 they will be pushed about two steps down the scale of water availability. Countries that now have adequate water availability will be pushed to the very low level. Countries that are now on the low level will be pushed to the catastrophically low level. By 2030 there will be a large increase in the number of people suffering the catastrophically low level of water availability.

Regional development of the water crisis.[52] In the Asia-Pacific region about 500,000 infants die each year because they have no access to clean water and sanitation facilities. Water pollution in this region is high, water availability is decreasing, and the withdrawal of groundwater is excessive.[53]

Africa suffers a high degree of water stress: up to 25 countries are expected to have severe problems by 2025. Groundwater is being depleted at a rapid rate. Africa also suffers severe economic loss through floods and droughts.

In Europe and Central Asia, water withdrawal is increasing rapidly. Between 1950 and 1990 the demand rose nearly six-fold. The withdrawal of groundwater is high in Europe and in parts of Asia. Many countries in these regions have high levels of groundwater pollution.

In Latin America-Caribbean, the depletion of groundwater and pollution levels are high. Sewage treatment in Latin America is deplorable and the associated health risks are high.

In North America, groundwater reservoirs are being depleted rapidly. North America also has high levels of pollution from agriculture and industry.

Access to safe water and sanitation. In 2000, 44 per cent of the global population (2.64 billion people)[54] did not have access to adequate sanitation facilities, and 19 per cent (1.14 billion people) did not have access to improved water sources. In the least-developed countries, 55 per cent of the population did not have access to sanitation services, and 37 per cent did not have access to safe water. About half of sub-Saharan Africa's population did not have access to sanitation services or safe water. In all developing countries, 48 per cent did not have access to sanitation facilities, and 22 per cent did not have access to safe water.[55]

The most affected regions are Asia and Africa. In Asia, 50 per cent of the population (1800 million people[56]) have no access to sanitation services, and 20 per cent (730 million people) have no access to pure water. In Africa, the respective figures are 40 per cent (320 million people) and 38 per cent (304 million people).

The percentage of the population with access to clean water or sanitation is lower in rural areas than in urban areas. Worldwide, about 84 per cent of the population without access to clean water live in rural areas.[57]

There is a strong correlation between water quality and child survival. In industrialised countries, where there is a good access to safe water, the survival rate of young children is nearly 100 per cent. In developing countries, where only 20 to 50 per cent of the population have access to safe water, the survival rate is 80 per cent. On average, for every 100 children below the age of five who live in developing countries, about 20 are destined to die just because they do not have access to clean water, which is taken for granted in many countries.[58]

Water-related conflicts. Water conflicts occur within and between countries. Conflicts within countries occur when various economic sectors compete for scarce water resources. Conflicts between countries occur over shared rivers or lakes. International disputes also erupt over groundwater (see Table A4-1 in Appendix A).

For instance, Israel, Jordan, Lebanon, Syria, and the West Bank are in conflict over the Jordan River. Turkey is in conflict with Syria and Iraq because of its Grand Anatolia Project, which includes establishing 22 dams on the Euphrates River, 19 hydropower plants, and a network of 1.7 million hectares of irrigated land. The project is scheduled for completion in 2006, and by then it will reduce the water flow to Syria by 40 per cent and to Iraq by 80 per cent.[59]

The Nile, Euphrates, Ganges, and Mekong are just a few of the world's contested rivers. The water basin of the Nile is shared by 10 countries. The basins of the Euphrates, Ganges, and Mekong are each shared by six countries. The Danube's basin is shared by 14 countries, and the Rhine basin by seven. Of the 261 international river basins, 71 are in Europe, 53 in Asia, 38 in South America, 39 in North and Central America, and 60 in Africa.[60] In many countries, such as Turkmenistan, Egypt, Hungry, Mauritania, Botswana, Bulgaria, and Uzbekistan, dependence on an external supply of water is exceptionally high (see Table 4-12). This creates a strong need for cooperation, but it is also a source of water-related conflicts.

Table 4-12 Dependence on the external supply of surface water

Country	Dependence (per cent)	Country	Dependence (per cent)
Turkmenistan	98	Syria	80
Egypt	97	Congo	77
Hungary	95	Sudan	77
Mauritania	95	Paraguay	70
Botswana	94	Niger	68
Bulgaria	91	Iraq	66
Uzbekistan	91	Albania	53
Netherlands	89	Uruguay	52
Gambia	86	Germany	51
Cambodia	82	Portugal	48
Romania	82	Bangladesh	42
Luxembourg	80	Thailand	39

Example: In Turkmenistan, 98 per cent of surface fresh water comes from external sources.

Source: Björklund et al 2000

So far, it has been possible to resolve international conflicts by agreements and cooperation. However, with the worsening water shortage it is becoming more difficult to find peaceful solutions to the problem of sharing water.

The future. Water shortages will increase everywhere, but they will be felt more severely in developing countries because their populations increase much faster than those of industrialised countries. Technological problems, such as wasteful irrigation systems and the lack of efficient water purification, also contribute to the water crisis in these countries.

Estimating the global maximum amount of accessible water. What is the maximum volume of renewable fresh water we can make available by better control of runoff water? The global volume of accessible runoff is 12,500 cubic kilometres per year [km³/y]. Can we boost it by building more dams?

It has been estimated that the maximum volume we can recover using present and future dams is about one-fifth of runoff water.[61] Considering that the average annual volume of runoff in geographically accessible places is 34,400 km³/y, we can see that the maximum volume we can recover by using all possible dams is 7000 km³/y. (This includes the estimated 3500 km³/y which is already being recovered by existing dams.) Consequently, we can still boost the availability of accessible water by 3500 km³/y—from the present 12,500 km³/y to 16,000 km³/y; that is, by 28 per cent.

Now we should compare this estimated increase of accessible runoff water with the estimated increase of the global population. If we take my low-level projections of the global population we shall find that by 2030 it will increase from the current 6.2 billion to 8.3 billion—that is by 34 per cent. By 2050 it will be 9.2 billion, which corresponds to a 48 per cent increase. Between now and the projected maximum of 9.4 billion in 2063, global population will increase by 52 per cent. We can therefore see that there is a significant mismatch between the projected increase of global population and the expected maximum increase of accessible runoff water.

This simple exercise illustrates two points: first, global runoff water resources are small; and, second, we are approaching the limit of available runoff water.

Maybe we could improve the technology for extracting fresh water from seawater and make it more affordable. Examples of recent developments in this field include seawater greenhouse technology[62] and the rapid-spray evaporation system.[63] The seawater technology relies on wind. It also needs large land areas which, however, could be used for growing food. The rapid-spray evaporation system requires substantial energy input. It is unlikely that poor countries will be able to pay for desalination technology and for the costs of water distribution to consumers. Richer countries will have to help them.

How many people will live with severe water stress? By 2025, people living in 48 countries will suffer severe water shortage, and water availability will be catastrophically low for 2.6 billion people.[64] Global demand for water will increase by 30 per cent for agriculture, 55 per cent for industry, and 70 per cent for public consumption. The global irrigation area will expand from 253 million hectares in 1995 to 330 million hectares in 2025.[65]

UNEP projections[66] also do not leave much room for optimism (see Table 4-13). Even if we assume that from now on we will all work together harmoniously for a sustainable future (the Sustainability First development), the percentage of the global population suffering severe water stress will be almost constant. This means that the number of people suffering severe water stress will increase from 2.4 billion in

2002 to 3.2 billion in 2032. However, with business as usual (the Market First development), the global number of people suffering severe water stress in 2032 will be 4.5 billion. The largest number of people suffering severe water stress is in Asia, where 1700 million people are now in this category. This number will increase to at least 2000 million, but more likely to 3000 million.

Table 4-13 Number of people suffering severe water stress, 2002 and 2032 (millions)

Region	2002	2032 Market First	2032 Sustainability First
Africa	172	520	340
Asia-Pacific	1 693	2 910	2 240
Europe	239	310	80
LAC	112	140	190
North America	124	140	80
West Asia	74	320	290
World	**2 414**	**4 450**	**3 220**

LAC – Latin America and Caribbean
Market First – Continuation of the status quo (business as usual)
Sustainability First – Complete, and unlikely, change of direction involving a wide range of programs for a sustainable future.

Example: In 2002, 172 million people in Africa suffered severe water stress. With business as usual, the number of people suffering severe water stress in Africa is projected to increase to 520 million. In the most favourable case of the Sustainability First scenario, the number of people suffering severe water stress in Africa will increase to 340 million.

Source: The numbers for 2002 were taken from UNEP 2002. The numbers for 2032 were calculated using the percentage values listed in UNEP 2002, and the populations calculated using the data of PRB 2002.

Economic water scarcity. It has been suggested that, in addition to physical water scarcity we should also consider economic water scarcity.[67] The point is that so-called physically available water might not be available in practice unless a country is able to invest enough money to make it available through a suitable technological development.

By definition, a country will experience economic water scarcity if it has to invest enough to increase its water availability by at least 25 per cent by 2025.

Countries with physical water scarcity face critical water restrictions. Even if they could invest in developing renewable water resources, they do not have enough water to meet their current or future needs. In this group of countries are Saudi Arabia, Pakistan, Jordan, Iran, Syria, Tunisia, Egypt, Iraq, Israel, South Africa, and Algeria.[68]

Examples of countries that will suffer economic, but not physical water scarcity include Brazil, Turkey, Mexico, the Philippines, Thailand, Ethiopia, Australia, Nigeria, Bangladesh, Argentina, Viet Nam, Sudan, and Morocco. Their future will depend on investment in water resources.

This division by countries represents only an approximate situation. A better division would be by geographical regions. This would help to describe the degree of water scarcity in various countries. For instance, most of China will suffer only economic water scarcity, which means that, by developing its water resources, the nation will be able to meet most of its needs. On the other hand, most of India suffers physical water scarcity. Consequently, even if it develops its renewable water resources it will not be able to meet most of the country's requirements.

Other expectations. The growing shortage of water will have a negative influence on many activities and will limit our progress. It will affect industrial and agricultural production, and will increase the cost of living. Less water in industry will mean lower industrial output and lower availability of industrial goods. Less water for agriculture will mean less food on the market.

Water shortages will have an increasing effect on rapidly growing urban populations, particularly in developing countries where there are already supply problems. The problem of dividing water between various sectors (industry, agriculture, and municipal) will be more severe. Water shortage will increase the number of people suffering hunger, disease, and other forms of hardship.[69]

International disputes over water are likely to intensify. The global community is now more integrated, and local disputes can easily involve a wider circle of nations. It has been suggested, for instance, that conflict over scarce water resources in the Middle and Near East could escalate into a global conflict.[70]

Chapter 5
The Destruction of the Atmosphere

'Some recent occurrences such as the BSE disaster and even perhaps—dare I mention it—the present severe weather conditions in our country are, I have no doubt, the consequences of mankind's arrogant disregard of the delicate balance of nature.'

—Prince Charles (1948–), British Medical Association
Millennium Festival of Medicine, London, November 2000

We know that greenhouse gases have the capacity to create a warm blanket around the Earth. We know that the emission of these gases is increasing, yet we put off decisive action. Humans are endowed with enormous privilege, but also carry a huge burden of responsibility. We have power, knowledge, and technology. We can now 'boldly go where no man has gone before' and do what no one has done before. And through our 'arrogant disregard of the delicate balance of nature', we can destroy the environment.

It is hard to tell which human interference with nature is the most dangerous, but the destruction of the atmosphere is probably at the top of the list. Maybe we shall create the climate of Mars or Venus on Earth. We have the power and technology to make our world as hostile as other planets in the solar system.

For the first time in human history, our influence on the atmosphere is so great that we are changing substantially and critically its composition and function. We have managed to create two crucial atmospheric problems: global warming and the depletion of the stratospheric ozone layer.

Global warming deserves special attention because it affects climate, which is likely to have the most severe consequences for the future. In addition, we have two other types of human-made atmospheric problems that directly affect our health: the ground-level ozone problem and problems with emissions of various other toxic and harmful gases that hover close to where we live.

The atmosphere and global warming.[1] The earth's atmosphere is made up of layers characterised by their specific distributions of temperature (see Box A5-1, Appendix A). The lowest is the troposphere, the layer that is most affected by our activities and that in turn affects us most strongly. This is where the changes influencing global and local climates are taking place. The Earth's surface absorbs about 50 per cent of the incoming solar radiation and the atmosphere about 20 per cent. The rest reflect back to space.

From this point on, there is a complex exchange of energy between the atmosphere and the surface. The amount of the energy fed back to the Earth by the atmosphere depends on the composition of gases in the atmosphere. At this stage also, the original short-wave solar radiation, such as visible light, is converted to long-wave heat radiation.

Eventually, solar energy absorbed by the surface of the Earth and by the atmosphere escapes back to space to make room for the constant inflow of solar energy. It is a dynamic process, with a continuous circulation of heat between the surface and the atmosphere.

This circulation keeps us warm, and thus is a desirable process. However, by our intensified emissions of greenhouse gases we are disturbing the natural exchange of heat and increasing the surface temperature. The result is excessive global warming, leading to undesirable and potentially disastrous climate change.

The term 'global warming' is unfortunate, because it does not reflect the true nature of the associated problem. It also does not distinguish between the natural phenomenon and the human-caused problem. A better term would be 'global climate instability', because excessive warming influences global and local climates.

Climate change usually occurs slowly, but now human-induced changes are too fast and they do not give ecosystems enough time to adapt. The climate may also undergo a sudden change. After reaching a critical threshold, the slow process can become dramatic, with potentially disastrous outcomes.[2]

A suggested mechanism of sudden climate change is associated with the oceanic conveyor-belt system—that is, with the system of currents and pumps.[3] As an example, excessive melting of the Arctic ice cover caused by global warming could change the flow of water in the conveyor belt system suddenly. Europe, which is now kept warm by the Gulf Stream/North Atlantic current, could become excessively cold.

Greenhouse gases. These gases can absorb and emit radiation. They include carbon dioxide, water vapour, methane, tropospheric ozone,[4] nitrous oxide, hydrofluorocarbons, perfluorocarbons, and sulphur hexafluoride.[5] The Kyoto Protocol[6] deals with six of them: carbon dioxide, methane, nitrous oxide, hydrofluorocarbons, perfluorocarbons, and sulphur hexafluoride.

Some gases are removed from the atmosphere relatively quickly, but others remain for hundreds or even thousands of years. More than one mechanism can be also involved in the removal of a gas.

A gas is characterised by its lifetime in the atmosphere. However, a lifetime does not mean that all the gas will be removed in a specific time. Typically, if the lifetime is 100 years, 37 per cent of the gas will remain in the atmosphere after 100 years, 14 per cent after 200 years, 5 per cent after 300 years, and 2 per cent after 400 years.[7]

The lifetime of carbon dioxide is more than 100 years. Even if we reduced our carbon dioxide emissions it would be many years before we could observe a substantial drop in the atmospheric concentration of this gas. The approximate lifetime of methane is 10 years, nitrous oxide 120 years, hydrofluorocarbons up to 1700 years, perfluorocarbons more than 1000 years, and sulphur hexafluoride 3200 years.[8]

Not all greenhouse gases have the same influence on climate. Their effect is described using radiative forcing (also known as climate forcing), which is defined as the ability to change the energy balance of the Earth and thus influence climate. Radiative forcing, measured in watts per square metre [W/m^2], specifies the amount of energy forced back to Earth by the blanket of greenhouse gases in the atmosphere. It is like having a heater with so many watts per square metre hanging (and working) over our heads. If we consider the total surface area of the atmosphere, even small values of radiative forcing can have a large combined effect.

Radiative forcing depends on the concentration of a given gas in the atmosphere. The highest radiative forcing is for carbon dioxide, which is now 1.5 W/m^2. Radiative forcing for methane is 0.5 W/m^2 and for nitrous oxide 0.15 W/m^2. The most harmful greenhouse gas is carbon dioxide—not because it is the most efficient in trapping heat energy, but because its atmospheric concentration is so high.[9]

Records show that radiative forcing started to increase from about 1750. Between 1750 and 1870, the increase was only 0.02 W/m^2 per decade, but between 1870 and 1950 it was 0.07 W/m^2 per decade. From about 1950 it accelerated again, and by 2000 it was 0.32 W/m^2

per decade, or 16 times faster than between 1750 and 1870.[10] If the current rate of increase continues, we can expect radiative forcing at the end of this century to be at least twice the 2000 level.

Even though we have identified a series of greenhouse gases, we might be still surprised by new discoveries, because we release so many pollutants into the atmosphere. For instance, a mysterious new pollutant (trifluoromethyl sulphur pentafluoride) has been identified recently that turns out to be 18,000 times more efficient in trapping solar energy than carbon dioxide.[11]

This pollutant has a long lifetime. It can stay in the atmosphere for thousands of years, and its concentration is increasing at the rate of 6 per cent per year. Its contribution to global warming is still negligible, but it has the potential to do much damage. We do not know where it is coming from, but a possible source could be sulphur hexafluoride, which is used as an insulator in large power stations.

Carbon emissions. Global annual emissions of carbon, mainly in the form of carbon dioxide, increased more than four times in the second half of the 20th century—from 1.6 billion tonnes per year [Gt/y] in 1950 to about 6.6 Gt/y in 2000.[12] The weight of carbon emissions should not be confused with the weight of carbon dioxide emissions. The molecular weight of carbon dioxide is 3.67 times greater than the molecular weight of carbon. Consequently, each tonne of emitted carbon corresponds to 3.67 tonnes of emitted carbon dioxide.

Carbon emissions per person vary over a large range of values between countries (see Table A5-1, Appendix A), from zero in Chad to 18 tonnes per person per year [t/p/y] in Qatar. The emissions in the US are 5.6 t/p/y and in Australia 4.7 t/p/y. The average emissions in industrialised countries are 3.4 t/p/y and in developing countries only 0.5 t/p/y. Average global emissions are 1.1 t/p/y.[13]

The Intergovernmental Panel on Climate Change (IPCC) report shows a linear correlation between gross domestic product (GDP) and the emission of carbon per person per year.[14] According to this correlation, the annual emission per person increases at the rate of about 0.16 t/p/y for each $1000 of GDP. Countries with a GDP of $1000 per person per year emit 0.2 tonnes of carbon per person per year; richer countries with a GDP of $10,000 emit 2 tonnes of carbon per person per year; and affluent countries such as Australia and the US, with a GDP between $20,000 and $30,000, emit 3–5 tonnes of carbon per person per year.

If we analyse the data of global carbon emissions,[15] we shall find that between 1850 and 2000 we have released 280 billion tonnes [Gt] of carbon into the atmosphere, with nearly half of it in the last 25 years of the 20th century. In the early 1900s, global carbon emissions were increasing at 6 Gt per decade, but by the end of the 20th century they were increasing at the rate of 67 Gt per decade.

In the 125 years after 1850, we released the first 140 Gt of carbon into the atmosphere. The second 140 Gt was released in 25 years. If the trend continues, we shall release the third 140 Gt in 15 years. Global carbon emission can be described by an exponential function with a doubling time of about 26 years.

Projected emissions of carbon. It is hard to predict future emissions of carbon because it is hard to predict how countries will react to the danger of global warming. It is also hard to predict to what extent developing countries will be able to develop their technology. However, we can still consider the most likely development based on current trends.

In 1990, global carbon emissions were 5836 million tonnes per year [Mt/y], and increased to 6600 Mt/y in 2000 (see Table 5-1). The largest national emissions of carbon in 2000 were in the US. That country accounts for nearly a quarter of global carbon emissions. The largest country in the world, China, accounts for only 15 per cent, which is almost the same as the combined emissions of Western Europe. The US emissions are nearly the same as the combined emissions of all Asia.

It can be expected that Asia and other developing countries will add substantially to the future emission of carbon. Indeed, emissions in the US are projected to increase 27 per cent between 2000 and 2020, and in China by 118 per cent. In the whole of Asia, carbon emissions are projected to increase 96 per cent during that period. By 2020, China and the US will be emitting about the same amount of carbon per year.

The combined emission by industrialised countries in 2000 was 4062 Mt/y, and by developing countries 2538 Mt/y (see Table 5-2).[16] The contribution to carbon emissions by developing countries is still relatively small, but their emissions are increasing much faster than in industrialised countries. By 2020, carbon emissions by developing countries will be nearly the same as emissions by industrialised countries. From about 2020, developing countries will be emitting more carbon into the atmosphere than industrialised countries.

Table 5-1 Carbon emissions by country and region, 1990–2020 (million tonnes per year)

Country/Region	1990	2000	(1)	2005	2010	2015	2020	(2)	(3)
USA	1 345	1 556	24	1 683	1 787	1 893	1 979	423	27
China	620	959	15	1 186	1 457	1 685	2 091	1 132	118
FSU	1 034	654	10	668	728	781	875	221	34
Japan	274	305	5	318	331	344	354	49	16
India	153	271	4	328	385	425	487	216	80
Germany	267	239	4	248	257	267	274	35	15
UK	166	160	2	167	175	183	189	29	18
North America	1 553	1 807	27	1 959	2 090	2 227	2 344	537	30
Asia	1 067	1 728	26	2 071	2 479	2 812	3 380	1 652	96
Western Europe	934	941	14	979	1 016	1 056	1 094	153	16
EE/FSU	1 337	896	14	927	992	1 050	1 151	255	28
Middle East	229	321	5	362	422	473	552	231	72
CSA	174	257	4	310	399	479	617	360	140
Africa	180	231	4	260	292	329	380	149	64
Australasia	90	113	2	120	126	131	137	24	22
World	**5 836**	**6 600**	**100**	**7 308**	**8146**	**8 901**	**10 009**	**3 409**	**52**

Countries and regions are arranged in descending order of carbon emissions for 2000

FSU – Former Soviet Union

EE/FSU – Eastern Europe/Former Soviet Union

CSA – Central and South America

(1) – Percentage of global emissions in 2000

(2) – Increase (in million tonnes) per year in carbon emissions between 2000 and 2020

(3) – Percentage increase in carbon emissions between 2000 and 2020

Example: Carbon emissions in the USA in 1990 were 1345 million tonnes per year. They increased to 1556 million tonnes per year in 2000, which represented a 24 per cent of global emissions in that year. Carbon emissions in the US are projected to increase to 1979 million tonnes per year in 2020, or by 423 million tonnes per year—that is, by 27 per cent, as compared with emissions in 2000.

Source: Based on EIA 2000

Projected global emissions for 2020 are 10,009 Mt/y, or 52 per cent higher than in 2000. By 2020, the gap between developing and industrial countries for total emissions of carbon will almost close—it will narrow from 1500 Mt/y in 2000 to only 150 Mt/y. However, the gap for the emissions per person per year will increase by 11 per cent. Emissions per person are increasing everywhere, but much faster in industrialised countries. The race is on, and the future of the atmosphere does not look promising. If these projections are realised, we can forget about the Kyoto Protocol and any hope of saving the planet.

Table 5-2 Carbon emissions by level of development and globally, 1990–2020

	1990	2000	(1)	2005	2010	2015	2020	(2)	(3)
Industrial countries									
Total [Mt/y]	4 187	4 062	62	4 304	4 555	4 808	5 079	1 017	25/7
Per person [t/p/y]	3.8	3.4		3.5	3.7	3.8	4.0	0.6	17/7
Developing countries									
Total [Mt/y]	1 649	2 538	38	3 004	3 591	4 093	4 930	2 392	94/30
Per person [t/p/y]	0.4	0.5		0.6	0.6	0.7	0.8	0.3	49/30
Total world									
Total [Mt/y]	5 836	6 600	100	7 308	8 146	8 901	10 009	3 409	52/26
Per person [t/p/y]	1.1	1.1		1.1	1.2	1.2	1.3	0.2	20/26
Gap									
Total [Mt/y]	2 538	1 524		1 300	964	715	149	-1 375	-90
Per person [t/p/y]	3.4	2.9		3.0	3.0	3.1	3.2	0.3	11

Mt/y – megatonne (million tonnes) per year

t/p/y – tonne per person per year

(1) – Percentage of global emissions in 2000

(2) – Increase (in million tonnes) per year in carbon emissions between 2000 and 2020. For instance, industrialised countries will be emitting 1017 million tonnes more carbon per year in 2020 than in 2000—that is, 0.6 tonnes more per person per year

(3) – Percentage increase in carbon emissions between 2000 and 2020 compared with the percentage increase in the population. For instance, in industrialised countries total carbon emissions will increase by 25 per cent between 2000 and 2020, but the population will increase by only 7 per cent

Gap – The gap in carbon emissions (total and per person) per year between industrialised and developing countries. The gap for total emissions is projected to grow smaller rapidly, but the gap in emissions per person per year will be increasing. Between 2000 and 2020 the gap in total emissions will decrease by 90 per cent. However, the gap in emissions per person per year will increase by 11 per cent between 2000 and 2020.

Source: Based on EIA 2000

Carbon storage and exchange. Carbon is stored almost exclusively in the ocean, but smaller amounts are distributed between the land and atmosphere (see Box 5-1) Terrestrial carbon is mainly in the soil. The estimated amount of carbon in soil is 2011 billion tonnes. This does not include the large deposits of carbon locked in fossil fuels, which we are pumping and burning. Thus we are moving large amounts of carbon from the earth's crust into the atmosphere.

Box 5-1
Distribution of the world's carbon

Global carbon is mainly located in oceans. The atmosphere contains only a small fraction of global carbon, but the amount is increasing.

Location	Amount [Gt]	Per cent of global
Oceans	40 000	92
Land	2 477	6
Atmosphere	800	2
Total	**43 277**	**100**

Terrestrial storage of carbon

The following inventory of the terrestrial storage of carbon includes vegetation and a one-metre-thick layer of the earth's surface. It does not include the large deposits of carbon in the form of fossil fuels deeper in the earth's crust. A substantial proportion of these fossil deposits of carbon are being shifted into the atmosphere.

	Vegetation [Gt]	Soil [Gt]	Total [Gt]	Per cent of land total
Tropical forests	212	216	248	17.3
Temperate forests	59	100	159	6.4
Boreal forest	88	471	559	22.6
Tropical savannas	66	264	330	13.3
Temperate grasslands	9	295	304	12.3
Deserts and semideserts	8	191	199	8.0
Tundra	6	121	127	5.1
Wetlands	15	225	240	9.7
Croplands	3	128	131	5.3
Total	**466**	**2 011**	**2 477**	**100.0**

Gt – gigatonne (billion tonnes)

Source: FAO 2000b; UNDP et al 2000b

Terrestrial vegetation contains 466 billion tonnes of carbon. The amount of carbon stored in all the forests of the world is only 359 billion tonnes, which is about a half of the amount of carbon stored in the atmosphere.

Between 1989 and 1998, global industrial activities were releasing about 6.3 billion tonnes of carbon per year, and land use was releasing 1.6 billion tonnes per year (see Table 5-3). Oceans and terrestrial vegetation absorb a fraction of the released carbon, but the net flow of carbon into the atmosphere is still substantial—3 billion tonnes per year and increasing. Using the projections of carbon emission from industrial activities, we can estimate that by 2020 the net flow of carbon into the atmosphere will increase to 7 billion tonnes per year.

Table 5-3 Global flow of carbon, 1989–1998

Type of emission or absorption	Flow [Gt/y]
Industrial activity	6.3
Land use	1.6
Total emission	**7.9**
Absorption by oceans	2.3
Absorption by terrestrial vegetation	2.3
Total absorption	**4.6**
Net flow into the atmosphere	**3.3**

Gt/y – gigatonne (billion tonnes) per year

Source: UNDP et al 2000b

Atmospheric concentrations of greenhouse gases.[17] The atmosphere contains only a small fraction of global carbon (about 2 per cent), and it would be better to keep it at this level. The world is warm enough, and we do not want to increase the thickness of the atmospheric blanket covering us. However, carbon concentration in the atmosphere is increasing. It is estimated that from 1850 to 1998 the amount of carbon in the atmosphere increased by 176 billion tonnes.[18]

We can study carbon dioxide concentrations in the atmosphere in the past by analysing the air trapped in ice. The records show (see Table 5-4) that for 900 years of the last millennium, carbon dioxide concentration in the atmosphere was nearly constant at 280 parts per million (ppm). However, from 1900, it increased rapidly to reach 370 ppm at the end of the 20th century, representing a 32 per cent increase.

The atmospheric concentrations of other greenhouse gases show a similar time-dependence. The concentration of methane was nearly constant at 700 parts per billion [ppb] before 1900, but increased to 780 ppb at the end of the 20th century—that is, by 154 per cent. The concentration of nitrous oxide was nearly constant at 270 ppb before 1900, but increased to 310 ppb between 1900 and 2000, or by 15 per cent.

Table 5-4 Atmospheric concentrations of carbon dioxide, methane, and nitrous dioxide, 1000–2000

Year	Carbon Dioxide [ppm]	Methane [ppb]	Nitrous Oxide [ppb]
1000	280	660	270
1200	280	690	270
1400	280	730	270
1600	280	690	270
1700	280	710	270
1800	280	740	270
1850	285	790	275
1900	290	860	280
1950	310	1 110	290
2000	370	1 780	310
Increase	**32**	**154**	**15**

ppm – part per million; ppb – part per billion

Increase – Percentage increase towards the end of the 20th century as compared with nearly constant values over hundreds of years in the last millennium

Example: The carbon dioxide concentration in the atmosphere was 280 ppm in AD 1000, but it increased to 370 ppm in 2000, which represents a 32 per cent increase.

Source: Houghton et al 2001

Records extending much further back show variations in the concentration of carbon in the atmosphere. Sometimes the concentration was as low as 180 ppm, but over thousands of years it was never as high as it is now.[19] Another interesting feature of these long-range studies is a remarkably close correlation between the concentration of carbon dioxide in the atmosphere and the average global surface temperature: the higher the concentration of carbon, the higher the average surface temperature.[20]

Industrialisation is supported by fossil fuels, and as long as we rely heavily on them the concentration of carbon in the atmosphere will continue to increase. However, there is also a possibility that the process of industrialisation and industrial production might be slowed by the physical limitations of the planet.

In order to predict future concentrations of carbon dioxide in the atmosphere I have analysed the data for the years 1960–2000.[21] The calculations show that the concentration will continue to increase, and by the end of this century it might be 90 per cent higher than in 2000 (see Table 5-5). My calculations agree well with the IPCC calculations,[22] and with values given by UNEP.[23]

Table 5-5 Projected concentrations of carbon dioxide in the atmosphere and changes in global average temperature, 2010–2100

Year	Author [ppm]	IPCC [ppm]	UNEP[a] [ppm]	UNEP[b] [ppm]	IPCC [°C]
2010	380–390	380–390	400	400	0.3–0.9
2020	400–410	410–420	430	410	0.4–1.0
2030	420–440	420–460	470	430	0.5–1.3
2040	430–460	450–510	510	440	0.7–2.0
2050	450–500	480–580	550	450	0.8–2.6
2060	470–530	500–640			1.1–3.7
2070	490–560	520–740			1.2–4.5
2080	510–600	540–810			1.2–5.1
2090	530–640	550–900			1.3–5.5
2100	550–690	560–980			1.4–5.8

Author – Projections are based on an analysis of the data (Dunn 2001b) for concentrations of carbon dioxide in the atmosphere between 1960 and 2000

IPCC – Houghton et al 2001

UNEP – UNEP 2002

[a]) Projections based on assumption of the Market First scenario (continuation of the status quo, or business as usual).

[b]) Projections based on assumption of the Sustainability First scenario (a desirable, but less likely change of direction aimed at developing a sustainable future).

ppm – parts per million

°C – A change in the average global temperature by a given year.

Example: My projections show that by 2010 the carbon dioxide concentrations in the atmosphere will increase from 370 ppm in 2000 to 380–390 ppm. By 2100 these concentrations will increase to 550–690 ppm.

Are global temperatures rising?[24] A well-established global network of meteorological stations, which has been working since 1880, gives accurate measurements of the earth's surface temperatures. We also have less accurate, but still reliable records of surface temperatures going back to 1861. All these data show that the average global surface temperature has been increasing.

Between 1861 and 1915, the mean global surface temperature was about 0.3°C below the reference temperature,[25] but in the last 85 years of the 20th century it climbed to 0.35°C above the reference temperature. This represents an increase of 0.65°C. The accepted value by the IPCC[26] is 0.6±0.2°C—that is, between 0.4 and 0.8°C.

In Australia, the mean surface temperature between 1910 and 1950 was 0.4°C below the reference temperature, but increased to 0.3°C above the reference temperature at the end of the last century. Thus, the average increase in the surface temperature in Australia between 1910 and 2000 was 0.7°C, about the same as the increase in global average temperature. In some regions of Australia (parts of Queensland and Western Australia) the average temperature increased by 2°C, and around Kalgoorlie by up to 2.5°C. In two regions of New South Wales the average temperature decreased by 0.5–1°C.[27]

The small increase of 0.6°C in the average global surface temperature may sound negligible, but it does not reflect local changes which can be significantly larger. It also seems that even this small average increase has an undesirable influence on global and local events: ice covers are melting faster than before, the average temperature of oceans is rising,[28] coral reefs are dying, the frequency of extreme weather is increasing, and the costs associated with weather-related losses are rising rapidly.

The Northern and Southern Hemispheres show a similar pattern in the trends of surface temperature. However, the overall increase in the Northern Hemisphere of 0.75°C is slightly higher than the 0.55°C increase in the Southern Hemisphere.

If we look as far back as the year 1000, we shall find that the only period marked by a rapid increase in the surface temperature is after 1900. Records of the surface temperature before 1860 are in the form of so-called proxy data. They are gathered from palaeoclimatic studies of organic and inorganic sedimentary deposits, ice covers, tree rings, pollens, insects, oxygen isotopes, and trace elements.

The data for the Northern Hemisphere show that about 1000 AD the mean surface temperature was 0.2°C below the reference temperature, and that it was decreasing slowly to reach the average of 0.4°C below the reference temperature in 1900. However, from that year on, there was a rapid increase.

Thus the Northern Hemisphere data for the last millennium show that we have had a drop of 0.2°C in the mean surface temperature in the first 900 years, followed by an increase of 0.8°C in the last

100 years. Records for the Southern Hemisphere display a similar time dependence.[29] The question therefore is not whether the mean surface temperature is increasing, but how fast—and how much more it will increase in the near future.

Projected increase in global and local temperatures. In 1995 the IPCC estimated that by the end of this century the average global surface temperature would increase 1.3–2.5°C.[30] The range of projections depended on the estimated population growth, global economic growth, and the associated growth in greenhouse gas emissions. Considering that the average increase of 0.6°C is already causing climate problems, any further increase by even the lowest estimate of about 1°C is undesirable. New projections by the IPCC give higher values for the expected increase: 1.4–5.8°C by 2100 (see Table 5-5).[31] The global climate will continue to change—perhaps even faster than in the past.

Calculations carried out at the University of East Anglia in the UK give a comprehensive overview of expected changes in the mean surface temperature of 188 countries during this century.[32] The calculations were carried out using five climate models (two from a modelling centre in UK, one from Germany, one from Canada, and one from Japan). For most countries, the mean values of expected temperature increases are between 4°C and 5°C. However, increases in excess of 6°C are expected in Canada, Kyrgyzstan, and Russia (see Box 5-2).

A team of scientists at the Hadley Centre for Climate Predictions in Bracknell, UK, recently published results of calculations based on a general circulation model. The model includes for the first time all conceivable factors, natural and anthropogenic, that have an influence on the change of global climate.[33]

Two opposite phenomena influence the degree of global warming: the warming-up effect of greenhouse gases and a net cooling down effect of aerosols. If all factors are included, the calculations show that the increase in the mean global surface temperature during this century will be 5.5°C, which is close to the maximum increase predicted by the IPCC.

Global warming can also turn into a self-accelerating process. As we have seen earlier (see Box 5-1), the soil stores large quantities of carbon. Rising surface temperature will increase the evaporation rate of soil moisture, thus increasing the rate of carbon release, which in turn will speed up global warming and increase surface temperature. This will release more carbon from the soil, and so on. Such runaway warming could also lead to a sudden change of climate.

Box 5-2
Global survey of the projected increase in surface temperatures during the 21st century

Increase of less than 3°C. Only three countries are listed in this category: New Zealand (2.1°C), Uruguay (2.6°C), and Ireland (2.9°C). For the remaining countries, the projected increases are higher.

Increase of 3°C to 3.9°C. Chile, Argentina, parts of the Middle America, West Indies, Greenland, Belgium, England, Indonesia, Malaysia, Vietnam, New Guinea, the Philippines, and Japan.

Increase of 4°C to 4.9°C. Most countries of the world, including most countries of Europe, the north countries of South America, Mexico, the US (4.9°C) most countries of Africa, Turkey, Syria, Israel, India, Pakistan, Nepal (4.8°C), and Australia (4.1°C).

Increase of 5°C to 5.9°C. Saudi Arabia, Iraq, Iran, Afghanistan, Turkmenistan, Kazakhstan, Mongolia, China (5.3°C), and a number of countries of Africa.

Increase above 5.9°C. Canada (6.3°C), Kyrgyzstan (6.2°C), and Russia (6.7°C).

Example: The projected increase in the average surface temperature in the course of the current century is only 2.1°C in New Zealand, but it is 6.7°C in Russia.

Source: Mitchell and Hulme 2000

International contributions to climate change. Between 1900 and 2000, industrialised nations contributed an estimated 63 per cent to the build-up of carbon dioxide in the atmosphere from industrial activities (see Table 5-6). The US and Canada contributed 25 per cent, and Europe contributed 20 per cent, about the same as Asia.[34]

The average carbon emission per person in industrialised countries is about seven times higher than in developing countries; however, industrialised countries will not be affected as strongly by the change of climate.[35] This conclusion is based on the relation between gross domestic product (GDP) and expected increases in surface temperature.

Table 5-6 Burden of responsibility for climate change

Country/Region	Contribution (per cent)
USA and Canada	25
Europe	20
FSU	12
Japan/Australia	6
Total industrial countries	**63**
Asia	19
SCA	10
MENA	4
Sub-Saharan Africa	4
Total developing countries	**37**

FSU – Former Soviet Union
SCA – South and Central America
MENA – Middle East and North Africa
Contribution – Relative contribution to climate change from industrial activity between 1900 and 2000

Example: The USA and Canada have contributed 25 per cent to climate change.

Source: Baumert and Kete 2002

For instance, Australia and Ethiopia can expect a similar increase in the mean surface temperature during this century. However, a person living in Australia will have incomparably more money to cope with climate change than a person living in Ethiopia. An Australian will be able to install an air-conditioning unit at home, but an Ethiopian will roast in an arid country. An Australian will have more money to deal with droughts and floods, to feed livestock, to irrigate crops, and do many other things that a poor Ethiopian will not be able to do. Heavy polluters will suffer less than those who pollute less.

Slow progress in tackling environmental problems.[36] Rapid technological progress does not match progress towards a sustainable future. The following time scale of events shows how reluctant we are to change. We have been busy travelling, meeting, debating, and writing reports, but there has been little action. The desirable course of action would be to take a radically new direction and break the spell of critical global trends.

Of the four future pathways,[37] the least likely seems to be the Sustainability First development. We shall probably follow the Market

First development (business as usual) or, even worse, the Fortress World development (global apartheid). The signs of the Fortress World future can be seen already in the increasing division between rich and poor, and in the global distribution of military power (see Chapters 7 and 8).

In 1972, an international conference, the United Nations Conference on Human Environment, was held in Stockholm, with only two heads of states in attendance. The result of the conference was the formation of the United Nations Environmental Program (UNEP). The conference identified areas of concern about the biophysical environment, brought environmental issues to the attention of governments, and launched an establishment of environmental ministries in various countries.

In 1983, the UN set up the World Commission of Environment and Development (WCED), also known as the Brundtland Commission. The aim was to examine global environmental issues and produce a report. This commission prepared the way for the now well-known concept of sustainable development.

In 1987, the Brundtland Commission published its report under the title of *Our Common Future*.[38] This report served later as the basis for Agenda 21. In the same year, 24 participating countries signed the Montreal Protocol on the reduction of ozone-depleting substances. The protocol went into force in 1989 and was followed by two reinforcing amendments: the London Protocol, signed in 1990 by 93 nations; and the Copenhagen Protocol, signed two years later by 87 nations. The Montreal agreement is probably the only big step forward we have made, and it seems to be producing desirable results.

In 1988, UNEP and the World Meteorological Organisation (WMO) established the IPCC, which in 1990 published its first assessment report in three volumes.[39]

In 1992, the UN Conference on Environment and Development, also known as the Earth Summit, was held in Rio de Janeiro. It attracted representatives of 176 governments. More than 100 heads of state were present, 10,000 delegates, 1400 representatives of non-government organisations, and 9000 journalists. Conference results were published as Agenda 21,[40] which contained recommendations for sustainable development. The conference also established the UN Commission for Sustainable Development and resulted in two international conventions, the UN Framework Convention on Climate Change, and the Convention on Biological Diversity. It also issued the Statement of Principles for the Sustainable Management of Forests.

In 1995, the first Conference of the Parties (COP1)[41] was held in Berlin, and the IPCC presented its second assessment reports.[42]

In March 1997, a Rio+5 (five years after the Rio Summit) international forum was held in New York. Concerns were expressed at slow progress with the implementation of Agenda 21.

In December 1997, the third Conference of the Parties (COP3) was held in in Kyoto, Japan, and it attracted representatives of 170 nations. This conference is famous for its first attempt to set up targets for the reduction of carbon emissions by industrialised countries. The debates were too close for comfort for many delegates. The conference was poised to end in a fiasco, but after last-minute negotiations on the introduction of allowances, concessions, and avoidance schemes, it produced the now well-known and controversial Kyoto Protocol.

In 2000, COP6 was held in The Hague. The purpose was to work out details for the implementation of the Kyoto Protocol, and delegates from 185 nations attended. It was a disaster, and negotiations broke down without reaching any agreement. An extra day or two probably would have helped to resolve the differences—but the venue was already booked for an oil convention.

In 2001, a series of events took place. In January, the IPCC released its third assessment reports,[43] which were scrutinised and accepted by all governments. In March, President George W. Bush made an official announcement that the US would have nothing to do with the Kyoto Protocol. Almost immediately, Australia followed suit. In July, representatives of 180 countries met in Bonn to try to put life back into the Kyoto Protocol. About 4500 participants, including 88 ministers, attended.

The meeting ended with the acceptance of the so-called Bonn Agreement,[44] a compromise about pollution allowances and trading rules[45] incorporated hastily in 1997 into the Kyoto Protocol.[46] The agreement was hailed as a breakthrough: without it the Kyoto Protocol was practically dead and buried.

According to the Bonn Agreement, industrialised countries will get pollution credits for revegetation, and for the management of forests, cropland, and pastures. Plants absorb carbon dioxide, so if you plant a tree or look after a forest in your country you will help to remove carbon from the atmosphere. To compensate for the good work nature is doing, you will be allowed to increase your carbon emissions.

Countries with vast areas of forests will have the right to take advantage of this allowance. For instance, the Kyoto target for Canada was a 6 per cent reduction in carbon emissions., but now Canada will be able to increase carbon emissions by 5 per cent. Russia will be able to increase emissions by 4 per cent.

The Bonn Agreement also includes rules for the so-called Clean Development Mechanism (CDM). If you help a developing country to develop cleaner technology you will be eligible to pollute more in your own country.

An extension to the CDM is known as joint implementation. The idea is the same, but it applies only to a collaboration between industrialised countries.

The Bonn Agreement made an attempt to define rules for pollution trading between developed countries. A country that is polluting less than allowed for in the Kyoto Protocol can sell pollution rights to a country that wants to pollute more. For the process of climate change it does not matter who pollutes the atmosphere, but it matters to us because the loss of pollution rights means the loss of profit. Our short-sighted actions are likely to result in huge economic loss for future generations.

In November 2001, COP7 was held in Marrakech, Morocco, resulting in the so-called Marrakech Accords,[47] which define more closely the rules for Kyoto Protocol implementation. The document redefines targets and gives additional credits to such countries as Canada, Japan, and Russia. With all these allowances, avoidance schemes, modifications, and escape rules, one wonders whether the Kyoto Protocol still has any meaning.

In 2002, the World Summit on Sustainable Development (also known as Rio+10) was held in Johannesburg, South Africa, from 26 August to 4 September. It attracted 40,000 delegates, including more than 100 heads of state and heads of government. The topics discussed included water, health, agriculture, energy, and biodiversity. The aim was to revive the decisions of the 1992 Earth Summit and find a way of implementing the recommendations of Agenda 21.

Some important points about the Kyoto Protocol. The primary aim is to set up targets for the reduction of carbon emissions. However, there are certain points we have to understand to appreciate this objective:

1. Developing countries, whose contribution to global carbon emissions is increasing rapidly, are not included.

2. The proposed carbon-emissions targets represent- only a minor reduction below the 1990 level. European Union countries are expected to reduce average emissions by 8 per cent, but they have their own internal arrangement that allows some to increase emissions at the cost of substantial reductions in other countries. The internal targets range from a 28 per cent reduction (Luxembourg) to a 27 per cent increase (Portugal).[48] The combined reduction by all industrialised countries listed in Annex I[49] is proposed to be only 5.2 per cent below the 1990 level.

3. It is not proposed to reduce emissions immediately, but to reduce them gradually by 2012.

4. This gradual reduction is not proposed to start immediately, but only after the year 2008.

5. The protocol can come into force if it were ratified by at least 55 countries responsible for at least 55 per cent of global carbon emissions in 1990. Without the participation of the US and Australia, the future of the Kyoto Protocol depended on Russia (see Table 5-7). Fortunately, Russia ratified the Protocol in November 2004 and it entered into force on 16 February 2005.

Table 5-7 Greenhouse gas emissions by developed countries in 1990

Country	Per cent of 1990 emissions
USA	33.6
European Union	23.9
Russia	16.9
Japan	6.6
Canada	3.4
Australia	2.4
Other	13.2
Total	100.0
Total[a]	64.0
Total[b]	47.1

[a] Total without the USA and Australia

[b] Total without the USA, Australia, and Russia

Example: Without the USA, Australia, and Russia, the combined contribution to greenhouse gas emissions in 1990 was only 47.1 per cent, or less than the 55 per cent required for bringing the Kyoto Protocol into force.

Source: POST 2002

What needs to be done about carbon emissions? Ideally, to heal the atmosphere we should stop emitting carbon. However, we continue to depend heavily on fossil fuels. Can we have a less radical yet effective solution to global warming? Will the proposed 5.2 per cent reduction by industrialised countries be enough to slow down substantially global climate change?

Calculations carried out in 1996 show that if we wanted to stabilise the atmospheric concentration of carbon dioxide at 450 parts per million [ppm], industrialised countries would have to cut emissions to about 10 per cent below the 1990 level before 2005, and to 30 per cent below before 2010.[50] Even then, atmospheric carbon dioxide—and the resulting increase in mean global surface temperatures—would still be too high. The elevated temperatures could still lead to sudden climate change. If we wanted to stabilise carbon dioxide at 350 ppm, industrialised countries would have to reduce carbon emissions to about 25 per cent below the 1990 level before 2005, and to 45 per cent below before 2010.

Consequently, to begin healing the atmosphere, the average reduction in carbon emissions by industrialised countries would have to be about nine times greater than the average reduction recommended by the Kyoto Protocol. However, there would still be no guarantee that the atmosphere could heal itself.

More recent calculations by Blanchard also show that the reduction of carbon emissions by industrialised countries would have to be substantially greater than the 5.2 per cent required by the Kyoto Protocol.[51] He considers three outcomes, which differ only in the relative share of carbon emissions between industrialised and developing countries. His calculations show that in order to achieve stabilisation of carbon dioxide at 450–550 ppm (that is, at 61–96 per cent above the pre-industrial level), emissions in 2030 by industrialised countries would have to be 18–38 per cent lower than in 1990.

He also shows that, with business as usual, carbon emissions by industrialised countries in 2030 will be 25 per cent higher than in 1990, which is the opposite to what we should be doing to heal the atmosphere.

Calculations by the IPCC show that even if we reached a maximum in global carbon emissions by the middle of this century, the damage already done would haunt us for a long time.[52] Carbon dioxide in the atmosphere and the associated global mean surface temperature would continue to increase for more than 100 years. Thermal expansion would mean a continued rise in sea levels for more than 1000 years;

the melting of ice would extend the process for thousands of years.

Are we already experiencing the effects of climate change?

A study of climate change is not easy, as records must be taken over a long period and it is necessary to distinguish between usual and unusual weather events. It is also necessary to study the frequency as well as the intensity of weather-related events. This is a long and laborious process, but we do have sufficient evidence of climate change.[53]

For instance, nine of the 10 warmest years on record occurred after 1990. The warmest was 1998, but 2001 and 2002 have also been declared exceptionally warm. These years had a now-familiar progression of weather-related disasters—droughts, heavy rainfalls, tropical storms, and floods (see Box A5-2, Appendix A).[54]

The number of tropical cyclones in Australia varied between one and eight per year before 1940, and between eight and 17 after 1960. The annual average before 1940 was five, and after 1960 it was 12. The maximum number of cyclones per year before 1940 was about the same as the minimum number after 1960.[55]

Records of the Swiss Re insurance group show that the number of natural catastrophes in the world increased from 30 per year in the 1970s to 140 per year at the end of the 20th century. Of the worst 40 natural disasters, only six were not related to the weather.[56]

The probability of global bankruptcy. Perhaps the best and the most convincing short-cut to the problems associated with studying the slow process of climate change and extreme weather lies in weather-related economic losses. The relevant records are not only well documented and scrutinised, but are also expressed in terms of a single quantity we can easily understand and appreciate—the money we have to pay for weather-induced damage. These records are maintained by insurance companies, and it is in their interest to make them reliable.

One such company is the Munich Re group, which has clients in more than 150 countries. Before being made available to clients, records of weather-related losses are checked and verified several times. They involve large sums of money, and many companies rely on their accuracy.

According to Munich Re, global weather-related economic losses increased from $3 billion per year in 1980 to $80 billion per year at the end of the 20th century. Losses per decade increased from $86 billion for 1980–89 to $474 billion for 1990–99.[57]

Only a small percentage of the losses are covered by insurance, but someone has to pay for them. Only 34 per cent of Australia's weather-related losses in 1998 were covered. In that year only 29 per cent were insured on the continent of America, 27 per cent in Europe, 7 per cent in Africa, and 4 per cent in Asia.[58]

Global weather-related losses covered by insurance increased from $26.2 billion for the 1980–89 decade to $123.5 billion for 1990–99.[59] These data show that, on average, only 26 per cent of weather-related losses were insured. Accurate records of uninsured losses are also important for insurance companies, because they contribute to an understanding of what is insurable.

How long can we cope with weather-related economic losses? If global income is substantially greater than the losses, and if it increases at least as fast as the losses, we have nothing to worry about. There will always be enough money to repair the damage. If global income increases more slowly than the losses, it is worthwhile to calculate how long the money will last.

To estimate this period I have analysed the data for weather-related economic losses[60] and for gross world product (GWP),[61] both expressed in 2001 US dollars.

Preliminary examination of the data shows that the prospects are not encouraging, because the losses are increasing much faster than income. As we have seen, global weather-related losses per decade increased from $86 billion to $474 billion, or 450 per cent, in the last two decades of the 20th century. However, GWP increased from $291 trillion per decade to $386 trillion, or 33 per cent, during the same period. GWP is still greater than the weather-related losses, but the losses are increasing much faster, and in time they might match global income. That would mean global bankruptcy.

Weather-related economic losses can be fitted by using exponential function. The best fit corresponds to a doubling time of 4.42 years. GWP can be fitted using a polynomial function, which increases slowly and has no doubling time. The two calculated curves cross in 2045. If about that time we decide to repair the damage there will be no money left for anything else.

Great natural disasters.[62] If the calculations are correct, global bankruptcy could occur earlier than 2045, for two reasons. First, the costs of weather-related damages might drain global resources to a critical level, even if they are lower than global income. Second, we also have to consider the cost of other natural catastrophes.

A subgroup of natural catastrophes is made up of the so-called great natural disasters. They do not happen often, but they are responsible for enormous economic losses, and their number has increased from about two per year in the 1950s to nine per year in the 1990s.

A great natural disaster is defined as one in which thousands of people are killed and hundreds of thousands made homeless, or when economic loss is so great that the affected region cannot recover by itself. An analysis of economic losses caused by great natural disasters is less reliable because the data vary over a large range of values.[63] However, tentative analysis shows that the costs of such events could alone lead to global bankruptcy about 2135.

Considering that these costs represent only a fraction of the total cost of all natural disasters, including weather-related events, the analysis shows again that global bankruptcy might occur well before the end of this century.

How reliable is the prediction of global bankruptcy? The projections of global income (GWP) can be accepted with confidence because of the high quality of the data. However, the projections of weather-related economic losses are less reliable, because only a relatively smaller number of data points are available and the data quality is not as high. The substantial increase in the cost of weather-related economic losses occurred only in the last decade of the 20th century, and the data points do not follow a smooth line.

An important parameter in predicting global bankruptcy is the doubling time of weather-related economic losses. This parameter is determined by fitting the data with an exponential function, so it depends on the range and the quality of the data.

With the current set of data, the analysis gives a doubling time of 4.42 years and global bankruptcy in 2045. If future weather-related losses are lower than expected, global bankruptcy will be delayed. For instance, with a doubling time of 6.6 years, global bankruptcy will be delayed until the end of this century. If future weather-related losses are higher than expected, global bankruptcy will occur earlier.

The first 10–20 years of this century will be crucial in determining whether weather-related economic losses will lead to global bankruptcy. The cost of damage is increasing rapidly, and the additional sets of data will make the analysis more reliable.

It is too early to argue about the probability of weather-related global bankruptcy: it is not the most important point at this stage. The point is that the risk of global bankruptcy is real, and we might have to deal with it in the not-too-distant future.

Insurance companies are worried about the effects of global warming and the increasing cost of weather-related natural disasters. This cost is increasing so rapidly that properties might become uninsurable. For instance, in the 1990s, the loss covered by insurance was 15 times greater than in the 1960s.[64] The Chartered Insurance Institute estimates that a rapid increase in weather-related economic losses could mean global bankruptcy before 2065.[65] It is interesting that this is close to the date I determined by direct mathematical analysis of the data. There also seems to be a high likelihood of national bankruptcies caused by local, weather-related economic losses.

Climate change and the oceans. Global warming has an undesirable effect on the oceans. Snow and ice covers are melting, the oceans are getting warmer, and sea levels are rising.[66] These changes affect not only aquatic life, but also life on land, including the life of humans.

Projections by the IPCC show sea levels rising by up to 30 centimetres by 2050, and by up to 90 centimetres by the beginning of the 22nd century. Coastal regions support the bulk of the world's population, and many will be affected by rising sea levels. Millions of people will be displaced, and millions of hectares of arable land lost. Some of the most affected areas will be the coastal regions of Bangladesh, India, Indonesia, Vietnam, Malaysia, Egypt, Nigeria, Côte d'Ivoire, Senegal, Gambia, Argentina, Belize, Guyana, Uruguay, and Venezuela. People in the Philippines will lose their properties; most of the Maldives islands and hundreds of Pacific atolls will disappear.[67]

Many densely populated cities are threatened by rising sea levels, including Shanghai, Manila, Dhaka, Bangkok, and Jakarta.

Coral reefs all over the world are dying. They provide shelter for a rich variety of life, but this shelter is being destroyed and with it the species that depend on it.[68] The estimated global value of coral reef goods and services is about $375 billion per year. It is projected that up to 60 per cent of reefs in the world will be destroyed by 2030.[69]

About 60 per cent of coral reefs have been destroyed in the Indian Ocean, 40 per cent in the Middle East, 35 per cent in South-East and East Asia, 20 per cent in the Caribbean, and 10 per cent in the Pacific. Millions of people depend on ocean resources for their livelihood.

Air pollution and human health.[70] The continuing use of fossil fuels causes air pollution, which has harmful health effects for an increasing number of people. Urban areas are most affected, and their populations are increasing rapidly. Air pollution exists in various forms: suspended particulate matter, oxides of nitrogen, ground-level ozone, polycyclic aromatic hydrocarbons, sulphur dioxide, carbon monoxide and carbon dioxide, heavy metals (for example, lead and arsenic), methane, and acid mists.

Suspended particulate matter. Suspended particulate matter (SPM) includes soot, ash, smoke, and dust from the burning of fossil fuels, and from industrial discharge. The associated health risks depend on particle size. Very small particles (called PM-2.5, with a diameter of less than 2.5 micrometres)[71] are more dangerous than larger particles (PM-10, with a diameter of 2.5-10 micrometres). PM-2.5 particles can be inhaled deep into the lungs, where they can enter the bloodstream. Vehicle exhaust fumes and industrial discharge contain large quantities of PM-2.5 particles. The less harmful PM-10 particles are mainly in the form of sand and dust blown by the wind.

Health effects associated with SPM of various sizes include runny or stuffy noses, sinusitis, sore throat, hay fever, eye irritation, wheezing, coughing, phlegm, shortness of breath, chest pain, aggravation of asthma attacks, acute or chronic bronchitis, cardiovascular disease, chronic obstructive pulmonary disease, and premature death. About 1.9 million people in developing countries die each year from exposure to SPM.

In 1995, in the group of cities each with a population of more than 9 million, the highest concentration of SPM was in Delhi. Concentrations there were seven times an acceptable level. Close behind were Beijing and Calcutta. Other strongly polluted cities were Tianjin, Mexico City, Shanghai, Bombay, Manila, and Rio de Janeiro.

Almost all cities are polluted with SPM. In many of them the pollution level is higher than the guideline level of 90 micrograms per cubic metre [$\mu g/m^3$], but still below 200 $\mu g/m^3$. However, in many large cities the pollution levels are higher than 200 $\mu g/m^3$. These cities are mainly in China and India.[72]

Nitrogen oxides. About 80 per cent of emitted nitrogen oxides comes from car engines. The rest is from domestic, commercial, and industrial combustion. Nitrogen oxides irritate and damage the respiratory tract, increase the likelihood of bacterial infections, and lead to chronic wheezing, coughing, prolonged respiratory illness, and even death. High concentrations of nitrogen dioxide in the air, and heat, contribute to a risk of cerebral infarction and cerebral ischemia in males over 65 years of age.[73]

In 1995, the heaviest nitrogen oxide pollution was in Mexico City and Beijing, with concentrations close to three times an international health limit. Other cities with high levels were Sao Paulo, New York, Shanghai, Los Angeles, Tokyo, Osaka, and Seoul. In 1990, the highest emission of nitrogen oxides was in the US, followed by the former Soviet Union, European OECD countries, South America, and East Asia.[74]

Ground-level ozone. Ground-level ozone is present in urban smog and is produced by an interaction of nitrogen oxides, ultraviolet (UV) radiation and volatile organic compounds (VOCs), which come from motor vehicles, and industrial sources. Ozone is a noxious gas because it is a powerful oxidising agent. It easily destroys cell membranes and thus the cells. It causes inflammations of the airways and scarring, and increases susceptibility to bacterial infection. Some studies show that exposure to ground-level ozone increases the risk of cancer. The most vulnerable members of society are children and elderly people. All cities with urban smog are exposed to varying levels of ground-level ozone.

Volatile organic compounds. VOCs include such compounds as propane, butanes, benzene, toluene, esters, ethers, formaldehyde (a compound used for preserving human cadavers), and ketones. Many of them are toxic; they may lead to headaches, nausea, sensory irritation, and cancer.

The highest concentration of VOCs is in heavily populated and industrialised regions of the world. About 95 per cent of VOC emission is in the Northern Hemisphere.[75]

Polycyclic aromatic hydrocarbons.[76] The group of polycyclic aromatic hydrocarbons (PAHs) contains about 10,000 compounds. They are produced not only by burning wood or coal, but also by burning oil, gas, petrol, diesel, garbage, and other organic matter such as tobacco or meat. They are also present in tars, dyes, plastics, and pesticides.

PAHs have an attractive, even pleasant, smell, but many of them are carcinogenic. They can cause damage to skin, lungs, liver, and kidneys. They can also cause reproductive problems and interfere with the development of unborn babies.

Sulphur dioxide. This compound is emitted by car engines, and by power plants, smelters, oil refineries, and other industrial sources. It causes swelling of airway tissue, aggravates asthma and other lung diseases, and irritates the respiratory tract. Exposure to sulphur dioxide can be lethal. Sulphur dioxide in the atmosphere reacts with water to form sulphuric acid, which falls as acid rain.

In 1995, the heaviest sulphur dioxide pollution was in Rio de Janeiro, with concentrations three times an international health limit. Other cities with strong sulphur dioxide pollution included Moscow, Beijing, Tianjin, Mexico City, and Cairo. In 1990, the highest emissions of sulphur dioxide were in South Asia, followed by the former Soviet Union, European OECD countries, and the US.[77]

Carbon monoxide. The main source of carbon monoxide is the burning of petrol in motor vehicles. It is a toxic gas that cannot be seen, smelled, or tasted. It combines with haemoglobin about 220 times more readily than oxygen and thus deprives the body of oxygen. It affects cerebral and heart functions. It can cause headaches, heart damage, and death. Carbon monoxide combines preferentially with foetal haemoglobin and can interfere with the development of the foetus. It can also interfere with the growth and development of young children.

Lead. Because of physical and metabolic differences, young children absorb up to eight times more lead than adults. Lead has a strong influence on children's physical and intellectual development and performance.

After entering the body, lead is distributed everywhere, in soft tissue, such as the brain, kidneys, and liver, and in mineral tissue, such as bones and teeth. It can cause neurotoxicity, chronic kidney toxicity, and sterility. It affects bone function (growth and stature) and the metabolism of vitamin D. Lead contributes to stillbirths, and to neonatal morbidity and mortality.[78]

Worldwide, 85 per cent of petrol is now lead-free. However, it has been estimated that 120 million people have elevated levels of lead in their bodies. About 97 per cent of children affected by lead exposure live in developing countries. Lead is also responsible for 234,000 deaths each year and for the loss of 12.9 million productive life years.[79]

A summary of expected outcomes of climate change. The effects of climate change have been discussed by many authors, but a good outline may be found in the recent IPCC reports (see Box 5-3).[80]

Climate change will affect agriculture and food security. A higher concentration of carbon dioxide in the atmosphere is likely to increase the yield of crops in certain parts of the world, but it will also help weeds to grow. Droughts and floods are likely to offset any gain from increased yields.

Climate change will affect the hydrologic cycle and thus water availability in various parts of the world. Model calculations show that by 2050 the global pattern of annual runoff water will be substantially changed. Many areas that have enough runoff water will become dry. Other areas will be flooded.

The rising temperature will affect terrestrial and aquatic ecosystems. The risk of extinction will increase for many species. The rich habitat of coral reefs is in danger of being destroyed.

Climate change will have a harmful effect on human health. There will be an increase in the number of people suffering infectious diseases. The number of people suffering diarrhoeal and respiratory diseases, particularly in developing countries, is expected to increase. Heatwaves are expected to affect a great number of people, particularly those who live in urban areas. People living in coastal areas will experience flooding caused by rising sea levels and by coastal storm surges. Weather-related economic losses will increase

The IPCC also makes projections for individual regions of the world. Africa will experience frequent flooding, and droughts. The food supply will diminish and the incidence of infectious diseases will increase. Asia will also experience more floods, droughts, and fires. There will be more tropical storms. Water availability will decrease in arid areas, but it will increase in North Asia.

Australia is expected to experience more frequent cases of extreme weather in the form of severe storms, wind, heavy rain, tropical cyclones, and droughts. Water is going to be the main problem in Australia. Many unique ecosystems, the coral reef, wetlands, and coastal zones are going to be affected.

In Europe, 50 per cent of alpine glaciers will disappear during this century. The frequency of river flooding will increase. The north is likely to become warmer; the south, drier.

Box 5-3
A brief list of continuing and projected effects of global warming

Basic changes
    Increasing global and local surface temperatures
    Changes in precipitation patterns
    Rising sea levels

Health
    Increase in weather-related mortality
    Spread of infectious diseases
    Increase in respiratory problems

Agriculture
    Changes in crop yields
    Increased demand of water for irrigation
    Crop loss through floods and droughts

Forests
    Changes in forest cover
    Changes in forest composition

Water
    Changes in water availability
    Changes in water quality
    Increase in demand
    Increase in water-related conflict

Weather
    Change in weather patterns
    Increase in extreme weather events
    Increase in economic loss
    Possibility of global bankruptcy

Coastal areas
    Beach erosion
    Flooding of many coastal areas
    Displacement of people from coastal areas

Biodiversity
    Displacement of species
    Loss of species

Source: Effects of climate change are discussed in a wide range of sources. See, for instance, McCarthy et al 2001 and IPCC 2001.

In Latin America, rapid melting of glaciers will affect water supply. The frequency of extreme weather events (heavy rainfalls, strong wind, tropical cyclones, storms, and flooding) will increase. The incidence of infectious diseases such as malaria, dengue fever, and cholera is expected to increase. The level of poverty is likely to rise.

All these projections do not consider the effects of a possible sudden climate change, which are in an entirely different category and could be even more disastrous.

The stratospheric ozone problem.[81] Stratospheric ozone serves as a shield against harmful ultraviolet (UV) radiation. Ozone in the stratosphere is produced by an interaction of UV light with oxygen.

Ozone is an unstable molecule that occurs in relatively small amounts, even in the stratosphere. In its densest part, about 35 kilometres above ground level, the stratosphere contains eight particles of ozone per million particles of air. If we could bring the whole stratospheric ozone layer to ground level it would be a maximum of 3 millimetres thick.

For countless ages, the stratosphere was beyond our reach. Now, for the first time in human history, technology has allowed us to extend our damaging influence to this remote environment.

The destruction of the stratospheric ozone layer is caused by ozone-depleting substances, mainly those containing chlorine atoms, which are released from man-made chlorofluorocarbons (CFCs). The chlorine atom is for ozone molecules like a nuclear bomb is for us. One chlorine atom can destroy up to 100,000 ozone molecules. Chlorine flows to the atmosphere not only from human activities, but also from natural sources (see Table 5-8). However, 84 per cent of the destruction of ozone by chlorine is caused by human activities.

When they were invented in 1928, CFCs were hailed as wonderful compounds. Chemically, they do not readily interact with other substances, so they were considered to be safe. Since about 1950, CFCs have been produced in large quantities and have been used as refrigerants, aerosol propellants, foam-blowing agents, and as solvents in the electronic industry.

It is because they are so stable that they have turned out to be so destructive. They can remain in the atmosphere for a long time, eventually drifting into the stratosphere. There, their fate is entirely different from the situation at ground level—they are no longer stable. Attacked by UV radiation, they break down and release chlorine atoms, which in turn destroy ozone molecules. The moral of the CFC story is that what we regard as safe may turn out to be damaging if we overlook even one link in the complex chains of ecological interactions.

Table 5-8 Relative contributions of chlorine and bromine substances to the depletion of stratospheric ozone

Chlorine-containing substances	
Natural source	Per cent
Methyl chloride (CH$_3$Cl)	16
Human-made sources	
CFC-12 (CCl$_2$F$_2$)	32
CFC-11 (CCl$_3$F)	23
Carbon tetrachloride (CCl$_4$)	12
CFC-113 (CCl$_2$FCClF$_2$)	7
HCFCs	5
Methyl chloroform (CH$_3$CCl$_3$)	4
Other gases	1
Total from natural sources	**16**
Total from human-made sources	**84**
Bromine-containing substances	
Natural sources	Per cent
Very short-living gases	16
Methyl bromide (CH$_3$Br)	27–42
Human-made sources	
Methyl bromide (CH$_3$Br)	5–20
Halon-1211 (CBrClF$_2$)	20
Halon-1301 (CBrF$_3$)	14
Other halons	4
Total from natural sources	**43–58**
Total from human-made sources	**43–58**

Example: Human-made sources of chlorine contribute 84 per cent to the destruction of the stratospheric ozone layer.

Source: WMO and UNEP 2003

It takes 5–10 years for CFCs to drift to the stratosphere, where they form a reservoir. Under the influence of UV radiation this reservoir leaks chlorine atoms into the atmosphere for decades. Some CFCs break up faster; some, slower. In general, their lifetime varies between 50 and 200 years. Even if we halt the production of CFCs their damaging influence will continue for a long time.

UV light can wound or kill living cells. It can destroy intricate and delicate cellular structures, including the code-bearing DNA molecules. In humans, UV radiation interacts via the skin and eyes, causing skin cancer, premature skin ageing, and cataracts. It also interferes with the immune system and suppresses its activity.

UV radiation retards the growth of plants. It penetrates easily through water and destroys plankton, an important food for water-living organisms. Plankton is at the lowest end of the food chain. Increased intensity of UV radiation upsets the balance of nature and affects the more complex organisms, including humans. Much of the animal protein consumed by humans comes from oceans, lakes, and rivers.

UV radiation inhibits photosynthesis. All plants, including phyto-plankton (plant-like plankton), use photosynthesis to convert carbon dioxide into oxygen and sugars. Thus they give partial protection against global warming. Phytoplankton plays an important role in this process because, as we have seen earlier, almost all carbon on the Earth is stored in oceans.

The use of ozone-destroying substances triggers an undesirable chain of events. Damaging the ozone layer in the stratosphere brings about an increase in the intensity of UV radiation, which damages or destroys living cells. One of the outcomes is that UV radiation impedes photosynthesis, and slows the conversion of carbon dioxide to oxygen. This contributes to the acceleration of global warming, which causes climate change with all its harmful effects. It is a domino effect—one change triggers a series of events. Global warming and depletion of the ozone layer are in principle different phenomena, but indirectly the depletion of the ozone layer contributes also to global warming and resultant environmental damage.

The destruction of the ozone layer occurs mainly over Antarctica, and to a lesser degree over the North Pole.

Why is ozone depleted mainly around the poles? First, it is because ice crystals help in the destruction of ozone. They absorb CFCs and provide the surface for destructive chemical reactions. Second, a vortex around each pole traps ozone and the CFCs in a kind of corral or cauldron. Nothing spectacular happens during winter, but in spring, with the first rays of light shining into the corral, UV radiation begins to break down CFC molecules, releasing the deadly chlorine atoms. They attack and destroy large numbers of trapped ozone molecules. The ozone-depletion process is more efficient above the South Pole than above the North Pole because the more uniform distribution of land and sea in the south makes the corral much tighter.

The depletion of the ozone layer in the stratosphere in spring is 50 per cent in the Antarctic region and 15 per cent in the Arctic. The springtime (September and October) thickness of the South Pole ozone layer, which used to be 300 Dobson Units,[82] has been reduced by half.

Global production of CFCs reached a peak of about 1 million tonnes per year in the late 1980s and fell to 50,000 tonnes per year by the end of the 20th century. This is a welcome change, but the influence of CFCs will remain with us for a good part of this century. Currently, the largest producers of CFCs are China and India, but China has promised to phase out production by 2010.

The future of the ozone layer is uncertain: not only CFCs, but also bromine-containing substances are contributing to its destruction (see Table 5-8). One of the most efficient ozone-killers is halon-1211. Emission of this substance was thought to have peaked in 1988. However, research carried out at CSIRO in Australia shows that emissions are increasing at 200 tonnes per year. About 90 per cent of global production of halon-1211 is in China.[83]

The continuing use of ozone-destroying substances does not seem to augur well for the ozone layer; however, opinions on this topic are divided. For instance, earlier model calculations[84] indicated that increasing concentrations of harmful gases in the atmosphere will delay the healing of the ozone layer. More recent calculations show that the Antarctic ozone hole is likely to heal itself by about 2040.[85]

In the recent four-yearly report, the Scientific Assessment Panel of the Montreal Protocol on Substances that Deplete the Ozone Layer says chlorine concentration in the atmosphere has reached a maximum and will probably start to decrease.[86] If this is confirmed, it should encourage us to work more vigorously on reducing greenhouse gas emissions.

The future. Global emissions of carbon will continue to increase. The projected increase is from 6 billion tonnes per year [Gt/y] in 2000 to 10 Gt/y in 2020.[87] It is also possible that carbon emissions will be even higher. If we adopted sustainable development we could reduce carbon emissions to 7 Gt/y by 2030, after an initial increase to 9 Gt/y. In the more likely case of business as usual, global carbon emissions will rise to 15 Gt/y by that year.[88]

With increasing carbon emissions, the atmospheric concentration of carbon dioxide will also increase. My calculations show that, by

2050, carbon dioxide concentration is likely to reach 450–500 ppm. This prediction is close to the IPCC predictions of 480–580 ppm.[89] Thus, by 2050, carbon dioxide concentration in the atmosphere is expected to be 20–60 per cent higher than in 2000 and 60–110 per cent higher than in pre-industrial times.

Even if we consider the most optimistic UNEP projections based on the assumption of the Sustainability First development, we shall find that the carbon dioxide concentration will rise to 450 ppm in 2050.[90] This will be 20 per cent higher than in 2000 and 60 per cent higher than in pre-industrial times. If we consider the more likely prediction based on the Market First development, we shall find that carbon dioxide concentration will rise to 550 ppm by 2050. In this case, the concentration will be 50 per cent higher than in 2000 and almost twice as high as in pre-industrial times.

Model calculations by Blanchard (2002) show that, with some effort, we might be able to stabilise the carbon dioxide concentration in the atmosphere at 450–550 ppm by 2030. But with the more likely business-as-usual scenario, the concentration will continue to rise.[91]

All these calculations show consistently that, unless we take positive and radical steps to reduce the emission of carbon, radiative forcing will increase, global mean temperature will rise, and the climate will continue to change. This will lead to severe and far-reaching outcomes.[92]

Large areas of coral reefs will be destroyed. Many coastal wetlands will be lost, many species will become extinct, the frequency of bush-fires will increase, ice covers will continue to melt, and water demand for irrigation will increase, but water supply will decrease in many areas. The frequency and intensity of extreme weather events will increase.

Small islands will be inundated and coastal areas will suffer losses from coastal storm surges. Climate change will increase the suffering and deprivation of people in poor countries. It will have a harmful effect on their health, and will reduce their access to food and water.

The only hopeful prognosis is that the stratospheric ozone layer might mend itself. However, the IPCC report does not share this optimistic outlook.[93] Humanity's interference with nature unleashes forces that are beyond our control. Extreme weather events inflict rapidly increasing economic losses, which could lead to global bankruptcy during this century.

Chapter 6
The Approaching Energy Crisis

'We are approaching a crisis of great proportions because oil
production capacity is reaching its limit.'

—Ali Rodriguez (1937–), OPEC president and Venezuelan oil minister,
12 September 2000

Human progress relies on an abundant supply of energy, but do we
have sufficient energy resources on our planet to cope with increasing
demand? Our future depends not only on the availability of various
energy sources, but also on our decisions. What choices do we have
and what decisions are we likely to make on future sources of energy?

Units of energy. Energy production, consumption, and resources are
measured using a variety of units such as joules, calories, British thermal units, watt-hours, barrels oil equivalent, tonnes oil equivalent,
cubic metres, and cubic feet. Navigation between the numerical values
of energy is often hindered by having to convert the various units. To
complicate the matter further, publications might use an unspecified
production-efficiency factor. However, this factor can be unravelled if
one is aware of its presence and knows the standard conversions.[1]

The internationally accepted unit of energy is the joule, but sometimes it is better to use more familiar units: for crude oil, these are
barrels and tonnes; for natural gas, cubic metres; and for coal, tonnes.
For electricity, the familiar unit is the watt-hour or its multiples, such
as kilowatt-hour. I use these units in preference to joules when discussing specific sources of energy.

Energy trends and projections.[2] Global consumption of energy
(total and per person) increased substantially during the 20th century
(see Table 6-1). Total consumption increased from about 30 exajoules
[EJ][3] in 1900 to 413 EJ in 2000.

In the first half, energy consumption increased a little more than
two-fold and in the second half nearly six-fold.

Table 6-1   Global annual consumption of energy, 1900–2020

Year	Total energy [EJ]	[GJ/p]	Fossil fuels [EJ]	Per cent
1900	30	20	15	50
1925	53	27	37	70
1950	73	29	60	82
1975	253	62	187	74
2000	413	68	351	85
2005	463	71	395	85
2010	520	75	447	86
2015	582	80	504	87
2020	645	84	562	87

EJ – exajoule (billion billion joules)
GJ/p – gigajoule (billion joules) per person
Per cent – Percentage of total energy consumption

Example: In 2000, the world consumed 413 exajoules of energy (68 billion joules per person), of which 85 per cent (351 exajoules) was in the form of fossil fuels.

Source: Based on the data of EIA 2002a and Meadows et al 1992

Global consumption of energy from all sources increased at the rate of 8.6 EJ per decade between 1900 and 1950, and by 68 EJ per decade, or close to eight times faster, from 1950 to 2000. Over the entire 20th century, global consumption of energy increased nearly 14-fold.

Global population is increasing, so it is not surprising that the consumption of energy is also increasing. However, the consumption per person per year has also increased, from 20 gigajoules in 1900 to 68 gigajoules in 2000. By 2020, our demand for energy will be 84 gigajoules per person per year.

Global consumption of energy per person per year rose only 45 per cent in the first half of the 20th century, but increased by 134 per cent in the second half. Consumption per person in 2020 will be more than four times higher than in 1900.

Fossil fuels are our main source of energy. Unlike renewable sources of energy, they are concentrated in small areas, which can be accessed and exploited relatively easily, but they are already causing great damage to the environment and the cost will only increase in the future.

Our reliance on fossil fuels increased from 50 per cent of energy requirements in 1900 to 85 per cent in 2000. It is projected to be 87 per cent in 2015. The consumption of fossil fuels increased more than 23-fold during the 20th century. In the first half, the consumption of fossil fuels increased at the rate of 9 EJ per decade, but speeded up to 58 EJ per decade in the second half.

Until at least 2020, fossil fuels will be our main source of energy (see Table 6-2). Between 2000 and 2020, global demand for energy from this source will rise about 60 per cent. In particular, global demand for crude oil will increase by 55 per cent (see Table 6-3). Global natural gas consumption will increase 87 per cent and coal consumption 41 per cent. This prognosis does not bode well for the health of our planet.

Table 6-2 Global consumption of energy by fuel, 1990–2020

Year	Fossil		Nuclear		Other		Total
	[EJ]	Per cent	[EJ]	Per cent	[EJ]	Per cent	[EJ]
1990	315.9	86	21.5	6	28.0	8	365.4
2000	350.6	85	27.0	6	35.7	9	413.3
2005	394.8	85	28.4	6	39.7	9	462.9
2010	446.8	86	29.0	6	43.9	8	519.7
2015	503.9	87	29.2	5	49.0	8	582.1
2020	562.1	87	29.5	5	53.5	8	645.1
Increase	60		10		50		56

EJ – exajoule (billion billion joules)
Per cent – Per cent of the total consumption of energy
Increase – Increase (in per cent) between 2000 and 2020

Example: In 2000, 85 per cent of global consumption of energy (350.6 exajoules) was in the form of fossil fuels, 7 per cent (27 exajoules) in the form of nuclear energy, and 9 per cent (35.7 exajoules) in the from of all other sources of energy. The total consumption of energy in 2000 was 413.3 exajoules. Between 2000 and 2020, the consumption of fossil fuels is projected to increase by 60 per cent.

Source: Based on the data of EIA 2002a

In 2000, crude oil accounted for 40 per cent of global energy consumption; natural gas, 23 per cent; and coal, 22 per cent. Nuclear energy accounted for only 6 per cent, and all other primary resources (such as hydroelectricity, geothermal, solar, and wind) for only 9 per cent.

Between 2000 and 2020, global consumption of nuclear energy will increase by only 10 per cent, but the consumption of all other energy forms (that is, excluding fossil fuels and nuclear energy) will increase by 50 per cent. However, the relative contribution of nuclear energy to total consumption will fall from 6 per cent in 2000 to 5 per cent in 2020, and the contribution of all other sources from 9 per cent to 8 per cent.

Table 6-3 Global consumption of fossil fuels, 1990–2020

Year	Oil [EJ]	Oil Per cent	Natural gas [EJ]	Natural gas Per cent	Coal [EJ]	Coal Per cent
1990	142.3	39	78.6	22	95.0	26
2000	164.3	40	94.9	23	91.4	22
2005	183.0	40	111.0	24	100.9	22
2010	206.2	40	130.2	25	110.5	21
2015	231.1	40	153.8	26	119.0	20
2020	255.1	40	177.9	28	129.0	20
Increase	55		87		41	

EJ – exajoule (billion billions joules)
Per cent – Percentage of the total energy consumption
Increase – Increase (in per cent) between 2000 and 2020

Example: In 2000, 40 per cent (164.3 exajoules) of the global consumption of fossil fuel energy was in the form of crude oil, 23 per cent (94.9 exajoules) in natural gas, and 22 per cent (91.4 exajoules) in coal. Between 2000 and 2020, the consumption of crude oil is projected to increase by 55 per cent.

Source: Based on the data of EIA 2002a. See also EIA 2000, 2001 and BP 2001, 2002a. The numerical data in the EIA 2000, 2001, 2000a are not exactly the same as in BP 2001, 2002a.

Consumption of energy by regions, countries, and levels of development.[4] The United States has the highest national consumption of energy (see Table 6-4). In 2000, it accounted for a quarter of the global total. China consumed 9 per cent. The highest regional consumption is in North America. In 2000, it was higher than in Asia and the same as for all the countries of Europe and the former Soviet Union.

By 2020, energy consumption in the US will increase by 33 per cent and in China by 150 per cent. However, even then, the total consumption of energy in China will be much lower than in the US. The largest increase (92.5 EJ) in energy consumption between 2000 and 2020 will be in Asia. By 2020, the region's consumption will be about the same as in North America.

In 2000, energy consumption in all the developed countries accounted for 68 per cent of global consumption (see Table 6-5). A person living in a developed country consumes, on average, nine times more energy than a person living in a developing country.

Table 6-4 Consumption of energy by country and region, 1990–2020 (exajoules per year)

Country/Region	1990	2000	(1)	2005	2010	2015	2020	(2)	(3)
USA	88.8	104.2	25	113.5	122.0	130.4	138.1	33.9	33
FSU	64.0	42.2	10	46.5	50.6	56.0	60.2	18.0	43
China	28.5	35.6	9	45.3	58.1	72.6	89.1	53.5	150
Japan	18.9	23.1	6	24.2	25.5	26.8	28.1	5.0	21
Germany	15.6	15.0	4	16.1	16.8	17.3	17.9	2.9	20
Canada	11.5	13.4	3	14.5	15.6	16.7	17.6	4.2	31
UK	9.8	10.6	3	11.3	11.8	12.3	12.9	2.3	22
France	9.3	11.0	3	11.8	12.3	13.0	13.7	2.7	24
India	8.2	13.4	3	16.0	19.2	23.0	26.8	13.4	100
North America	105.6	124.5	30	136.4	148.2	159.6	170.2	45.7	37
Asia	53.8	78.6	19	97.6	120.2	144.7	171.1	92.5	118
Western Europe	63.1	70.6	17	75.4	78.8	82.0	86.0	15.4	22
EE/FSU	80.5	54.3	13	59.9	65.2	72.0	77.4	23.1	43
CSA	14.5	21.4	5	24.0	29.9	37.6	45.5	24.1	112
Middle East	13.8	20.8	5	23.2	27.7	32.2	36.7	15.9	76
Africa	9.8	12.8	3	14.8	16.6	19.1	21.4	8.6	67
Australasia	5.1	6.6	2	7.2	7.7	8.2	8.8	2.2	32
World	365.3	412.9	100	462.8	519.7	582.1	645.2	232.3	56

Countries and regions are arranged in descending order of energy consumption for 2000.

(1) – Percentage of global energy consumption in 2000. For instance, in 2000, 25 per cent (104.2 exajoules) was in just one country, the USA.

(2) – The increase in the annual consumption of energy (in exajoules) between 2000 and 2020. For instance, the annual consumption of energy in the USA in 2020 will be 33.9 exajoules higher than in 2000.

(3) – Percentage increase between 2000 and 2020. For instance, energy consumption in the USA will increase by 33 per cent.

exajoule – billion billion joules

FSU – Former Soviet Union

EE/FSU – Eastern Europe/Former Soviet Union

CSA – Central and South America

Source: Based on the data of EIA 2002a

By 2020, energy consumption in industrialised countries will increase by 91.4 EJ and in developing countries by 141 EJ. The gap between total energy consumption in industrialised and developing countries, which is now about 145 EJ, will narrow to 96 EJ in 2020. However, the gap in energy consumption per person will increase from 210 gigajoules to 250 gigajoules. It seems therefore that, with energy consumption per person, developing countries do not have a promising future. Their consumption of energy—total and per person—may be increasing, but so is the gap in energy consumption per person between them and developed countries.

Table 6-5 Consumption of energy by level of development and globally, 1990–2020

	1990	2000	(1)	2005	2010	2015	2020	(2)	(3)
Industrialised									
Total [EJ/y]	273.3	279.2	68	303.2	325.2	348.6	370.6	91.4	33/7
Per person [GJ/p/y]	248.7	236.4		249.4	262.3	277.1	292.5	56.1	24/7
Developing									
Total [EJ/y]	92.0	133.7	32	159.5	194.2	233.5	274.6	140.9	105/30
Per person [GJ/p/y]	22.0	27.5		30.4	34.6	39.0	43.2	15.7	57/30
World									
Total [EJ/y]	365.3	412.9	100	462.8	519.7	582.1	645.2	232.3	56/26
Per person	69.2	68.3		71.6	75.7	80.3	84.7	16.4	24/26
Gap									
Total [EJ/y]	181.3	145.6		143.7	131.3	115.1	95.9	-49.7	-34
Per person [GJ/p/y]	226.7	209.0		219.0	227.7	238.1	249.2	40.2	19

(1) – Percentage of global consumption in 2000. For instance, industrialised countries were consuming 68 per cent.

(2) – The increase in annual energy consumption between 2000 and 2020. For instance, total energy consumption in industrialised countries in 2020 will be 91.4 exajoules per year higher than in 2000. Consumption per person will be 56.1 billion joules higher per year than in 2000.

(3) – Percentage change in annual energy consumption between 2000 and 2020. For instance, for industrialised countries, the total consumption of energy will increase by 33 per cent, but the population will increase by only 7 per cent. The annual consumption of energy per person will increase by 24 per cent.

Gap – The difference between energy consumption by industrialised and developing countries for total consumption and consumption per person. For instance, the gap for the total consumption is projected to decrease from 145.6 exajoules per year in 2000 to 95.9 exajoules in 2020 (49.7 exajoules), or by 34 per cent. However, the gap in consumption per person is projected to increase by 19 per cent.

EJ/y – exajoule (billion billion joules) per year; GJ/p/y – gigajoule (billion joules) per person per year
Source: Based on the data of EIA 2002a

The complex network of energy flows.[5] In order to understand global trends in the production and consumption of energy, and in order to appreciate the possible problems with future demand and supply, it is necessary to understand the complex network of energy flows (see Tables A6-1 and A6-2 in the Appendix A). An explanation has to be in two parts: energy flows at the source and at the sink—that is, at an area of application.

Energy can flow not only directly from a primary source to a sink, but also indirectly via the generation of grid electricity. This intermediate step acts as sink and source.

The sources to consider are: crude oil, coal, natural gas, nuclear energy, hydroenergy, all other primary sources of energy, and grid electricity. The sinks are: transport, industry, residential and commercial applications, non-energy applications, losses, and grid electricity.

The values listed in Tables A6-1 and A6-2 are approximate, as energy flows change over time. There are also differences in flow estimates in various publications. However, the values in these tables should be a sufficiently reliable guide to the energy network.

Energy flow at the source

Crude oil. Transport consumes an estimated 44–60 per cent of global crude oil production. The next application is industry, followed by residential and commercial. The use of crude oil in generating electricity is low, and about 6 per cent of oil is lost in refinery operations.

Coal. Industry uses an estimated 32–75 per cent of global production. The next application is electricity generation, which uses 30–49 per cent.

Natural gas. The use of natural gas is almost equally divided between industry, residential and commercial, and electricity generation. A large fraction of natural gas is lost in transmission.

Hydroenergy and nuclear energy. These energy sources are used exclusively in electricity generation.

Other primary sources of energy. These sources support mainly residential and commercial applications, but they also support industry.

Electricity. This source is used mainly by industry and by the residential and commercial sectors. Transport uses only about 2 per cent of electricity.

The production and consumption of electricity involves substantial loss of energy. In addition, 10 per cent of electricity is lost in transmission and distribution. These additional losses vary between countries and regions. In 1996, the loss in South Asia was 18.7 per cent, in East Asia 10.1 per cent, sub-Saharan Africa 9.6 per cent, and North America 7.6 per cent. The average transmission loss in member countries of the Organisation for Economic Cooperation and Development (OECD) was 6.4 per cent and in the least-developed countries 20.9 per cent.[6]

Energy flow at the sinks

Transport. Crude oil supports 96 per cent of transport. This is an undesirable and potentially critical situation. Coal supports almost entirely the remaining 4 per cent. The contributions of natural gas and electricity are relatively small.

Industry. This sector depends more on coal than on any other energy source. However, the contributions of crude oil and natural gas are also relatively high. Unless compensated by coal, the shortage of crude oil might slow industrial production. During this century we shall also have to deal with the shortage of natural gas. A gradual replacement of the two sources by coal might help to solve the problem of industrial production, but not the problem of atmospheric pollution. The expected increase in industrial development in richer and poorer countries will promote global warming and associated changes in global and local climate.

Residential and commercial. These sectors depend heavily on fossil fuels. However, the degree of their dependence is nearly the same for crude oil, coal, and natural gas. As these sectors also depend on other primary sources of energy, the development of alternative sources could reduce their consumption of fossil fuels.

Electricity generation. This activity is mainly supported by coal, which is both a reassuring and worrying aspect. It is reassuring because the expected oil crisis is not likely to affect this sector. However, it is worrying because the oil shortage can probably be easily compensated by coal. One way or another, the production of grid electricity will continue to contribute to carbon accumulation in the atmosphere, and there will be also no pressing need to develop alternative sources of energy for this sector. Furthermore, as we shall see later (see Table 6-14), the relative contributions of hydroelectricity and nuclear energy will gradually decrease.

Non-energy use of crude oil. Crude oil supports about 73 per cent of non-energy applications. The shortage of crude oil is not going to have a strong direct effect on industrial output, but as a source of raw material it will. Thousands of products depend heavily on crude oil: asphalt, plastics, fertilisers, medicines, insulation, glues, solvents, detergents, antiseptics, tyres, deodorants, shoes, dresses, pillows, boats, insect repellents, paint brushes, linoleum, wood filler, rubbish bags, TV cabinets, videotapes, CDs, film, loudspeakers, credit cards, telephones, sweaters, pantyhose, cassettes, toilet seats, life jackets, toys, contact lenses, car battery cases, hand lotion, fishing rods, water pipes, oil filters, and many more.[7]

Crude oil reserves. Demand for this energy source will continue to increase (see Table 6-3). To appreciate its future role we need to understand not only how it is used, but also estimate the reserves and know how they are distributed. It is necessary to identify the main users and producers of crude oil.

Table 6-6 Deposits of fossil fuels, 2001

Region	Crude oil [Gb]	Per cent	Natural gas [Tm³]	Per cent	Coal [Gt]	Per cent
Middle East	685.6	65.3	55.91	36.1	1.7	0.2
SCA	96.0	9.1	7.16	4.6	21.8	2.2
Africa	76.4	7.3	11.18	7.2	55.4	5.6
FSU	65.4	6.2	56.14	36.2	230.0	23.4
North America	63.9	6.1	7.55	4.9	257.8	26.2
Asia-Pacific	43.8	4.2	12.27	7.9	292.5	29.7
Europe	18.7	1.8	4.86	3.1	125.4	12.7
World	**1 050.1**	**100.0**	**155.07**	**100.0**	**984.6**	**100.0**
OECD[a)]	85.0	8.1	14.87	9.6	445.8	45.3

Regions are arranged in the descending order of crude oil reserves.

[a)] Members of Central Europe are not included.

Gb – gigabarrel (billion barrels);

Tm³ – tera cubic metre (trillion cubic metres);

Gt – gigatonne (billion tonnes).

FSU – Former Soviet Union. SCA – South and Central America.

Example: The Middle East holds 65.3 per cent (685.6 billion barrels) of global deposits of crude oil, 36.1 per cent (55.91 trillion cubic metres) of global deposits of natural gas, and only 0.2 per cent (1.7 billion tonnes) of global deposits of coal.

Source: BP 2002a

The largest deposits of crude oil are in the Middle East (see Table 6-6). The reserves in other parts of the world are relatively small, but there are also significant deposits in South and Central America, Africa, the former Soviet Union, and North America. Europe has the smallest deposits of crude oil. The deposits in all OECD countries, the main users of crude oil, account for only 8 per cent of global deposits.[8]

Since 1962, annual discoveries of new oil fields have been decreasing rapidly.[9] It is therefore doubtful whether future discoveries will boost substantially the known reserves. Oil fields are also not equally productive, and mining them is not always economically viable.

Oil companies, governments, and private corporations are interested in estimating how much oil might still be hidden from us. Numerous attempts have been made, but estimate is difficult and the results might well depend on who is producing the figures—buyers or sellers.

Estimated ultimately recoverable (EUR) reserves vary from as low as 1000 billion barrels to 3600 billion barrels. Most of them indicate EUR reserves of 2200 billion barrels. By the end of 1999, the world had used 857 billion barrels of these reserves.[10]

The production of crude oil usually follows the well-known, bell-shaped Gaussian distribution. The area under the Gaussian shape is equal to the total volume of reserves. Consequently, a peak in production is expected to occur when about half of the reserves are extracted. Even if we assume that EUR reserves are 2200 billion barrels, we are now not far from the half-way mark for extracted volume and therefore not far from a maximum in the global production of crude oil.

Ultimately recoverable reserves will be discussed in the last chapter of this book. This chapter is based on *proved* reserves.

Users and producers of crude oil.[11] The main users and producers of crude happen to be divided into two distinct groups: the OECD countries and OPEC.[12] OPEC holds about 78 per cent of global proved crude oil reserves, and OECD countries 8 per cent. In particular, the US has only about 3 per cent of global reserves, Australia, 0.3 per cent, North America 6 per cent, and Europe 2 per cent.

OECD countries account for about 63 per cent of global consumption, but only 29 per cent of global production (see Table 6-7). On the other hand, OPEC members account for 6 per cent of consumption, but about 40 per cent of production.[13] OECD countries have to buy crude oil: OPEC sells it.

Table 6-7 Producers and consumers of crude oil, 2001

Region/Group	Reserves [Gb]	Production [Gb/y]	Production Per cent	Production R/P	Consumption [Gb/y]	Consumption Per cent	Consumption R/C	Balance [Gb/y]
Middle East	685.6	8.1	30	84	1.6	6	436	6.5
SCA	96.0	2.6	9	38	1.7	6	56	0.8
Africa	76.7	2.9	10	27	0.9	3	84	1.9
FSU	65.4	3.2	12	21	1.2	5	53	1.9
North America	63.9	5.1	19	12	8.5	31	7	-3.4
Asia-Pacific	43.8	2.9	11	15	7.6	28	6	-4.7
Europe	18.7	2.5	9	8	5.9	21	3	-3.4
World	1 050.1	27.2	100	39	27.5	100	38	-0.3
OPEC	818.8	11.0	41	74	1.6	6	497	9.4
Non-OPEC	231.2	16.2	59	14	25.8	94	9	-9.7
OECD	85.0	7.8	29	11	17.3	63	5	-9.5

Regions and groups of countries have been arranged in descending order of crude oil reserves.

Gb – gigabarrel (billion barrels); Gb/y – Gigabarrel per year

Per cent – Percentage of global production or consumption. For instance, the Middle East accounts for 30 per cent of global production of crude oil, but only 6 per cent of global consumption.

R/P – Reserves to production ratio (in years). For instance, at the current rate of production of crude oil in the Middle East, the proven reserves in this region will be used up in 84 years.

R/C – Reserves to consumption ratio (in years). For instance, if the Middle East countries produced crude oil to meet only their own consumption, they would have sufficiently large reserves to last 436 years.

Balance – The annual difference between consumption and production. Negative numbers indicate the annual deficit. For instance, in North America the consumption of crude oil in 2001 was 3.4 billion barrels greater than local production.

Source: Based on the data of BP 2002a

The US accounts for nearly 26 per cent of global consumption and 11 per cent of global production; Europe consumes 21 per cent, but produces less than 9 per cent; and the Asia-Pacific region consumes 28 per cent, but produces only 11 per cent.

North America consumes 3 billion barrels of crude oil per year in excess of what it produces. The annual deficit in the OECD is 9.5 billion barrels, a shortfall that is made up by OPEC.

Regional dependence on crude oil. The global volume of proved reserves is not constant. It rose from 679 billion barrels in 1981 to 1001 billion barrels in 1991, and to 1050 billion barrels in 2001. However, it seems that this gradual increase is now coming to an end. More detailed, year-by-year records of proved reserves extending back to

1973 show that from about 1990 the global volume of proved reserves has been levelling off.[14]

Constant production. If we divide the volume of reserves by the volume of production we get an estimate of how long the reserves will last at a given level of production. The reserves-to-production ratio (R/P) is expressed in years.[15]

For crude oil, the global R/P ratio is 39 years (see Table 6-7). With the current proved global reserves and level of global production, the world has enough oil to last 39 years.

For OECD countries, the R/P ratio is 11 years, but for OPEC it is 74 years. This again illustrates the vulnerable position of OECD countries and the strength of the OPEC club. The R/P ratio for North America is 12 years, and for Europe eight years. For the US it is 11 years, and for Australia 14 years.

Constant consumption. The reserves-to-consumption ratios (R/C) are not calculated or published, but they seem to be as important, or even more so (see Table 6-7). Here, the question asked is slightly different: how long will current reserves last if we assume that countries will rely solely on their own reserves of crude oil?

The results show that the R/C ratio for OECD countries is only five years, but for OPEC 497 years, which is a huge and undesirable difference. If OECD countries had to rely on their own reserves to support their excessive consumption, their reserves would run out in only about five years, but OPEC would have enough oil for nearly 500 years.

It seems, therefore, that OPEC has a choice: to sell or not to sell. OECD countries do not have a choice: they must buy. OPEC could afford to reduce production, relax, sit back and wait for the lights to go out all over the world, or it could use oil as a highly effective bargaining chip.

Any local monopoly is undesirable; global monopoly can be disastrous. If OPEC decided to stop all exports of crude oil, it could paralyse OECD countries in less than a decade.

The R/C ratio for North America is only seven years, and for Europe only three. The R/C ratio for the US is four years, and for Australia twelve.

Crude oil and economic balance. Mutual interests and needs dictate the economic balance between countries. The world relies heavily on fossil fuels, and the balance of economic power is already in favour of OPEC. A factor that seems to be mitigating the bias against OECD countries is food. However, even here the balance is swinging in favour of OPEC.

OPEC needs to buy food; OECD countries need to buy crude oil. Between 1950 and 1973, the exchange rate between food and oil was one bushel of grain for one barrel of crude oil. This increased to two bushels per barrel in 1974 and to five bushels between 1975 and 1998. In 1999, it rose to six bushels per barrel and a year later to ten.[16] However, crude oil is also needed for producing and transporting food. The circle is closed, and it is in favour of OPEC.

Arab countries have enough money to desalinate water. They are also interested in building seawater greenhouses.[17] This new technology offers not only a cheap method of desalination, but also efficient production of food.[18] If Arab countries implement this technology on a large scale, they might be less dependent on food from industrialised countries.

Natural gas reserves.[19] The largest proved reserves of natural gas are in the former Soviet Union and in the Middle East (see Table 6-6). Reserves in other regions are relatively small. The smallest reserves are in Europe. Reserves of natural gas are distributed less critically over the globe than reserves of crude oil.

Proved reserves of natural gas have been increasing, from 82.44 trillion cubic metres in 1981 to 123.97 trillion cubic metres in 1991, and to 155.08 trillion cubic metres in 2001. Unlike crude oil reserves, proved reserves of natural gas keep increasing at a steady pace. To estimate the current strength of natural gas reserves and our dependence on them we can again calculate the R/P and R/C ratios.

Constant production. The current global R/P ratio for natural gas is 62 years. At the current rates of production, global reserves of natural gas will last only about 20 years longer than oil reserves. For North America, the R/P ratio is 10 years; for South and Central America, 71.6 years; Europe, 16.1 years; the former Soviet Union, 78.5 years; the Middle East, 245.2 years; Africa, 90.2 years; and the Asia-Pacific region, 43.8 years.

Reserves in the Middle East and the former Soviet Union are almost the same, but their production rates are different, which is reflected in the R/P ratios. The average R/P ratio for OECD countries is 13.7 years; for the US, 9.2 years; and Australia, 77.9 years.

Constant consumption. The striking feature of natural gas reserves is that for a number of regions—North America, South and Central America, the Middle East, and Asia-Pacific—the R/C ratios are almost the same as the R/P ratios. The dependence on external supply, which is so critical for crude oil, is much less pronounced for natural gas. These regions are fairly independent and they can rely on their own reserves sufficiently well to meet demand. For OECD countries, and for the US, the R/P and R/C ratios are also nearly identical.

However, for Europe the R/C ratio is 10.3 years; for the former Soviet Union, 102.4 years; Africa, 185.7 years; and for Australia, 113.3 years. In Africa and Australia, the domestic consumption of natural gas is much lower than the production. Africa produces about twice as much as it consumes and Australia about 50 per cent more. Former Soviet Union countries produce about 20 per cent more than they consume. On the other hand, Europe consumes about 70 per cent more than it produces.

Coal reserves.[20] Coal reserves are distributed over an even larger area than crude oil and natural gas (see Table 6-6). Another interesting feature is that the Middle East has almost no coal reserves. The largest reserves are in the Asia-Pacific region—mainly in China (39.2 per cent of Asia-Pacific reserves), India (28.9 per cent), and Australia (28.1 per cent).

The current global R/P ratio for coal is 216 years. We have huge proved reserves of coal in the world. Even if we do not discover new deposits, current reserves have the potential to support global production for more than 200 years. The R/P ratio for North America is 234 years; for South and Central America, 381 years; Europe, 167 years; Africa, 246 years; and the Asia-Pacific region, 147 years.

Production and consumption of electricity.[21] The largest installed capacity for electricity generation is in North America, followed by the Asia-Pacific region and Western Europe (see Table A6-3 in Appendix A).[22] The installed capacities for electricity generation in Africa and the Middle East are relatively small.

Worldwide, fossil fuels support two-thirds of installed electricity generation capacity. In particular, North America and the Asia-Pacific region rely heavily on this energy source. The largest nuclear electricity generation capacity is in Western Europe and North America. Hydroelectric capacity is high in North America, the Asia-Pacific region, Western Europe, Central America, and South America.

Nationally, the largest installed capacity for electricity generation is in the US (see Table A6-4 in Appendix A). About a quarter of global installed capacity is in this one country. Only four countries—the US, China, Japan, and Russia—account for nearly half of global installed capacity.

Table 6-8 Consumption of electricity by country and region, 1990–2020 (terawatt-hours per year)

Country/Region	1990	2000	(1)	2005	2010	2015	2020	(2)	(3)
USA	2 817	3 329	25	3 793	4 170	4 556	4 916	1 587	48
China	551	1 157	9	1 523	2 031	2 631	3 349	2 192	127
FSU	1 488	1 099	8	1 219	1 331	1 479	1 600	501	46
Japan	765	962	7	1 036	1 117	1 194	1 275	313	33
Germany	489	507	4	567	609	645	699	192	38
Canada	438	508	4	558	612	661	711	203	40
France	326	408	3	450	486	523	527	119	29
UK	287	338	3	365	389	413	438	100	29
India	257	443	3	537	649	784	923	480	108
Brazil	229	361	3	398	494	613	748	387	107
South Korea	93	246	2	309	348	392	429	183	75
North America	3 362	4 018	30	4 585	5 093	5 605	6 100	2 082	52
Western Europe	2 077	2 486	19	2 743	2 959	3 183	3 455	969	19
Asia	1 259	2 448	19	3 092	3 900	4 819	5 858	3 410	139
EE/FSU	1 906	1 485	11	1 651	1 807	2 006	2 173	688	46
CSA	449	701	5	788	988	1 249	1 517	816	116
Middle East	263	507	4	572	690	808	932	425	84
Australasia	181	235	2	255	278	299	322	87	2
Africa	287	383	3	460	550	671	776	393	103
World	**10 549**	**13 225**	**100**	**15 182**	**17 380**	**19 835**	**22 407**	**9 182**	**69**

Countries and regions are arranged in the descending order of the consumption of electricity in 2000.

(1) – Percentage of global electricity consumption for 2000. For instance, 25 per cent (3329 trillion watt-hours) was in just one country, the US.

(2) – The increase in annual electricity consumption between 2000 and 2020. For instance, consumption in the US in 2020 is projected to be 1587 TWh per year greater than in 2000.

(3) – Percentage change between 2000 and 2020. For instance, annual electricity consumption in the US is projected to increase by 48 per cent between 2000 and 2020.

TWh – terawatt-hour (trillion watt-hours)

FSU – Former Soviet Union

EE/FSU – Eastern Europe/Former Soviet Union

CSA – Central and South America

Source: Based on the data of EIA 2002a

Installed capacity in the US is nearly three times greater than in China, but per person it is about 12 times greater. Installed capacity in Germany is the same as in India, but per person it is 12 times greater. Installed capacity per person in the US is twice that in Germany.

Global electricity consumption in 2000 was 13,225 terawatt-hours per year [TWh/y], and is projected to be 22,407 TWh/y in 2020 (see Table 6-8). The US accounts for one-quarter of global electricity consumption. The largest regional consumption of electricity is in North America.

Table 6-9 Consumption of electricity by level of development and globally, 1990–2020

	1990	2000	(1)	2005	2010	2015	2020	(2)	(3)
Industrialised									
Total [TWh/y]	8 291	9 187	69	10 270	11 253	12 287	13 325	4 138	45/7
Per person [kWh/p/y]	7 544	7 779		8 446	9 068	9 767	10 517	2 738	35/7
Developing									
Total [TWh/y]	2 258	4 038	31	4 912	6 127	7 548	9 082	5 044	125/30
Per person [kWh/p/y]	540	829		937	1 090	1 260	1 430	601	72/30
Total world									
Total [TWh/y]	10 549	13 225	100	15 182	17 380	19 835	22 407	9 182	69/26
Per person [kWh/p/y]	1 997	2 186		2 350	2 533	2 736	2 941	755	35/26
Gap									
Total [TWh/y]	6 033	5 149		5 358	5 126	4 739	4 243	-906	-18
Per person [kWh/p/y]	7 004	6 949		7 509	7 978	8 507	9 087	2 138	31

(1) – Percentage of global consumption in 2000. For instance, 69 per cent of global celectricity onsumption in 2000 was by industrialised countries.

(2) – Increase between 2000 and 2020. For instance, total electricity consumption in industrialised countries in 2020 is projected to be 4138 trillion watt-hours per year higher than in 2000. Consumption per person is projected to be 2738 kilowatt-hours per year higher.

(3) – Percentage change between 2000 and 2020 in annual consumption compared with percentage change in population. For instance, electricity consumption in industrialised countries will increase by 45 per cent between 2000 and 2020, but the population will increase by only 7 per cent.

Gap – Difference between annual electricity consumption by industrialised and developing countries in total consumption and in consumption per person. For instance, the gap in total electricity consumption is projected to decrease from 5149 trillion watt-hours per year in 2000 to 4243 trillion watt-hours per year in 2020, which represents an 18 per cent reduction. However, the gap in consumption per person is projected to increase by 31 per cent.

TWh/y – terawatt-hour (trillion watt-hours) per year

kWh/p/y – kilowatt-hour per person per year

Source: Based on the data of EIA 2002a

Global electricity consumption in 2020 is projected to be 9000 TWh/y higher than in 2000. The largest national increase will be in the US and China, and the largest regional increase in Asia and North America. In 2020, the highest national electricity consumption will still be in the US, and the highest regional consumption will be in North America.

If we look at electricity consumption according to development levels, we shall see that close to 70 per cent of global consumption is by industrialised countries (see Table 6-9). The consumption per person is also high in industrialised countries, where it is more than nine times higher than in developing countries. However, electricity consumption in developing countries is increasing faster than in developed countries. In 2020, industrialised countries will be consuming 4000 TWh/y more than in 2000, and developing countries about 5000 TWh/y more.

The rapid increase in electricity consumption in developing countries will help them to narrow the gap with industrialised countries, but only for total consumption. The gap for consumption per person will increase. Here again, as with energy consumption, developing countries will be left further behind.

The gap in total electricity consumption will decrease from 5000 TWh/y in 2000 to 4000 TWh/y in 2020. However, the gap in consumption per person will widen from about 7000 kilowatt-hours per year [kWh/y] to about 9000 kWh/y.

Nuclear energy. We can produce nuclear energy by fission (splitting heavy nuclei) or by fusion (joining light nuclei). In both cases, minute differences in mass, arising from a rearrangement of the packing order of individual nucleons, are released in the form of enormous amounts of energy according to the famous energy-mass equation of Albert Einstein

Nuclear fusion is cleaner and more efficient than nuclear fission, but we do not yet know how to control it. One of the problems is that the colliding light nuclei have to move fast to overcome their repelling force before they can fuse. The collision energy corresponds to a temperature of about 100,000,000°C, which is about five times higher than the temperature at the centre of the sun.

Progress in fusion research has been made, but the work is hampered by technical problems and a lack of funding. We know how to unleash fusion energy in thermonuclear explosions, but we do not know how to harness it for our benefit.

The Group of Eight (G8) countries[23] are showing renewed interest in this energy source. These countries account for about 60 per cent of global energy consumption, including more than 70 per cent of nuclear energy consumption, close to 60 per cent of electricity, more than 40 per cent of coal, and close to 70 per cent of natural gas.[24]

The fission process is incomparably easier to handle, and we know how to do it. However, it is a dirty source of energy because of the production of radioactive fragments.

Nuclear fission consists in splitting heavy nuclei such as uranium 235 (U-235, which is made of 235 nucleons) by neutrons. The splitting of heavy nuclei leaves behind various lighter, radioactive fragments. Some of these newly created lighter nuclei last a short time and cause no real harm. However, others continue, causing problems for thousands, even millions, of years.

We generate large quantities of useful energy, but also produce large quantities of harmful radioactive waste—and we do not know what to do with it. The process starts with one type of a radioactive isotope that is converted into a multitude of smaller radioactive fragments.

Potential danger of nuclear power plants. Nuclear power plants are opposed by communities because of the potential for accidents, and for military or terrorist attacks. The plants can be damaged in various ways, ranging from brute force such as bombing or a suicide attack to computer hacking or jamming. An attack on a nuclear reactor cannot result in a nuclear explosion, but it can release large quantities of radioactive material.

Depending on the degree of damage, radioactive contamination around a nuclear facility could make the land unsuitable for crops, factories, and habitation. Radioactive waste released into the atmosphere can drift long distances and contaminate large areas of land.

Any serious damage to a nuclear power plant is demoralising. People feel helpless when they are subjected to radioactive contamination and suffer incurable diseases.

The disposal of radioactive waste is an unsolved problem.[25] One obvious response is to bury it; underground repositories are the preferred option, but are they safe? The design must result in a safe storage not just for hundreds of years, but for a much longer period, because of the long half-lives of some isotopes created in nuclear fission.

It is claimed that geological repositories can be made safe.[26] However, nature is full of surprises, and what we think as being safe may turn out to be harmful. A good example is the use of CFCs, which was considered safe until the stratospheric ozone layer began to deteriorate.

Unwanted radioactive material has been piling up in the US since 1940. It is stored in nuclear reactor facilities, naval bases, and research facilities. The proposed disposal site is Yucca Mountain in Nevada.[27]

A computer simulation is being used to study deposits of nuclear wastes. The program incorporates 120 parameters and is used to anticipate natural processes in the radioactive dump.[28]

The calculations show that, even if radioactive material is placed in well-sealed containers deep in a tunnel drilled into the mountain, such storage is not safe over hundreds of years. Radioactive decay of nuclear waste will generate enough heat to dislodge water molecules locked in the rock. By interacting with rock minerals, the water will be converted to a cocktail of corrosive compounds, damaging containers and allowing the leakage of radioactive material.

The best way to get rid of nuclear waste would be to shoot it to the sun. This is where nuclear reactions take place and where nuclear waste belongs.

If not for radioactive waste, security, and the potential for serious accidents, nuclear fission energy could be a good solution to at least some of our energy problems. However, we cannot blame the public if there is opposition to this energy source.

The first application of nuclear fission was to kill and destroy, which did not help in attracting people to the idea of nuclear energy. The fear of nuclear accidents adds to its bad reputation.

Consumption of nuclear energy.[29] At the end of the 20th century, there were 438 nuclear power plants operating in the world. The greatest number was in the US (104), followed by France (59), and Japan (53). Close to 30 per cent of global installed nuclear generating capacity was in the US (see Table A6-5, Appendix A). Just two countries, the US and France, accounted for almost half the total.

Almost all nuclear generating capacity is in industrialised countries, but the situation is slowly changing. By 2020, the installed nuclear generating capacity in industrialised countries will decrease by about 8 per cent, but in developing countries it will more than double—it will increase from 27 gigawatts [GW] in 2000 to 62 GW in 2020. The gap in installed nuclear generating capacity between developing and industrialised countries will slowly close.

Global production capacity of nuclear reactors rose from only 1 GW in 1960 to 350 GW in 2000. Production capacity is projected to increase to only 360 GW in 2020.

Close to 50 per cent of global consumption of nuclear energy takes place in two countries, the US and France, with 70 per cent in two regions, Europe and North America (see Table 6-10). OECD countries consume about 87 per cent of global nuclear energy. The lowest regional consumption is in Africa, and in South and Central America. The consumption of nuclear energy in the Asia-Pacific region is less than 20 per cent of the global total.

Table 6-10 Consumption of nuclear energy by country, region, and country group, 2000

Country	Consumption [TWh]	Share (1)	Share (2)	Region	Consumption [TWh]	Share (1)	Share (2)
USA	793.7	30.7	30.7	Europe	965.6	37.3	37.3
France	415.4	16.1	46.8	North America	875.0	33.8	71.2
Japan	319.5	12.4	59.1	Asia-Pacific	501.2	19.4	90.6
Germany	169.7	6.6	65.7	FSU	217.9	8.4	99.0
Russia	130.4	5.0	70.7	Africa	13.7	0.5	99.5
South Korea	109.2	4.2	74.9	SCA	11.9	0.5	100.0
UK	85.3	3.3	78.2	**World**	**2,585.3**	**100.0**	
Ukraine	77.3	3.0	81.2	OECD	2,239.3	86.6	
Canada	72.9	2.8	84.1	EU	864.9	33.5	
Spain	62.3	2.4	86.5				
Sweden	57.5	2.2	88.7				

Countries and regions are arranged in the descending order of the consumption of nuclear energy.

(1) – Percentage of global consumption of nuclear energy in 2000. For instance, 30.7 per cent (793.7 trillion watt-hours) of global consumption of nuclear energy in 2000 was in just one country, the USA.

(2) – The cumulative percentage of global consumption of nuclear energy in 2000. For instance, close to half (46.8 per cent) of global consumption of nuclear energy in 2000 was in just two countries, the USA and France.

TWh – terawatt-hour (trillion watt-hours)

SCA – South and Central America; FSU – Former Soviet Union. EU – European Union

Source: Based on BP 2002a. The values listed in million tonnes oil equivalent have been converted to terawatt-hours using the production-efficiency factor of 38 per cent.

Even though the population of industrialised countries is relatively small, they consume more than 11 times more nuclear energy than developing countries (see Table 6-11). The consumption of nuclear energy per person in industrialised countries is about 47 times higher than in developing countries.

Table 6-11 Consumption of nuclear energy by level of development and globally, 1990–2020

	1990	2000	(1)	2005	2010	2015	2020	(2)	(3)
Industrialised countries									
Total [TWh/y]	1 800	2 226	92	2 295	2 285	2 253	2 189	-37	-2/7
Per person [kWh/p/y]	1 638	1 885		1 887	1 841	1 791	1 728	-157	-8/7
Developing countries									
Total [TWh/y]	105	197	8	260	325	384	478	281	143/30
Per person [kWh/p/y]	25	40		50	58	64	75	35	86/30
World									
Total [TWh/y]	1 905	2 423	100	2 555	2 610	2 637	2 667	244	10/26
Per person [kWh/p/y]	361	400		395	380	364	350	-50	-13/26
Gap									
Total [TWh/y]	1 695	2 029		2 035	1 960	1 869	1 711	-318	-16
Per person [kWh/p/y]	1 613	1 844		1 838	1 783	1 727	1 653	-191	-10

(1) – Percentage of global consumption of nuclear energy in 2000. For instance, 92 per cent of global consumption of nuclear energy in 2000 was by industrialised countries.

(2) – The change between 2000 and 2020. For instance, the consumption of nuclear energy by industrialised countries in 2020 is projected to be 37 trillion watt-hours per year lower than in 2000, but the consumption in developing countries will be 281 trillion watt-hours per year higher.

(3) – Percentage change in the consumption of nuclear energy between 2000 and 2020 compared with the percentage increase in the population. For instance, the total consumption of nuclear energy in industrialised countries will decrease by 2 per cent between 2000 and 2020, but the population will increase by 7 per cent.

Gap – The difference between the consumption of nuclear energy by industrialised and developing countries for total consumption and for consumption per person. For instance, the gap for total consumption of nuclear energy is projected to fall from 2029 trillion watt-hours per year in 2000 to 1711 trillion watt-hours per year in 2020, which represents a 16 per cent reduction.

TWh/y – terawatt-hour (trillion watt-hours) per year; kWh/p/y – kilowatt-hour per person per year

Source: Based on the data of EIA 2002a

Some countries depend heavily on nuclear energy for electricity generation. Countries in which 30–50 per cent of national electricity is generated using nuclear fission are Ukraine, Bulgaria, South Korea, Hungary, Sweden, Switzerland, Slovenia, Japan, Armenia, Finland,

and Germany. Countries in which more than 50 per cent of national electricity is produced this way are France (76 per cent), Lithuania (74 per cent), Belgium (56 per cent), and Slovakia (54 per cent).

By 2020, the consumption of nuclear energy in industrialised countries will decrease slightly, but in developing countries it will increase by 143 per cent. Nuclear energy is the only form of energy for which the gap between developing and industrialised countries in consumption per person will be narrowing. Between 2000 and 2020, the gap in total consumption will decrease by 16 per cent and in consumption per person by 10 per cent.

Will the technology and management in developing countries be of sufficiently high standard to guarantee the safe operation of nuclear reactors? Will their nuclear technology be used only for peaceful purposes?

Hydroelectricity.[30] Of all forms of energy generation used on a large scale at present, hydroelectricity is regarded as the cleanest and the safest. What is the global potential for using this source? We can consider three types of potential: theoretical, technical, and economic (see Table A6-6).

Theoretical potential. The energy produced using a hydroelectric plant can be calculated using a simple equation that includes the flow rate of water (that is, how much is flowing through the turbines and how fast) and the vertical drop in the flow of water (called the head). The calculations take into account total runoff water (globally, nationally, or in any given area) and the elevation in places where water could be used for generating electricity.

The highest theoretical potential for hydroelectricity is in Latin America-Caribbean, followed by Central Asia and North America. The lowest is in the Middle East and North Africa.

Global theoretical potential is 40,784 trillion watt-hours of electricity per year [TWh/y], or 15 times global consumption of hydroelectricity in 2000. This is three times the global consumption of electricity from all sources in that year, and almost twice the projected global consumption of electricity for 2020.

Theoretically we have great potential for the production of hydroelectricity, but we must also consider technical and economic constraints. Environmental constraints are important, but they are usually not considered.

137

Technical potential. Technical constraints reduce substantially the hydrologic potential. The highest technical potential for hydroelectricity is again in Latin America-Caribbean, followed by Central Asia and Europe. The lowest is in the Middle East and North Africa.

Global technical potential for hydroelectricity is only 13,945 TWh/y, or about three times lower than the theoretical potential. It is also about the same as global consumption of electricity in 2000, but substantially lower than the projected consumption of electricity in 2020.

Economic potential. The economic potential is what determines the practical potential of hydroelectric resources. The greatest economic potential is in Central Asia, followed by sub-Saharan Africa and Latin America-Caribbean. More than 50 per cent of global economic potential is in these three regions. The lowest economic potential for hydroelectricity is in South Asia, which accounts for only 1.5 per cent of global economic potential. Hydroelectricity can therefore serve as a good source of energy only in a small number of regions.

Global economic potential for hydroelectricity generation is only 6964 TWh/y, which corresponds to about half the global consumption of electricity in 2000, and is more than three times lower than the projected global consumption of electricity for 2020.

It is expected that global production of hydroelectricity will reach 6000 TWh/y no earlier than 2050. Thus, the full practical potential of 6964 TWh/y might be developed no earlier than in the second half of this century. Hydroelectric resources will contribute to electricity generation, but their contribution will continue to be small. If we want to reduce the consumption of fossil fuels we should work harder on the development of alternative energy sources.

Consumption of hydroelectricity. Global consumption of hydroelectricity in 2000 was 2726 TWh/y (see Table 6-12). A quarter of this was in North America, and a half was shared by Canada, Brazil, the US, China, and Russia, which was about the same as the total consumption of all OECD countries.

In Norway, 99.5 per cent of electricity is generated using hydropower; in Brazil, it is 87 per cent; in Canada, 59 per cent; and in Sweden, 54 per cent.[31]

Collectively, industrialised countries consume about 65 per cent of globally generated hydroelectricity (see Table 6-13). By 2020, global consumption of hydroelectricity will increase to 4000 TWh/y. The increase will be a little higher in developing countries than in industrialised countries: 800 TWh/y and 600 TWh/y, respectively.

Table 6-12 Consumption of hydroelectricity by country, region, and country group, 2000

Country	Consumption [TWh]	Share (1)	(2)	Region	Consumption [TWh]	Share (1)	(2)
Canada	358.0	13.1	13.1	North America	666.9	24.5	24.5
Brazil	307.6	11.3	24.4	Europe	628.4	23.0	47.5
USA	275.8	10.1	34.5	Asia-Pacific	560.8	20.6	68.1
China	243.1	8.9	43.4	SCA	552.9	20.3	88.4
Russia	165.3	6.1	49.5	FSU	228.5	8.4	96.7
Norway	142.3	5.2	54.7	Africa	80.4	2.9	99.7
Japan	91.5	3.4	58.1	Middle East	8.4	0.3	100.0
Sweden	78.7	2.9	61.0	**World**	**2,726.3**	**100.0**	
India	76.9	2.8	63.8	OECD	1,385.5	50.8	
France	72.5	2.7	66.5	EU	351.8	12.9	
Venezuela	62.8	2.3	68.8				

Countries and regions are arranged in the descending order of their consumption of hydroelectricity in 2000.

(1) – Percentage of global consumption of hydroelectricity in 2000. For instance, 13.1 per cent (358 trillion watt-hours) of global consumption of hydroelectricity in 2000 was by Canada.

(2) – Cumulative percentage of global consumption of hydroelectricity in 2000. For instance, 24.4 per cent of global consumption of hydroelectricity in 2000 was by Canada and Brazil.

TWh – terawatt-hour (trillion watt-hours) per year

SCA – South and Central America; FSU – Former Soviet Union; EU – European Union

Source: Based on BP 2002a. The values listed in million tonnes oil equivalent have been converted to terawatt-hours using the production-efficiency factor of 38 per cent.

The gap in total consumption of hydroelectricity between industrialised and developing countries will narrow by 21 per cent from 2000 to 2020. However, the gap in consumption per person will increase by 23 per cent.

The most intensive development of hydroelectric facilities in 1997 was in Central Asia, where the total capacity under construction was 51,672 megawats [MW]. Other busy regions were Latin America-Caribbean (18,331 MW), sub-Saharan Africa (16,613 MW), and South Asia (13,003 MW). These regions accounted for about 80 per cent of global capacity under construction in that year.[32]

Table 6-13 Consumption of hydroelectricity by level of development and globally, 1990–2020

	1990	2000	(1)	2005	2010	2015	2020	(2)	(3)
Industrialised									
Total [TWh/y]	1 518	1 799	65	1 929	2 093	2 257	2 405	606	34/7
Per person [kWh/p/y]	1 381	1 523		1 586	1 686	1 794	1 898	375	25/7
Developing									
Total [TWh/y]	657	979	35	1 157	1 321	1 551	1 756	777	79/30
Per person [kWh/p/y]	157	201		221	235	259	276	75	37/30
World									
Total [TWh/y]	2 175	2 778	100	3 086	3 414	3 808	4 161	1 383	50/26
Per person [kWh/p/y]	412	459		478	498	525	546	87	19/26
Gap									
Total [TWh/y]	862	819		771	771	706	648	-171	-21
Per person [kWh/p/y]	1 225	1 322		1 365	1 451	1 535	1 621	299	23

(1) – Percentage of global consumption of hydroelectricity in 2000. For instance, 65 per cent (1799 trillion watt-hours) of hydroelectricity in 2000 was consumed by industrialised countries.

(2) – Change between 2000 and 2020. For instance, the consumption of hydroelectricity in industrialised countries in 2020 will be 606 trillion watt-hours per year (34 per cent) higher than in 2000.

(3) – Percentage change in the consumption of hydroelectricity between 2000 and 2020 compared with the percentage increase in the population. For instance, total consumption of hydroelectricity in industrialised countries will increase by 34 per cent between 2000 and 2020, but the population will increase by only 7 per cent.

Gap – Difference between the consumption of hydroelectricity by industrialised and developing countries for total consumption and for consumption per person. For instance, the gap for total consumption of hydroelectricity is projected to decrease from 819 trillion watt-hours per year in 2000 to 648 trillion watt-hours per year in 2020 (a 21 per cent reduction), but the gap in consumption per person will increase by 23 per cent.

TWh/y – terawatt-hour (trillion watt-hours) per year. kWh/p/y – kilowatt-hour per person per year.

Source: Based on EIA 2002a. The values listed in EIA 2002a in quadrillion British thermal units [PBtu] have been converted to trillion watt-hours [TWh] using the production-efficiency factor of 28 per cent. The factor has been calculated by comparing the EIA 2002a and the BP 2002a data for the years 1990, 1998, and 1999.

Global capacity under construction in 1997 was 124,161 MW (or 124 GW). Total installed capacity in that year was 655 GW.[33] Global installed capacity in 1999 was 740 GW.[34] Between 1997 and 1999, global capacity increased at an average rate of 42.5 GW per year.

Working at their full capacity of 655 GW in 1997, global hydro-electric facilities should have produced 5738 TWh of electricity in that year. However, global production was only 2600 TWh,[35] which shows that only 45 per cent of installed hydroelectric capacity has been of benefit to the user.

Summary and extended projections of global demand for electricity. A tentative projection of global demand for electricity in 2050 is 36,000 TWh/y, or almost three times global consumption in 2000 (see Table 6-14).

Table 6-14 Summary and extended projections of global demand for electricity, 2000–2050

Year	Fossil fuels [TWh]	Per cent (1)	Hydro electricity [TWh]	Per cent (1)	Per cent (2)	Nuclear energy [TWh]	Per cent (1)	Total [TWh]
2000	8 024	60.7	2 778	21.0	39.9	2 423	18.3	13 225
2005	9 541	62.8	3 086	20.3	44.3	2 555	16.8	15 182
2010	11 356	65.3	3 414	19.6	49.0	2 610	15.0	17 380
2015	13 390	67.5	3 808	19.2	54.7	2 637	13.3	19 835
2020	15 579	69.5	4 161	18.6	59.7	2 667	11.9	22 407
2025	17 266	70.4	4 496	18.3	64.6	2 749	11.2	24 511
2030	19 161	71.5	4 845	18.1	69.6	2 806	10.5	26 812
2035	21 057	72.3	5 194	17.8	74.6	2 863	9.8	29 114
2040	22 953	73.1	5 542	17.6	79.6	2 920	9.3	31 416
2045	24 849	73.7	5 891	17.5	84.6	2 977	8.8	33 717
2050	26 745	74.3	6 240	17.3	89.6	3 034	8.4	36 019
Increase								
2000–2020	94		50			10		69
2000–2050	233		125			25		172

(1) – Percentage of total production of electricity in a given year. For instance, 60.7 per cent (8024 trillion watt-hours) of global electricity in 2000 was produced using fossil fuels.

(2) – Percentage of the global economic potential of hydroelectricity production. For instance, the production of electricity in 2000 represented only 39.9 per cent of global potential.

Increase – Projected percentage increase in the dependence on a given source of electricity between 2000 and 2020, and between 2000 and 2050. For instance, the dependence on fossil fuels for electricity generation is projected to increase by 94 per cent (will nearly double) between 2000 and 2020, and by 233 per cent between 2000 and 2050.

TWh – terawatt-hour (trillion watt-hours) per year

Source: Projections up to 2020 are based on EIA 2002a. Projections after 2020 represent linear extrapolations.

Between 2000 and 2050, the contribution of fossil fuels to electricity generation is expected to rise from 61 per cent to 74 per cent. The contribution of hydroelectricity will fall from 21 per cent to 17 per cent and the contribution of nuclear energy from 18 per cent to 8 per cent. However, the production of electricity will increase for all three sources. For fossil fuels it will more than treble (from 8000 TWh/y to 27,000 TWh/y). The production of hydroelectricity will more than double, but the production from nuclear energy will increase by only 25 per cent.

These projections are based on current trends and do not consider the possibility of developing other energy sources and using them in the production of grid electricity.

By 2050, close to 90 per cent of the global economic potential for the generation of hydroelectricity will already have been developed. Unless we develop other sources of energy our global dependence on fossil fuels for electricity generation will rapidly increase. By 2050, about three-quarters of global electricity is expected to be generated using this environmentally unsafe energy source. Large deposits of coal might help us but it would be better to develop cleaner sources of electricity.

Alternative sources of energy.[36] Other sources of energy include wind electricity, solar photovoltaic electricity, solar thermal electricity, low-temperature solar heat, geothermal energy, and marine energy. Unfortunately, the combined contribution of all these sources is low, and is expected to remain so until at least 2020.

In 1998, wind energy was producing 18 terawatt-hours [TWh] of electricity, which corresponded to only 0.1 per cent of global electricity consumption in that year (see Table 6-15). Solar photovoltaic electricity accounted for a minuscule amount of only 0.5 TWh, and solar thermal electricity for only 1 TWh. The geothermal contribution was larger, but all alternative sources of energy were contributing only 66 TWh. If we include the generation of heat by alternative sources of energy we have a total of 120.2 TWh, which corresponded to only 1 per cent of global electricity consumption in 1998. We have to increase substantially the use of these sources.

Wind energy. Wind energy attracts a great deal of attention, and the development of this source is now exceptionally rapid. Global installed capacity of wind facilities grew from only 0.01 GW in 1980 to 10.2 GW in 1998 and to 18.1 GW in 2000.[37]

Working at full capacity of 10.2 GW, wind-energy facilities should produce 89.4 TWh of electricity per year. However, global production of electricity in that year was only 18 TWh, which shows that the capacity factor[38] for wind generators is only around 20 per cent. This is understandable because wind is an unstable source of energy.

Table 6-15 Contributions of alternative sources of energy to electricity production and to the generation of heat energy, 1998

Source	Operating capacity [GW]	Capacity factor Per cent	Energy produced [TWh]
	Generation of electricity		
Wind	10.8	20–30	18.0
Solar			
Photovoltaic	0.5	8–20	0.5
Thermal	0.4	20–35	1.0
Geothermal	8.0	45–90	46.0
Marine	0.3	20–30	0.6
Total			66.1
	Generation of heat energy		
Solar	18.0	8–20	14.0
Geothermal	11.0	20–70	40.0
Total			**56.0**

Capacity factor – A fraction of the operating capacity that is converted into energy. For instance, 10.8 GW of wind capacity should be able to generate 94.6 TWh of electricity per year. The energy produced in 1998 was 18.0 TWh, which means that only 19 per cent of the operating capacity was converted into electricity.

GW – gigawatt (billion watts); TWh – terawatt-hour (trillion watt-hours)

Source: Turkenburg 2000

Earlier analyses indicated that global installed wind-power capacity between 2030 and 2035 would be 1100–1900 GW.[39] These projections seem to have been underestimated. The analyses predicted 20 GW of installed capacity in 2002, whereas my extrapolation of the latest data[40] shows that installed capacity in that year must have reached 26–30 GW. Furthermore, if we analyse the available data for global installed capacity we shall find that it will reach 2000 GW well before 2035.

Between 1980 and 1987, the installation rate of wind-energy facilities was low. However, from about 1987 the installation rate increased from 9 per cent per year to 30 per cent. The average annual installation rate between 1987 and 2000 was 21 per cent, but the average in the last five years of the 20th century was 32 per cent. At this rate, installed wind-power capacity will double every 2.5 years.

Assuming that global installations of wind-power facilities will continue to increase at the rate of 32 per cent per year, we shall find that the total installed capacity will reach 1000 GW in 2015, and 2000

GW in 2017. Assuming that the annual installation rate will be 21 per cent, we find that the total installed capacity will reach 1000 GW by 2018, and 2000 GW by 2022.

The capacity of 2000 GW is about 100 times larger than installed capacity in the year 2000. We are clearly at the early stages of development for this energy source, and we can expect many more windmills sprouting from the ground or in offshore areas in future.

Table 6-16 The future of wind energy, 2000–2020

| Year | Annual installation rate: 21 per cent | | | | Annual installation rate: 32 per cent | | | | Global |
	Installed capacity [GW]	Electricity production [TWh/y]	Per cent (1)	Per cent (2)	Installed capacity [GW]	Electricity production [TWh/y]	Per cent (1)	Per cent (2)	consumption [TWh/y]
2000	18.1	32	0.2	0.2	18.1	32	0.2	0.2	13 225
2005	46.9	82	0.5	0.4	72.5	127	0.8	0.6	15 182
2010	121.8	213	1.2	1.1	290.7	509	2.9	2.5	17 380
2015	315.8	553	2.8	2.8	1,164.9	2,041	10.3	10.2	19 835
2020	819.2	1,435	6.4	7.2	4,668.3	8,179	36.5	40.9	22 407

Annual installation rate of 21 per cent – The average annual installation rate of global wind-power facilities between 1987 and 2000

Annual installation rate of 32 per cent – The average annual installation rate of global wind power facilities in the last five years of the 20th century

Installed capacity – The installed capacity in 2000 and the projected installed capacities at a given annual rate of installation

Electricity production – The annual production of electricity calculated using the capacity factor of 20 per cent

Global consumption – Global consumption (demand) of electricity in a given year

(1) – Percentage of global consumption of electricity in a given year. For instance, if we assume that installations of wind-power facilities will increase at 21 per cent per year, then by 2020 they will be able to contribute only 6.4 per cent to global electricity generation.

(2) – Percentage of the global economic potential of electricity production using wind power. For instance, if we assume that installations of wind-power facilities will increase at 21 per cent per year, then by 2020 global installed capacity will represent only 7.2 per cent of the ultimate economic potential.

GW – gigawatt (billion watts); TWh/y – terawatt-hour (trillion watt-hours) per year

Source: Based on an analysis of the data listed by Flavin 2001 for installed wind-power capacities between 1987 and 2000

Global installed capacity of wind energy is shared mainly by four countries: the US, Denmark, Germany, and Spain. In 1998, they accounted for 72 per cent of global installed capacity. The highest regional capacity was in Europe (6553 MW, or 65 per cent of global capacity), followed by North America (2292 MW, or 23 per cent), and Asia (1224 MW, or 12 per cent).

The theoretical potential for wind energy has been estimated at 500,000 TWh per year, and the economic potential at only 20,000 TWh, or three times the economic potential of hydroelectric resources. This value is also a little higher than the projected global consumption of electricity for 2015 (19,835 TWh) and a little lower than the projected consumption for 2020 (22,407 TWh).[41]

In principle, therefore, wind energy could make a substantial contribution to global electricity production, at least for a short period—but only if we develop it fast enough. Whether we want to have so many windmills around us is another question. However, we could perhaps build them in remote places, and conduct the electricity to where we need it.

At its full potential of 20,000 TWh, installed wind power capacity would have to be about 11,400 GW. To achieve this by 2020 we would have to increase the installation rate to 38 per cent per year. Assuming that installations will continue to increase at 32 per cent per year, the installed capacity in 2020 will be 4670 GW (see Table 6-16). Wind-power facilities would then be able to contribute 37 per cent to global electricity production in that year. However, if we assume a more conservative growth rate of 21 per cent per year, we can calculate that the installed wind-power capacity will be only 820 GW and will contribute only 6 per cent to global electricity production in 2020.

Solar energy. On average, the Earth receives about 340 watts of solar radiation per square metre. Solar devices aim at converting some of this radiation into useable energy. This can be done directly, by converting it into heat or electricity, or indirectly, by converting it into heat, which produces steam energy, which generates electricity.

Three groups of solar devices are in various stages of development: solar photovoltaic, solar thermal electricity, and low-temperature solar energy. Solar facilities have to spread their wings to catch sufficiently large amounts of solar radiation. They need to cover a certain surface area, and this is not always desirable or feasible.

Solar photovoltaic systems convert the radiation directly into electricity using photovoltaic solar cells. Global photovoltaic capacity in 1998 was 500 MW and was producing 0.5 TWh of electricity per year. This implies that the average capacity factor for these devices is only 11 per cent. Indeed, the recorded capacity factors for photovoltaic systems vary between 8 and 20 per cent.

We are still a long way from a practical application of photovoltaic technology on a larger scale. To meet only 10 per cent of projected global consumption of electricity for 2020, the installed capacity of photovoltaic facilities would have to make a sudden jump from 500 MW in 1998 to 2,326,000 MW in 2020. Photovoltaic technology will not play a big role in the supply of energy in the near future, and it is unlikely that it will play much of a role in the more distant future, unless we substantially step up its development.

One variation of photovoltaic technology is space-based energy generation. The idea is to install photovoltaic facilities in space and beam the energy to Earth, but this technology lies well into the future.

Solar thermal electricity systems are also in the early stages of development. Various technical solutions have been proposed or are being tried, including solar chimneys, towers, ponds, steam turbines, and Stirling engines.

The global capacity of thermal electricity systems in 1998 was 400 MW, and produced only 1 TWh of electricity in that year, which implies an average capacity factor of 29 per cent. The recorded capacity factors vary between 20 and 35 per cent. These devices are on average about three times more efficient in converting solar energy into electricity than photovoltaic devices.

The global capacity of thermal electricity systems is expected to grow to 2000 MW in 2010 and 12,000–18,000 MW in 2020. In 2020 they will be producing a maximum of 45 TWh of electricity per year. This will correspond to a maximum of only 0.2 per cent of projected global consumption for that year.

By 2050, the global capacity of solar thermal electricity systems is expected to increase to 800–1200 GW. They will then be able to generate only 2000 to 3000 TWh of electricity per year, contributing 6–8 per cent to projected global electricity consumption for that year.

The cost of the electricity produced by these systems is low. Unfortunately, they are not expected to play any great role in electricity generation.

Low-temperature solar energy devices convert solar energy into heat. They are used for space heating, water heating, cooking, and drying.

In 1994, the global area of low-temperature solar energy collectors was 22 million square metres [Mm^2]. By 1998, it had increased to 30 Mm^2. The combined installed capacity was 18 GW. These devices were generating 14 TWh of energy per year, which implies an average capacity factor of 9 per cent. The recorded capacity factors for these devices vary between 8 and 20 per cent. On average, these devices are about as efficient as photovoltaic devices.

It is expected that by 2010 the global area of low-temperature solar collectors will be 150 Mm^2. They will then be able to generate 70 TWh of energy per year—that is, about 0.25 exajoules, which corresponds to only 0.05 per cent of global energy demand for that year. The contribution of these devices to global energy production in the near future will be negligible.

Geothermal energy. Geothermal systems use the energy stored and produced (by radioactive decay) inside the Earth. The deeper we go the hotter it gets: the temperature increases about 3°C per 100 metres.

Geothermal energy was used for thousands of years as a source of heat energy. However, in the last few decades it has also been used for electricity generation.

In 1997, the greatest use of geothermal energy for heating was in Europe, where it accounted for 52 per cent of global generation of heat energy by this source, followed by Asia (32 per cent), and the Americas (10 per cent),[42] Oceania (5 per cent) and Africa (1 per cent). The total energy generated by these sources and used for heating in that year was 38.2 TWh. It increased to only 40 TWh per year in 1998—that is, by only 5 per cent.

The highest use of geothermal energy for the electricity generation in 1997 was in the Americas, where it accounted for 53 per cent of global electricity generated by this energy source, followed by Asia (30 per cent), Europe (10 per cent), Oceania (6 per cent), and Africa (1 per cent). The global contribution of geothermal energy to electricity generation in 1997 was 43.8 TWh. It increased to 46 TWh in 1998, which also represents only a 5 per cent increase. The annual generation of electricity by geothermal devices in 1998 corresponded to only 0.4 per cent of global electricity consumption in that year.

The largest increase in geothermal capacity between 1990 and 1998 was in the Philippines (957 MW), followed by Indonesia (445 MW), Japan (315 MW), Italy (224 MW), Costa Rica (120 MW), Iceland (62 MW), and Mexico (43 MW).

Global geothermal capacity in 1998 was 8239 MW. Considering that geothermal facilities were producing 46 TWh of electricity in that year, we can calculate that the average capacity factor for these facilities is 64 per cent, which is three times greater than the average capacity factor for wind-power facilities and six times higher than for photovoltaic devices. The recorded capacity factors vary between 45 and 90 per cent.

The highest installed capacity in 1998 was in the US (2850 MW), followed by the Philippines (1848 MW), Italy (769 MW), Indonesia (590 MW), Japan (530 MW), New Zealand (345 MW), Iceland (140 MW), Costa Rica (120 MW), and El Salvador (105 MW). Smaller installations were mainly in Nicaragua, Kenya, China, and Turkey.

The global potential for electricity generation by this source of energy has been estimated at 12,000 TWh per year,[43] which corresponds to 54 per cent of expected global demand in 2020. This source, therefore, could play an important role in meeting our energy needs. However, its development is slow and its contribution will remain low.

Marine energy. Oceans store large quantities of usable energy, which is in such forms as wave energy, the energy of daily cyclical tides (tidal energy), the energy of sea currents, thermal energy, and osmotic energy (the energy of variations in water salinity).

The tidal potential has been estimated at 22,000 TWh per year; the wave potential at 18,000 TWh per year; the thermal potential at 2,000,000 TWh per year; and the salt gradient potential at 23,000 TWh per year. The total potential of marine energy has been estimated at 2,063,000 TWh per year.[44]

The capacity factors for marine sources of energy vary from as low as 20–30 per cent (for tidal energy) to as high as 70–80 per cent (for thermal energy). Marine sources of energy seem to have huge potential for further development.

Biomass energy.[45] Biomass is the energy source of the poor. It is used mainly in household applications in developing countries, but does not play much of a role in industrialised countries.

In Africa and South Asia, biomass supports 90 per cent of household energy consumption; in South-East Asia, 70 per cent; in Latin America, 45 per cent; and in China, 70 per cent. The highest contribution of biomass energy to agriculture is in Latin America and Africa, where it supports 20 per cent of energy demands in this sector. The consumption of biomass energy for domestic purposes is projected to decrease in China, South-East Asia, and Latin America, and to increase in Africa and South Asia.

Hydrogen as an energy source.[46] Hydrogen is an attractive secondary source of clean energy because the product of combustion is water. In principle, it creates no pollution problems, but in practice this will depend on the source used for the production of hydrogen.

Hydrogen is an explosive material—as demonstrated dramatically by the sudden destruction of the airship Hindenburg and of the space shuttle Challenger. However, it can now be burned safely in fuel cells.

Hydrogen can be extracted from organic compounds, but the process of extraction creates pollution. The source of organic compounds could be fossil fuels or wastes such as wood chips, sawdust, and coconut shells. Hydrogen can also be extracted from water. The energy needed for hydrogen production can be supplied by renewable sources such as wind or solar radiation.

Prototypes of fuel-cell technology are already available and have been tested successfully in cars and buses by vehicle manufactures. Fuel-cell technology could be used to power the ever-increasing global fleet of motor vehicles and for many other applications, including domestic electricity generation. It could be used to make batteries for portable devices such as laptop computers or mobile phones. The technology is still in the early stages of development, but the results are most encouraging.

The future of transport.[47] About 20 per cent of global energy is now used for transport. This percentage is projected to remain fairly constant until at least 2020, but energy consumption in this sector will continue to increase.[48]

Transport is our Achilles heel because it depends heavily on an extensive supply of crude oil. Globally, the energy used by transport is shared 75 per cent by road transport and 11 percent by air transport. In industrialised countries, the division is 80 per cent and 12 per cent, respectively. This relative distribution is not expected to change in the near future.

Energy consumption by transport is projected to rise from 69 exajoules [EJ] in 1990 to 136 EJ in 2020. By 2020, this consumption will be twice the 1990 level, and 86 per cent higher than total global energy consumption in 1950.

The fleet, 75 per cent of which is made up of passenger cars,[49] is increasing. In 1950, global production of motor vehicles was 8 million per year, and the total world fleet was 53 million. On average, there were 20 motor vehicles per 1000 people in that year. In 2000, the production of motor vehicles had increased to 41 million per year and the total fleet to 532 million. In that year, the average was 90 motor vehicles per 1000 people.[50] About 25 per cent of the global fleet is in the US.[51]

In the last 50 years of the 20th century, the global population increased a little over two-fold, but the global production of motor vehicles increased five-fold and the global fleet 10-fold. The data show that in 2030 the fleet will be 1000 million—that is, 19 times larger than in 1950.[52]

With decreasing reserves of crude oil it will be more difficult to support this growing fleet of motor vehicles. There is therefore an urgent need to improve fuel consumption and to develop new technologies such as fuel cells and hybrid cars.

The above figures for the number of motor vehicles might be under-estimated. Other sources give consistently higher values.[53] Their range of data is less extensive, but they indicate that the values listed by Renner (2000a) might be underestimated by 40 per cent. This means that the projected 1000 million motor vehicles in the world will occur earlier. Indeed, if the projections of the US Department of Energy[54] are extrapolated, the global fleet will reach 1000 million in 2014.

In 1996, the number of motor vehicles per 1000 people was 479 in Australia and New Zealand, and only 44 in Central Asia, 41 in North-West and East Asia, 20 in South-East Asia, and five in South Asia.[55] In 1999, the number of motor vehicles per 1000 people was 777 in the US, 528 in Western Europe, 642 in Australasia, 57 in the Middle East, 26 in Africa, 12 in China, and 10 in India. The average for all indus-trialised countries was 614 per 1000; for developing countries, it was 32. The global average was 122 per 1000 people.[56] If motor-vehicle density everywhere in the world were the same as in industrialised countries, the total number would be 4000 million.

Motor vehicles contribute to global warming and urban air pollu-tion. Many cities of the world are overcrowded and heavily polluted, and the situation is growing worse.

We are becoming more dependent on transport, which depends on the availability of fossil fuels, in particular crude oil. The future of transport is uncertain—and, with it, our future.

Summary. The situation with energy resources is similar to the situation with water. We have huge resources of energy, but our access to them is limited or their use creates serious environmental problems.

We still have large deposits of fossil fuels, mainly in the form of coal. The most versatile form of fossil fuels—crude oil—will not last long and will have to be replaced by another source. Natural gas is more abundant, but its production peak is also expected to be reached during this century.

Fossil fuels are our main source of energy, but they create serious environmental and health problems. We are already paying a high price for their use, and the price will only increase. Much of the damage we are now causing to the environment will be costly—or even impossible—to repair.

We have huge amounts of energy stored in heavy atomic nuclei. If we could solve the problems of security and radioactive waste, we could have an abundant source of energy.

Nuclear fusion is an attractive, clean, safe, and practically inexhaustible source of energy, but its development is extremely difficult. Progress towards ignition[57] has been made in the past 30 years, and with much better financial support from governments this rich source of energy could eventually be made available. The development of fusion energy calls for strong international collaboration, but unfortunately we do not live on a peaceful planet and our priorities are elsewhere. We prefer quick and partial solutions to our energy problems.

Hydroelectricity is a useful source of energy, but it involves the construction of an undesirably large number of dams all over the world, creating environmental, social, and political problems. It is also a relatively small source of energy, and even at full practical capacity it is not expected to play a big role in the future.

We are surrounded with alternative and inexhaustible sources of energy (wind, solar, marine, and geothermal). Combined, they have significant potential for meeting future energy needs, but they are still in the early stages of development. Unless we increase substantially our commitment to the development of these sources, which seems unlikely, they will play only a marginal role in the near future.

Hydrogen is also a good and relatively clean energy source. The problem is to find a cheap way of producing it, and attempts to find a solution are being made. However, progress is slow, and it is not certain how soon we will be able to develop this technology.

The problem with energy is not the lack of options and possibilities, but our reluctance to move away from convenient and readily available fossil fuels. The future of the global energy supply is likely to be dictated by political, economic, and social attitudes rather than by its physical or technical potential. Future energy developments will be dictated by profit. As long as existing technology (with perhaps minor, but necessary modifications) and energy resources can produce a profit, companies will resist investing in radically new projects.

The future. Between now and at least 2020, we will rely strongly on fossil fuels for our energy needs. Our dependence on this energy source will increase from 85 per cent in 2000 to 87 per cent in 2020 (see Table 6-17).

Table 6-17 Future of global energy consumption, 2000–2020 (per cent)

Source of energy	Increase in consumption (demand) 2000–2020	Relative dependence		
		Increase/ decrease	From (in 2000)	To (in 2020)
All forms of energy	56			
Fossil fuels	60	Increase	85	87
Crude oil	55	No change	40	40
Natural gas	87	Increase	23	28
Coal	41	Decrease	22	20
Nuclear energy	10	Decrease	7	5
Other[a]	50	Decrease	9	8
Electricity (total consumption)	69			
Fossil fuels	94	Increase	61	70
Hydroelectricity	50	Decrease	21	18
Nuclear energy	10	Decrease	18	12

[a]) All forms of energy other than fossil fuels and nuclear energy.

Increase in consumption (demand) – The percentage increase in consumption (demand) of energy from a given source between 2000 and 2020. For instance, global energy consumption from all sources is expected to increase by 77 per cent between 2000 and 2020.

Relative dependence – Relative contribution of a given source of energy to global generation of energy or electricity in 2000 and 2020. For instance, global dependence on fossil fuels will increase from 85 per cent in 2000 to 87 per cent in 2020.

Increase/decrease – This column shows whether the relative contribution of a given source of energy is going to increase or decrease between 2000 and 2020. For instance, the relative contribution of fossil fuels to the generation of energy will increase between 2000 and 2020.

Source: Author's calculations based on the data of EIA 2002

The relative dependence on crude oil will not change, but the dependence on natural gas will increase. The relative dependence on coal will decrease slightly during this period. The relative dependence on nuclear energy and other energy sources will decrease. The share of fossil fuels in electricity generation will increase, but the share of hydroelectricity and nuclear energy will decrease.

However, we should notice that even though our relative dependence on crude oil (40 per cent) will not change between 2000 and 2020, global demand of crude oil will increase by 55 per cent. Consumption of coal and natural gas will also increase. By 2020, global demand of fossil fuels will be 60 per cent higher than in 2000. In contrast, the consumption of nuclear energy will increase by only 10 per cent. The consumption of all forms of energy will increase by 56 per cent.

Thus, for at least the next 20 years we will be burning larger quantities of fossil fuels and polluting the environment with ever-increasing vigour and determination. There is no end in sight to the accumulation of greenhouse gases in the atmosphere and to continuing climate change, with all its harmful effects.

Developing countries will increase rapidly their energy consumption (see Table 6-18), and in time they will contribute substantially to atmospheric pollution and climate change. The gap between industrialised and developing countries in energy production and consumption will decrease. The gap in consumption of all forms of energy is likely to close about 2050. The gap in consumption of nuclear energy and hydroelectricity is likely to close about the same time. However, the gap in consumption of electricity from all energy sources will not close earlier than next century.

Consumption of energy per person will increase in industrialised and developing countries. In industrialised countries, it will rise from 236 gigajoules per person per year [GJ/p/y] in 2000 to 293 GJ/p/y in 2020 (see Table 6-5). In developing countries, it will rise from 28 GJ/p/y to 43 GJ/p/y. This will mean greater stress for the environment. Energy consumption per person will increase faster in industrialised countries than in developing countries, and consequently the gap between these two groups will increase.

The consumption of nuclear energy will decrease in industrialised countries, but it will increase in developing countries (see Table 6-11). In 2020, industrialised countries will be consuming 37 terawatt-hours per year [TWh/y] less nuclear energy than in 2000, but developing countries will be consuming 281 TWh/y more.

Table 6-18 Changes in energy consumption by industrialised and developing countries, 2000–2020 (per cent)

Source of energy	Consumption			Gap		Years of equal consumption	
	Ind.	Develop.	Total	Per person		Total	Per person
All forms of energy	24	57	-34	19		2050	Never
Electricity	35	72	-18	31		2110	Never
Nuclear energy	-2	143	-16	-10		2042	2050
Hydroelectricity	34	79	-21	23		2058	Never

Consumption – Percentage increase or decrease (if negative) in total consumption of energy from a given source between 2000 and 2020 by industrialised (Ind.) and developing (Develop.) countries. For instance, the consumption of nuclear energy is projected to decrease by 2 per cent in industrial countries, but increase by 143 per cent in developing countries.

Gap – Percentage increase or decrease (if negative) in the gap of consumption from a given source of energy by industrialised and developing countries between 2000 and 2020 for total consumption and for consumption per person. For instance, the gap in total consumption of all forms of energy by industrialised and developing countries is projected to decrease by 34 per cent between 2000 and 2020, but the gap in the consumption per person is projected to increase by 19 per cent.

Years of equal consumption – Projected years when the total consumption of energy from a given source or the consumption per person will be the same in industrialised and developing countries. The projections are based on polynomial extrapolations. For instance, total consumption of nuclear energy is projected to be the same in industrialised and developing countries in 2042, and consumption per person to be the same in 2050.

Source: Based on an analysis of the data of EIA 2002a

The consumption of nuclear energy per person per year will decrease in industrialised countries, but it will increase in developing countries. Consequently the gap between these two groups will be growing smaller. This is the only form of energy for which the gap in consumption per person is decreasing.

If we look at global energy developments beyond 2020, almost anything is possible, except that our freedom to use crude oil, and perhaps natural gas, will be limited. The shift away from these two energy sources might even start earlier.

Future energy consumption can be described using three groups of resources: oil and gas, renewable and nuclear, and coal.[58] The most likely outcome will be dictated by our inertia. In physical systems, the inertia of moving objects helps them to move forward. Paradoxically, in our combined global behaviour, our inertia is likely to take us backwards—that is, back in time.

In the middle of the 1800s, consumption of renewable energy

(including biomass and animal power) was 80 per cent of the total, and the consumption of coal was 20 per cent. From about 1850 to 1900, there was a rapid shift from renewable sources to coal. In 1900, coal supported 70 per cent of global energy consumption; renewable sources, 28 per cent; and oil and gas, only 2 per cent.

In 1920, the contribution of coal was still 70 per cent, renewable resources fell to 20 per cent, and gas and oil rose to 10 per cent. After 1920, the contribution of renewable resources (which in time included nuclear energy) remained steady, coal decreased and gas and oil increased.

The most likely outcome in the future, from 2020 or earlier, will be a move back in time. Coal use will increase, oil and gas will decrease, and renewable and nuclear resources will be practically unchanged. By 2050, the contribution of coal is likely to rise from the current 22 per cent to 55 per cent; gas and oil will decrease from the current 63 per cent to 30 per cent.

There are also other paths for energy consumption after 2020, or maybe even earlier, but to follow them we would have to overcome our inertia. They involve the greater use of renewable and nuclear resources and the diminishing use of fossil fuels.

If we follow the most likely path, our existing energy sources will keep us going for at least 20 years. However, we shall continue to cause increasing damage to the environment.

Crude oil serves as a raw material in the manufacture of a wide range of products from asphalt to medicines. Finding a replacement quickly would be difficult, or even impossible. Products now in use every day might not be available in future, or they might be too expensive.

The increasing shortage of crude oil might also create a domino effect before 2020, as transport depends heavily on this energy source. Even if crude oil production is not going to decrease substantially in the next 15–20 years, the number of vehicles will increase and widen the gap between supply and demand. Unless we quickly find a viable replacement for crude oil, the fuel shortage will hinder transport and jeopardise many of our activities.

Chapter 7
Social Decline

'Civilisation is unbearable, but it is less unbearable at the top.'

—Timothy Francis Leary (1920–1996), US psychologist

'Civilisation degrades the many to exalt the few.'

—Amos Bronson Alcott (1799–1888), US educator, philosopher, and reformer,
Table-Talk – Pursuits

It is hard for the minority living in industrialised countries to see the extent of poverty and suffering in the world, the extent of social decline, and the developing crisis. We live in a microclimate of a profit-oriented society in which the standard of living is relatively high. If progress is measured by commercial output, life on Earth appears to be getting better. More products are available and they are becoming cheaper.

In 1992, computer simulations showed that with business as usual (a continuation of the present status quo) global production would continue to increase until about 2020.[1] After peaking about that time, it would steadily fall. However, they also showed that by changing our attitudes and behaviour we could have a more secure future. We could have economic growth combined with sustainable development. Our planet is not necessarily doomed, but the way we run our global business shapes our future.

With business as usual, adverse trends will get progressively worse from about 2020, and we can already see signs of a developing global crisis. Progress cannot be measured solely by production output, gross domestic product (GDP), and prices. Previous chapters have dealt with many areas which are not getting any better. This chapter will focus on the developing human crisis which we can see, for instance, in the widening inequality between and within nations, and in various forms of social decline.

How reliable are progress indicators? Progress indicators are useful, but they can be misleading, and we must be careful how we use them.

In the past 200 years, we have experienced unprecedented progress in science and technology. It has changed the way we live, and we can use many indicators to confirm and describe our achievements. For instance, many people have more access to education than in a distant past; they are healthier and they live longer.[2] However, who benefits most from scientific and technological progress, and who will benefit from it in future?

Who, for instance, will benefit from stem-cell research or progress in genetic engineering? Are we going to use these fields of science to improve the quality of life for everyone or only for selected groups? Will the free resources of nature benefit humanity generally or only a handful of rich corporations and individuals?

GDP is used for gauging economic progress and prosperity, but it does not describe internal levels of poverty, which can be present even in countries with high GDP, and does not reflect the environmental costs of progress.

Combined global wealth as measured by global GDP (known as gross world product, or GWP) is steadily increasing. The wealth of individual countries in GDP terms is also increasing, not only in the industrialised world, but also in the developing world. However, only a few countries achieve large increases; for the rest, GDP values remain low. Records extending back to 1820 show that for the poorest countries GDP has been close to constant, with a trend towards a slow decrease.[3]

Under the guise of progress, improvement, and development, civilisation paradoxically leads to degradation of the environment, society and quality of life. Behind the façade of progress lies a progressive decline.

The mortality rate of children under five years old in industrialised countries fell from 37 per 1000 live births in 1960 to six in 1998. The mortality in developing countries is higher, but it is also decreasing, from 216 per 1000 live births in 1960 to 95 in 1998. This is an encouraging indicator.[4]

However, closer examination shows that the 1998 mortality rate in developing countries was 2.6 times the 1960 rate in industrialised countries. Developing countries are still far behind, at least for this indicator. The mortality rate in sub-Saharan Africa and in the least-developed countries was five times higher. About 11 million children in developing countries die each year of preventable diseases.[5]

For this indicator, the developing countries are on average 30 years behind industrialised countries. The least-developed countries are 50 years behind, and sub-Saharan Africa is 70 years behind.

Individual countries are even further behind: for Afghanistan and Somalia, the gap is 90 years; for Niger, 140 years; for Sierra Leone, 160 years; and for Angola, 200 years.

Globally, primary school enrolment increased from 80 per cent of eligible children in 1990 to 84 per cent in 1998. However, 113 million children still do not have access to education. Furthermore, 97 per cent of those deprived of education live in developing countries.[6] From about 1980, enrolment rates have been falling in sub-Saharan Africa. Generally, primary enrolment favours boys: 60 per cent of the children not in primary school are girls. Of the 850 million illiterate people in the world, 540 million are women.[7] The enrolment indicator also says nothing about the number of children who finish their primary education.

About 500 million people in the world use the internet and the number is projected to increase to 1000 million by 2005. The internet is an efficient tool for sharing information. However, 72 per cent of users live in high-income OECD countries, which represent only 12 per cent of the global population. The rest of the world is left in a virtual darkness. About 164 million users (one-third of the global total) are in the United States.[8]

The proportion of people living on less than $1 per day[9] fell from 29 per cent of the global population to 22.7 per cent between 1990 and 1999, which looks like impressive progress. However, with the increase in global population the actual number of people living on less than $1 per day only fell from 1.3 billion in 1990 to 1.2 billion in 1999.[10]

Developing countries have a long way to go before they can move significantly closer to the prosperity of industrialised countries. Given the other limiting factors, such as land, water, and energy resources, they do not have a promising future. The standard of living in the world is improving, but only for a small number of people. The improvements are not only limited, but also unsustainable. They are supported by overexploitation of natural resources and are based on an increasing legacy of environmental degradation.

How to measure progress? GDP includes monetary transactions, but excludes many important factors that describe progress.[11] A significantly better way of measuring progress is by using the so-called genuine progress indicator (GPI). This is similar to GDP, but it includes components such as the value of free services from voluntary organisations and the costs of environmental damage associated with economic activities. These additional components are vital in measuring the true economic growth of a country. Some of them (such as the value of voluntary services) are added to the economic progress indicator; some (such as the cost of environmental damage) are subtracted. The GPI shows that many countries are not as economically strong as they seem to be.[12]

For instance, GDP for the US (expressed in 1996 US dollars) increased from $11,000 per person per year in 1950 to $33,000 per person per year in 2000. This appears to represent impressive growth. However, GPI rose from $5000 per person per year in 1950 to $10,000 per person per year in 1975. After that, GPI was decreasing until about 1994, when it reached a mere $7000. It then rose slightly to reach $9000 in 2000.[13] The gap between GDP and GPI in 2000 was $26,000 per person per year, or a total of $7400 billion per year according to the population at that time. The gap between GDP and GPI is increasing. Much of this gap represents the ecological deficit, the burden of environmental damage left for future generations.

The US is not an isolated example. Australia's GDP (expressed in 1999 US dollars) rose from $9000 per person per year in 1950 to $26,000 in 2000, but GPI increased from $7000 per person per year to only $13,000.[14] The difference between GDP and GPI in 2000 totalled $370 billion per year. The Australian economy is growing, but at a much lower rate than indicated by GDP. The gap between GDP and GPI is widening, and consequently the amassed debt of environmental damage is growing.

GDP is an illusion of progress, a disguise that hides the stark reality of an environmental overdraft. We borrow from the future to pay for the present, and we use GDP to make ourselves happy.

GPI is not yet listed in statistical tables, and GDP can still be used as an indicator of relative prosperity. We can also use it to measure absolute levels of prosperity, as long as we remember that much of the current prosperity is based on a growing environmental deficit that is likely to be reflected in steady economic decline—or even a rapid fall.

Social inequality. The distribution of wealth can be measured for people in a given country, between countries, or between all the people in the world. In principle, there is nothing wrong with an unequal distribution of wealth as long as those who are not rich have enough to support their life. Consequently, it is claimed that our aim should not be to reduce inequality, but to reduce poverty.[15]

The argument goes more or less like this: if the rich get richer without making poor people poorer, there is nothing wrong with an increase in income inequality.

However, what about increases in the cost of living? Rising prices will not harm the rich, but the poor will have the same income as before and will be able to buy less. The poorer sections of society will be poorer not only in a relative sense, but also absolutely. Even if the level of inequality were unchanged—if the rich did not get richer—it would be easier for them to absorb an increase in the cost of living.

Furthermore, the assumption that the rich are able to get richer without making the poor poorer can also be questioned.

Income inequalities can be regarded as a natural state of affairs. The problem is not so much with the existence of inequality, but with its magnitude. Increasing inequality is likely to lead to a gradual reduction in the number of people or countries in the middle-income range and to undesirable social polarisation. Excessive and increasing inequality is a sign of social decline.

Disappearance of the middle class. Several authors have indicated that socio-economic polarisation is becoming so strong that the middle class is disappearing.[16] Social stability depends on balanced distribution of wealth; in particular, on the existence of a middle class.[17] "Thus it is manifest that the best political community is formed by citizens of the middle class, and that those states are likely to be well-administered, in which the middle class is large ... where the middle class is large, there are least likely to be factions and dissension."[18]

Increased social polarisation, to the point of a gradual disappearance of the middle class in individual countries, increases social and economic instability within countries. Likewise, the gradually deepening division of the world between excessively rich and excessively poor countries increases the social, economic, and political instability of the world. "United we stand, divided we fall."[19]

A recent survey of 111 countries shows that the world is already divided into rich and poor countries, and that the number of people living in countries with middle-range national incomes is low.

In this survey, countries have been sorted into three groups: rich countries with incomes greater than the mean income in Italy; poor countries with incomes below the poverty line of Western countries;

and middle-class countries with incomes between the two extremes. The survey shows that 11 per cent of countries are rich, 78 per cent are poor, and 11 per cent are in the middle. Rich countries account for about 16 per cent of the global population; poor countries, 76 per cent; and middle-class countries, 8 per cent. The poor countries include India and China.[20]

How to measure income inequality? Several mathematical models have been proposed and used for describing observed wealth distributions.[21] However, it has been shown that no less than 11 models can be derived from one unifying mathematical formula called the five-parameter generalised beta distribution.[22] By using this grand formula and by carrying out straightforward simplifications one can create a family of simpler models. In this group of simpler models, three have been identified as giving the best results.[23] So, in analyses of wealth distribution we can use one complex model or one of three simpler models, all belonging to the same family of models.

If we are not interested in studying the details of wealth distribution, but only in measuring levels of inequality, we can simplify the problem even further by using a single number called the Gini index, also known as the Gini coefficient.[24] The Gini index measures the difference between an abstract situation when wealth is distributed evenly and the real situation when wealth is distributed unevenly.

The geometrical interpretation of the Gini index is an area between two Lorenz curves[25] (one corresponding to an equal distribution of wealth and the other to an unequal distribution) multiplied by two. This geometrical representation gives a convenient visual interpretation of the Gini index.[26] The greater the difference between the real distribution of wealth and the equal distribution, the greater the area between the two Lorenz curves and the larger the Gini index. The larger the Gini index, the greater the inequality.

A Gini index of zero per cent corresponds to an abstract situation in which all individuals have an equal share of wealth. A Gini index of 100 per cent corresponds to a situation in which one individual takes everything and leaves nothing to the rest of the population.

Inequality within countries is common, whether they are poor or rich. The big difference is that rich countries have a greater common wealth to share, so the poor have a better chance of living reasonably well than their counterparts in poor countries. In many poor countries, the level of inequality is exceptionally high, which means that a large proportion of the population is in abject poverty.

Statistical information for 1990–98 (see Table 7-1) shows that the lowest income inequality was in Slovakia and the highest in Swaziland. Inequality levels, as measured by the Gini index, were approaching 60 per cent in sub-Saharan Africa, Latin America-Caribbean, and in East Asia-Pacific. The widest range of inequality levels was in Eastern Europe and in the Commonwealth of Independent States.[27] In OECD countries, the lowest income inequality was in Denmark, and the highest in the US. It is interesting that, even in rich countries, income inequality varies over a wide range.[28]

Table 7-1 The range of inequalities within countries (per cent)

Groups of countries	Gini coefficient					
	Minimum		Maximum		Range
Sub-Saharan Africa	33.3	Burundi	60.9	Swaziland	27.6
LAC	36.4	Jamaica	60.3	Nicaragua	23.9
East Asia-Pacific	31.6	S. Korea	50.9	PNG	19.3
EE&CIS	19.5	Slovakia	48.7	Russia	29.2
Arab States	28.9	Egypt	41.7	Tunisia	12.8
OECD	24.7	Denmark	40.8	USA	16.1
South Asia	31.2	Pakistan	37.8	India	6.6

Groups of countries have been arranged in descending order of maximum levels of inequality as measured by the Gini coefficient.

Gini coefficients (in per cent) – Perfect equality is represented by Gini = 0 per cent, and perfect inequality by Gini = 100 per cent. For each group, a country with the lowest and the highest Gini values are listed. For instance, in sub-Saharan Africa, the lowest level of inequality (33.3 per cent) is in Burundi and the highest (60.9 per cent) in Swaziland.

Range – Difference between the highest and lowest Gini values for a given group of countries. For example, the widest range of inequality is in Eastern Europe and the Commonwealth of Independent States.

LAC – Latin America-Caribbean

EE&CIS – Eastern Europe and Commonwealth of Independent States

PNG – Papua New Guinea

Source: UNDP 2001. Statistical data for 1990–1998

More recent statistical information shows that the Gini index for individual countries varies between about 20 per cent and 70 per cent.[29] For the 20 countries characterised by the highest level of human development (as defined by the HDI ranking—see below), the lowest income inequality is in Sweden, Belgium, Japan, and Denmark, where the Gini index is 25 per cent. The highest level is in the US (41 per cent).

For the 20 countries at the lowest level of human development, the lowest inequality is in Rwanda (29 per cent) and the highest in Sierra Leone (63 per cent). The highest inequality in the world is in Namibia (71 per cent). In this country, the richest 20 per cent of society shares 79 per cent of the income, and the poorest 20 per cent shares only 1.4 per cent.[30]

A recent analysis of 82 sets of data for 23 countries indicated that the general trend is for inequalities to increase (see Table 7-2). For instance, inequality in Australia increased from 33 per cent in 1981 to 36 per cent in 1994, and in the US from 37 per cent in 1969 to 43 per cent in 1997. In this survey, income inequality eased only in France, Norway, Poland, and Spain.[31]

Table 7-2 Increasing income inequalities

Country	Years of survey		Change in Gini		Inequality
	From	To	From	To	
Australia	1981	1994	33.3	36.1	Increased
Belgium	1985	1997	25.5	32.7	Increased
Canada	1971	1997	37.9	39.2	Increased
Czech Republic	1992	1996	30.3	33.2	Increased
Denmark	1987	1997	38.6	47.8	Increased
Finland	1987	1995	40.1	43.3	Increased
France	1979	1994	50.4	38.7	Decreased
Germany	1973	1994	31.1	34.3	Increased
Israel	1979	1997	35.0	43.0	Increased
Italy	1986	1995	30.6	33.8	Increased
Mexico	1984	1996	46.6	54.8	Increased
Netherlands	1983	1994	27.7	34.0	Increased
Norway	1979	1995	52.8	37.3	Decreased
Poland	1986	1995	38.4	36.2	Decreased
Taiwan	1981	1995	29.3	34.1	Increased
Russia	1992	1995	46.6	53.7	Increased
Spain	1980	1990	35.6	34.8	Decreased
Sweden	1967	1995	39.9	44.6	Increased
Switzerland	1982	1992	36.0	37.3	Increased
USA	1969	1997	36.8	43.0	Increased

Gini – The Gini coefficient in per cent. Gini = 0 per cent represents perfect equality, and 100 per cent represents perfect inequality.

Example: Between 1981 and 1994 income inequality in Australia increased from 33.3 per cent to 36.1 per cent.

Source: Bandourian et al 2002

Another survey of 91 countries has shown a clear trend of increasing inequalities. These countries represented 85 per cent of the global population and 95 per cent of global income. The average Gini index for these countries rose from 63 per cent in 1988 to 66 per cent in 1993.[32] An increase in inequality has also been observed in a survey of transition economies (Bulgaria, Hungary, Latvia, Poland, Russia, and Slovenia) between 1987 and 1995.[33]

Why do we have income inequalities? In 1896, Vilfredo Pareto, a Paris-born engineer, discovered that the distribution of populations according to their levels of affluence can be described remarkably well by a simple mathematical function.[34] Pareto's law remained a mystery for more than 100 years until recently, when two physicists at the École Normale Supérieure in Paris explained its underlying mechanism and the reasons for unequal wealth distribution.[35] The Pareto distribution belongs to a group of two-parameter models that can be derived from a universal five-parameter formula.[36]

Simple and approximate descriptions are often useful because they help us to understand the basic and most prominent mechanisms of a process. The simple picture can be adorned later with modifications to give a better description of data. Such modifications and generalisations are likely to make mathematical formulae complex and unattractive. Complex models may better describe the real world, but the basic and most important features might get lost in the details. However, if we understand the basic underlying process, we should be able to see the simplicity in complexity.

It turns out that the basic mechanism of unequal wealth distribution is the network of financial transactions between people. A more complex, richer, more vigorous, and more evenly distributed network of interactions leads to a more equal distribution of wealth. Income distribution is guided primarily by chance,[37] but it can, and often is, moderated and modified by other factors such as history and policies. The effect of historical factors can be seen in the uneven distribution of wealth between developed and developing countries. The effects of policies can be seen in taxation rules, which might give an advantage to certain sections of society. Biased or restrictive trade arrangements also allow capital to flow mainly to certain groups of people or countries.

Globalisation. Globalisation is a transnational, global interaction between peoples and nations. It includes trade, technology, culture, science, political interaction, and governance. Globalisation is a natural process of integration that could have been expected to develop in a shrinking world. With the increasing global population, a strong network of trade connections and other links is being developed within and between countries.

In principle, there is nothing wrong with global interactions between people, and globalisation could be used to improve life on Earth.[38] However, globalisation includes trade, and the basic aim of trade is profit. History has shown that, while profit is necessary for businesses to operate, it is usually connected with greed, corruption, and deceit; with the desire for unlimited wealth; with the ruthless elimination of competitors; and with the brutal exploitation of individuals, countries, and resources. Vigorous and fair trade could help to wash out the differences between rich and poor nations. Regrettably, it tends to discriminate against the poor and favour the rich.

Globalisation is the Pareto process on a global scale, but it also influences local distributions. If uncontrolled, it will continue to deepen social and economic polarisation within and between countries.

Globalisation is a new phenomenon. It developed gradually in the 20th century, but exploded in the second half. Between 1950 and 1980, the global value of exports increased from $311 billion to $5400 billion, and the value of imports increased from $467 billion to $1300 billion. Foreign direct investment increased from $44 billion to $644 billion. The number of transnational corporations increased from 7000 to 53,600. Shipping increased six-fold, air transport 100-fold, tourism 25-fold, and international telephone lines six-fold.[39]

Globalisation leads to the formation of mega-corporations through mergers and acquisitions. In fact, mergers and acquisitions increased slowly from about 1980 then exploded from the mid-1990s. Their total annual value rose from $500 billion (in 1998 US dollars) in the early 1990s to $3400 billion in 1998. The latter figure is about 100 times higher than in 1980. Cross-boundary mergers increased from $150 billion in the early 1990s to $1100 billion in 1998.[40] The annual value of mergers in the field of computing increased from $21.4 billion in 1988 to $246.7 billion in 1998; in biotechnology from $9.3 billion to $172.4 billion; and in telecommunications from $6.8 billion to $265.8 billion.[41]

The quest for patents also increased substantially after 1980. The number of annual patent applications increased from a few thousand in 1980 to more than 50,000 in 1998.[42] The patenting of life forms, such as genetic resources and stem cells, is a worrying and controversial development because it slows research and development, and hampers progress. Patents restrict the exchange of information between scientists as well as affecting conventional freedoms in research and publishing. There is a growing and justifiable fear that such patents will limit the number of people benefiting from natural resources.[43]

Unequal share between countries.[44] At the end of the 20th century, the 20 per cent of global population living in industrialised countries shared 86 per cent of global wealth (GWP). The 20 per cent living in the poorest countries shared only 1 per cent of GWP. The remaining 60 per cent of global population shared 13 per cent of GWP. Each 20 per cent of the population in this larger group received 4.33 per cent of GWP (see Table 7-3). Five countries—China, India, Indonesia, Brazil, and Russia—accounted for 50 per cent of the global population and only 9 per cent of GWP.

Table 7-3 Unequal share in the benefits of progress

Share in	Top 20 per cent	Bottom 20 per cent	Remaining 60 per cent	
Gross world product	86.0	1.0	13.0	(4.33)
Global export market	82.0	1.0	17.0	(5.67)
Direct foreign investment	68.0	1.0	31.0	(10.33)
Telephone lines	74.0	1.5	24.5	(8.17)
Internet access	93.0	0.2	6.8	(2.27)

Top 20 per cent – Approximate 20 per cent of the global population living in the richest countries

Bottom 20 per cent – Approximate 20 per cent of the global population living in the poorest countries

Remaining 60 per cent – Approximate 60 per cent of the remaining global population

Gross world product – Global income

Figures in parentheses – Percentage share for each 20 per cent of the population in the last group

Example: The 20 per cent of the global population living in rich countries use 86 per cent of gross world product, the poorest 20 per cent use only 1 per cent, and each 20 per cent in the middle group use 4.33 per cent.

Source: UNDP 1999

Table 7-4 gives a summary of industrial outputs by region and level of development for 1985 and 1998. It seems that developing countries have little or no chance of a better future in terms of industrialisation.

Developing countries' share of industrial production in 1985 was

19.2 per cent, and only 21.7 per cent in 1998. The relative contribution to global industrial production by East Asia rose from 8.2 per cent to 11.5 per cent. However, the gap in manufacturing value between industrialised and developing countries increased from $1524 billion in 1985 to 3185 billion in 1998, and the gap in production per person increased from $2432 to $3802.[45] Even though their output is increasing, developing countries are being left further behind in industrial development.

Table 7-4 Unpromising future of industrialisation in developing countries

Group	1985			1998		
	Value	Value/p	Share	Value	Value/p	Share
East Asia	203.7	145	8.2	649.8	387	11.5
East Asia without China	98.0	278	4.0	293.3	668	5.2
China	105.7	101	4.3	356.5	288	6.3
South Asia	42.0	42	1.7	83.6	65	1.5
LAC	171.1	462	6.9	360.0	771	6.4
Sub-Saharan Africa	24.1	83	1.0	38.2	92	0.7
MENA and Turkey	35.8	202	1.4	94.1	392	1.7
Industrialised countries	2,003.3	2 579	80.8	4,410.3	4 102	78.3
Developing countries	476.6	147	19.2	1,225.8	300	21.7
World	**2,480.0**	**619**	**100.0**	**5,636.1**	**1,094**	**100.0**
Gap	**1,526.7**	**2 432**		**3,184.5**	**3 802**	

Value – Manufacturing value added in billion dollars

Value/p – Manufacturing value added per person in dollars

Share – The share in global industrial production (per cent)

Gap – The gap in total manufacturing output (billion dollars) and in manufacturing output per person (dollars) between industrialised and developing countries

LAC – Latin America-Caribbean; MENA – Middle East and North Africa

Example: The manufacturing value added in East Asia in 1985 was $203.7 billion, which corresponded to $145 per person. In 1985, East Asia was contributing 8.2 per cent to global industrial production. In 1985, the gap in manufacturing output between industrialised and developing countries was $1526.7 billion and the gap in manufacturing output per person was $2432.

Source: Based on the data of UNIDO 2002

In 1997, the G-7 Western economic powers—Canada, France, Germany, Italy, Japan, UK, and the US—with a combined population representing 11.8 per cent of the global total, shared 44 per cent of GWP. At the same time, the G-77 developing and transition countries, with a combined population representing 76 per cent of the global total, shared only 16.9 per cent of GWP.

The unequal share in global wealth distribution is reflected in large differences in the standard of living, and in particular in the severe shortage of such essential items as food. Children usually suffer most. The proportion of underweight children is only 1 per cent in rich countries, yet it is up to about 60 per cent in poor countries.

Rich countries are efficient at exploiting the natural resources of poor countries. For instance, export production in sub-Saharan Africa is high, but it is mainly in primary resources. Foreign investment in that region is also centred on the exploitation of minerals.

The share of the global market in finance and insurance is 90 per cent for the minority living in industrialised countries and 10 per cent for the majority living in developing countries.[46] Industrialised countries also own about 97 per cent of patents.[47]

The purchasing power available to people in various countries is spread over a large range.[48] In 1967, people living in the richest countries were spending $16,000 (in 1995 US dollars) per person per year on private consumption,[49] whereas those in the poorest countries were spending only $120 per person per year. Average spending on private consumption in the US was $21,680 per person per year; in Germany, $15,229; Poland, $5087; Indonesia, $1808; China, $1410; India, $1166; and Zambia $625.[50]

It is true that we have charitable individuals and institutions in the world, but the general trend of social interactions is towards ruthless exploitation of the poor—an attitude not confined to any particular group of people or countries. Class differences and contempt for the poor seem to be even greater in poor countries. We are not building a healthy and sustainable global society.

The richest individuals.[51] Of the 200 richest people in the world in 1998, 65 lived in North America, 55 in Europe, 30 in the Asia-Pacific region, 16 in Arab states, 17 in Latin America-Caribbean, and 17 in other parts of the world, including one in sub-Saharan Africa. Between 1994 and 1998, they increased their combined wealth from $440 billion to $1024 billion, which corresponds to an average increase of $2 million per person per day—clearly more than they can spend. An annual contribution of only 1 per cent of their assets could pay for the primary education of all the children in the world.

In 1998, their combined assets were the same as the combined gross national product (GNP) of the 48 poorest countries. The combined wealth of the richest three individuals in this exclusive group was larger than the combined GNP of 600 million people living in the least-developed countries.

Absolute poverty. Absolute poverty is defined by an income of less than $1 per person per day (corrected for purchasing power parity).[52] About 1.2 billion people now live in absolute poverty (see Table 7-5), mainly in South Asia, sub-Saharan Africa, and East Asia-Pacific.[53] These three regions account for close to 90 per cent of all people in absolute poverty. In 1999, nearly half the population in sub-Saharan Africa lived in absolute poverty. The number suffering absolute poverty fell only slightly in South Asia, but increased in sub-Saharan Africa.

Table 7-5 Little change in the number of people living in absolute poverty

Region	Change			
	Millions		Per cent	
	From 1990	To 1999	From 1990	To 1999
South Asia	495	490	44.0	36.9
Sub-Saharan Africa	242	300	47.7	46.7
East Asia-Pacific	452	260	27.6	14.2
Latin America-Caribbean	74	77	16.8	15.1
Eastern Europe and Central Asia	7	17	1.6	3.6
Middle East and North Africa	6	7	2.4	2.3
World	**1 276**	**1 151**	**29.0**	**22.7**

Regions are arranged in descending order of the number of people living in absolute poverty (that is, on income of less than $1 per person per day) in 1999.

Example: The number of people living in absolute poverty in South Asia changed from 495 million in 1990 to 490 million in 1999—that is, from 44 per cent of the population in that region in 1990 to 36.9 per cent in 1999.

Source: UNDP 2002

A country-by-country survey shows that, even though the largest number of people living on less than $1 per day are in Asia, countries with the largest percentage of the population in absolute poverty are mainly in sub-Saharan Africa (see Table 7-6). In many countries, over half the population suffers extreme deprivation.[54]

Table 7-6 The depths of absolute poverty

Region/Country Latin America-Caribbean	Per cent	Country Sub-Saharan Africa (continued)	Per cent
Ecuador	20.2	Zimbabwe	36.0
El Salvador	21.4	Mozambique	37.9
Honduras	23.8	Malawi	41.7
Nicaragua	82.3	Lesotho	43.1
South Asia		Ghana	44.8
India	34.7	Madagascar	49.1
Bangladesh	36.0	Sierra Leone	57.0
Nepal	37.7	Burundi	58.4
East Asia and Pacific		Gambia	59.3
Lao People's Dem. Rep.	26.3	Burkina Faso	61.2
Sub-Saharan Africa		Niger	61.4
Kenya	23.0	Zambia	63.7
Botswana	23.5	Central African Republic	66.6
Senegal	26.3	Nigeria	70.2
Mauritania	28.6	Mali	72.8
Cameroon	33.4	Ethiopia	81.9
Namibia	34.9	Uganda	82.2
Rwanda	35.7		

Countries are arranged in rising order of percentage of the population living in absolute poverty—that is, on incomes of less than $1 (PPP) per person per day. Only countries with absolute poverty of more than 20 per cent are listed here.

$1 (PPP) – $1 corrected for purchasing power parity—that is, expressed in term of real purchasing power in a given country.

Example: About 20 per cent of the population of Ecuador earn less than $1 per day.

Source: UNDP 2003; see also UNDP 2002. The latest statistical data are for years between 1990 and 2001.

The second level of poverty is defined by income of less than $2 per person per day (corrected for purchasing power parity).[55] The number of people on this income is about 3 billion, or about half the world's population.[56] In some countries, almost the entire population is on such an income (see Table A7-1, Appendix A). The worst situation is in Ethiopia, where 98 per cent of the population lives on less than $2 per person per day, but the percentage is also high in many other countries, particularly in sub-Saharan Africa.

In some countries the level of absolute poverty, as defined by the $1 limit, is low, but the level of poverty defined by the $2 limit is high.

For instance, in China, the figure for absolute poverty is 16.1 per cent, but 47.3 per cent of the population is on less than $2 per day. In Indonesia, the absolute poverty figure is 7.2 per cent, but 55.4 per cent of the population is on less than $2 per day.[57]

If you want to understand the extent of human suffering, learn more about these countries. If you want to help the poor, you have a list of countries (see Tables 7-6 and A7-1) in greatest need of help. If you want to understand the unfolding global crisis, pay close attention to these and similar countries.

About 170 million underweight children, mainly below the age of five, live in poor countries, and 3 million of them die each year of malnutrition. In the year 2000, 2 million children died of malnutrition in Africa and 1 million in Asia. Considering that each child could have had about 40 years of healthy, productive life, the deaths represent 120 million life-years lost annually. At the other end of the scale, about 1 billion people in the world are overweight, and 300 million are critically obese.[58]

There is a strong correlation between poverty and health-related risk factors. Children in families earning up to $2 per day are 15 times more likely to die before they reach the age of five than children in families earning more. The probability of dying between the ages of 15 and 59 is up to 12 times greater, and the probability of contracting tuberculosis is up to eight times greater. Children in families earning less than $1 per day have a two-to-three times greater risk of being underweight than children in families earning more than $2 per day.[59]

Access to safe water and sanitation is also strongly correlated with poverty. It is estimated that five million people die each year because they do not have such access. About 90 per cent are children, and virtually all of them are in developing countries.[60] This is like a quarter of Australia's population dying each year because they do not have access to clean water and sanitation.

Measuring levels of human development.[61] Levels of human development, which can be interpreted as levels of well-being, are measured using the Human Development Index (HDI), introduced in 1990. The index has three components: the average healthy lifespan, the level of knowledge (measured by the adult literacy ratio and the combined primary, secondary, and tertiary enrolment ratios), and the economic status of the country (measured by the GDP). The higher the HDI, the higher the level of human development (see Box 7-1).

Box 7-1
Levels of human development, 2001

The Human Development Index (HDI), which is used to describe levels of well-being, varies over a large range of values. To help understand this index, I give here the accepted scale of human development levels and a few examples of HDI values.

The scale of human development levels

Level of human development	Range of HDI values
High human development	0.800 and higher
Medium human development	0.500 – 0.799
Low human development	0.499 and lower

Examples of human development levels

High human development (55 countries)	HDI
Two highest	
Norway	0.944
Iceland	0.942
Two lowest	
Trinidad and Tobago	0.802
Mexico	0.800
Medium human development (86 countries)	
Two highest	
Antigua and Barbuda	0.798
Bulgaria	0.795
Two lowest	
Congo	0.502
Togo	0.501
Low human development (34 countries)	
Two highest	
Cameroon	0.499
Nepal	0.499
Two lowest	
Niger	0.292
Sierra Leone	0.275

Average values in 2001

High human development	0.908
Medium human development	0.684
Low human development	0.440
OECD	0.905
High-income OECD	0.929
Developing countries	0.655
Least-developed countries	0.448
World	0.722

Source: UNDP 2003

The values of the HDI components vary over a wide range. Life expectancy at birth varies from 47 years in sub-Saharan Africa to 78 years in high-income OECD countries. The average for developing countries is 64 years; for the least-developed countries, 50 years; and for the world, 67 years. The adult literacy rate is 56 per cent in South Asia and 99 per cent in high-income OECD countries, and in CEE/CIS.[62] The average for developing countries is 75 per cent and for the least-developed countries 53 per cent. The enrolment figure is 44 per cent for sub-Saharan Africa and 93 per cent in high-income OECD countries. The average for developing countries is 60 per cent and for the least-developed countries 43 per cent. GDP ranges from $500 to $54,000 (see section on GDPs below).[63]

The HDI is forced to be in the range of 0–1. The mathematical definition of HDI allows for values greater than one. However, in the calculations, the maximum allowed value for a healthy life span is 85 years, the maximum value for literacy and enrolment is 100 per cent, and the maximum value for GDP is $40,000 per person per year. Human development in high-income countries is now so high that the application of this index is already causing a problem. The GDP in Luxembourg in 2001 was $53,780 per person per year, but $40,000 had to be used in the calculation of HDI. Thus the calculated HDI for Luxembourg does not reflect the true level of human development in this country.

In almost all regions of the world there is a steady increase in human development (see Table 7-7), but in some countries there is a clear decline. Between 1980 and 1990, only four countries experienced a decline (Democratic Republic of Congo, Guyana, Rwanda, and Zambia). However, between 1990 and 2001, the number of such countries increased to 21 (Armenia, Belarus, Botswana, Burundi, Cameroon, Central African Republic, Congo, Democratic Republic of Congo, Côte d'Ivoire, Kazakhstan, Kenya, Lesotho, Mouldova, Russian Federation, South Africa, Swaziland, Tajikistan, Tanzania, Ukraine, Zambia, and Zimbabwe).[64]

Table 7-7 Changes in levels of human development, 1975–2001

Region/Group	HDI		Increase	
	1975	2001	(1)	(2)
CEE and CIS	0.773	0.787	0.014	1.8
LAC	0.668	0.777	0.109	16.3
East Asia-Pacific	0.643	0.722	0.079	12.3
Arab States	0.585	0.662	0.077	11.2
South Asia	0.416	0.582	0.166	39.9
Sub-Saharan Africa[a]	0.426	0.468	0.042	9.9
High-income OECD	0.846	0.929	0.083	9.8
Developing countries	0.517	0.655	0.138	26.7

Regions and groups are arranged in descending order of the HDI in 2001.

HDI – Human Development Index. It measures the average level of human development (well-being) in countries or regions. It does not measure levels of poverty. The higher the HDI the higher the average level of human development of the population. The HDI varies between 0 and 1.

(1) – Increase in the HDI between 1975 and 2001

(2) – Percentage increase in the HDI between 1975 and 2001

CEE and CIS – Central and Eastern Europe and the Commonwealth of Independent States. The HDI for this group of countries decreased between 1990 and 1995, but increased again between 1990 and 2001.

LAC – Latin America-Caribbean

[a] – The HDI for sub-Saharan Africa decreased between 1995 and 2001

Source: Based on UNDP 2003

The fastest progress in human development is in South Asia and the slowest in CEE/CIS. The lowest level is in sub-Saharan Africa. The progress in human development is now faster in developing countries than in high-income OECD countries, but the development levels are much higher in high-income OECD countries.

The HDI for 1975–2001 shows linear time dependence (see Table 7-8). Using linear extrapolation, we can find that high-income OECD countries will reach a maximum HDI value in 2024. This does not mean they will reach a maximum in human development: only that, for many of them, the calculated index will be meaningless.

The future of developing countries is not certain. Their GDP values are so low that they will probably never catch up with developed countries. For many countries of sub-Saharan Africa, GDP values have been falling since 1975. The GDP values for countries with low human development are close to $1000 per person per year, and they were almost constant between 1975 and 2001.

Table 7-8 Gaps in human development, 1975–2001

Group	1975	1980	1985	1990	1995	2001
Developing countries	0.517	0.546	0.572	0.600	0.633	0.655
High-income OECD	0.846	0.865	0.874	0.895	0.910	0.929
Gap	0.328	0.319	0.302	0.294	0.277	0.274

Gap – The difference between HDI values for high-income OECD countries and developing countries. For example in 1975, the HDI was 0.517 for developing countries and 0.846 for high-income OECD countries. The diffference (gap) was 0.328. Between 1975 and 2001, the gap decreased from 0.328 to 0.274—that is, by 16 per cent.

Source: Based on UNDP 2003

The gap in HDI values for high-income OECD countries and developing countries is slowly narrowing. It decreased from 0.328 in 1975 to 0.274 in 2001. Considering the environmental restrictions, the narrowing may slow down in future. However, if we use the optimistic assumption that the gap will continue to decrease at the rate of the past 25 years, we can calculate that it will not close earlier than 2120.

Linear extrapolations of HDI values for different regions indicate that Latin America-Caribbean might reach the maximum value of 1 about 2050. South Asia might reach the maximum value by 2065, East Asia-Pacific by 2090, the Arab states by 2140, and sub-Saharan Africa by 2190. However, considering the many obstacles that developing countries must surmount, projections of human development beyond 2050, or even earlier, are unreliable. These projections are also based on simple linear extrapolations, and the further in time they extend the less accurate they are.

The progress in human development in CEE/CIS is so slow that by 2040 HDI might increase to only 0.800 from the 1975 value of 0.773. At this rate, even sub-Saharan Africa would overtake CEE/CSI about the year 2130.

Improved way of measuring poverty levels.[65] Incomes of $1 or $2 per person per day give only approximate estimates of poverty levels. A better estimate is given by the so-called human poverty index (HPI). The HPI was introduced after the HDI, and is expressed as a percentage of the population living in poverty. The HPI is defined differently for developed and for developing countries.

Table 7-9 shows examples of poverty levels as measured by HPI. Even in rich countries large proportions of the population live in poverty.

Table 7-9 Large differences in levels of poverty (per cent)

Developing countries	HPI-1
Five countries with the lowest levels of poverty	
Barbados	2.5
Uruguay	3.6
Chile	4.1
Cost Rica	4.4
Cuba	5.0
Five countries with the highest levels of poverty	
Zimbabwe	52.0
Mali	55.1
Ethiopia	56.0
Burkina Faso	58.6
Niger	61.8

Developed countries	HPI-2
Five countries with the lowest levels of poverty	
Sweden	6.5
Norway	7.2
Finland	8.4
Netherlands	8.4
Denmark	9.1
Five countries with the highest levels of poverty	
Belgium	12.4
Australia	12.9
UK	14.8
Ireland	15.3
USA	15.8

HPI – The Human Poverty Index (per cent) includes income and other factors, such as life span and illiteracy. The HPI is defined differently for developing and developed countries.

Example: Only 2.5 per cent of the population in Barbados live in poverty, but the figure is 61.8 per cent in Niger. Only 6.5 per cent of the population in Sweden live in poverty, but 15.8 per cent in the USA.

Source: UNDP 2003

HPI-1. For developing countries, the HPI (labelled as HPI-1) includes the probability of dying below the age of 40, the adult illiteracy rate, the percentage of the population without access to clean water, and the percentage of underweight children under the age of five.

The probability of not surviving to the age of 40 is 3 per cent in Barbados and Kuwait, but 53.6 per cent in Zambia. Adult illiteracy varies between 2.2 per cent (Republic of Korea) and 84.1 per cent (Niger). The proportion of the population without access to clean water varies between zero (Cyprus, and Barbados) and 76 per cent (Ethiopia). The proportion of underweight children varies between 1 per cent (Chile) and 48 per cent (Bangladesh). The lowest proportion of the population living in poverty, as measured by the HPI-1, is in Barbados (2.5 per cent), and the highest in Niger (61.8 per cent).

HPI-2. For developed countries, the HPI (labelled as HPI-2) includes the probability of dying below the age of 60, the percentage of people with functional illiteracy, the percentage of people living below the poverty line (defined as having an income lower than 50 per cent of the median household income), and the rate of long-term unemployment (more than 12 months).

UNDP (2003) lists 51 countries for which the HPI-2 should have been calculated. However, it is calculated for only 17 countries, as the relevant data sets are incomplete for the remaining countries.

In this group of 17 countries, the lowest probability of dying before the age of 60 is in Sweden (7.3 per cent) and the highest is in the US (15.8 per cent). The lowest functional illiteracy is in Sweden (7.5 per cent), and the highest is in Ireland (22.6 per cent). The lowest unemployment level is in Norway and the US (0.2 per cent) and the highest is in Italy (6.1 per cent). The lowest proportion of the population living below the poverty line is in Luxembourg (3.9 per cent) and the highest is in the US (17 per cent).

The proportion of people living in poverty, by the standards of richer countries, ranges from 7 per cent to 16 per cent. The lowest proportion living in poverty, as measured using the HPI-2, is in Sweden (6.5 per cent) and the highest is in the US (15.8 per cent). About 45 million people in the US live in poverty, and about 2.6 million in Australia.

Finally, it is worth mentioning child poverty levels. The proportion of children living in poor households in industrialised countries varies between 2 per cent and 27 per cent. It is interesting that even in rich countries the proportion of children living in poor households is high.

Table 7-10 Child poverty in industrialised countries, 1999 (percentage of children living in poor households)

Country	Child poverty (Per cent)	Country	Child poverty (Per cent)
Russia	26.6	France	9.8
USA	26.3	Netherlands	8.4
UK	21.3	Luxembourg	6.3
Italy	21.2	Switzerland	6.3
Australia	17.1	Belgium	6.1
Canada	16.0	Denmark	5.9
Ireland	14.8	Austria	5.6
Israel	14.7	Norway	4.5
Poland	14.2	Sweden	3.7
Spain	13.1	Finland	3.4
Germany	11.6	Slovakia	2.2
Hungary	11.5	Czech Republic	1.9

Countries are arranged in descending order of child poverty.

Example: In 1999, 26.6 per cent of children in Russia lived in poor households, and 26.3 per cent in the US.

Source: UNICEF 2000b

Inequality in gross domestic product. Of the 175 countries surveyed by UNDP (2003), only 55 countries are in the so-called high human development group. Their average GDP in 2001 was $23,135 per person (see Table 7-11). For 23 countries in this exclusive group, GDP exceeded $20,000, and for only three countries (Luxembourg, the US, and Ireland) it exceeded $30,000. The lowest GDP in 2001 was in Sierra Leone ($470) and the highest in Luxembourg ($53,780).[66]

Between 1975 and 2000, GDP values increased everywhere, except in sub-Saharan Africa and CEE/CIS. The highest rate of increase was in high-income OECD countries, where GDP increased from $16,048 per person in 1975 to $27,843 per person in 2000—which corresponds to an average increase of $471 per person per year. The fastest increase in the rest of the world was in East Asia, from $1000 per person in 1975 to $4500 per person in 2000—which corresponds to an average increase of only $140 per person per year.[67] If we use the accepted way of measuring progress by GDP, we can see that rich countries are getting rapidly richer. With some exceptions, developing countries are also getting richer, but at a much slower pace. Even for fast-moving East Asia, progress is more than three times slower than for high-income OECD countries.

Table 7-11 A wide range of GDP values

High human development	GDP
Two countries with the highest GDP	
Luxembourg	53 780
USA	34 320
Two countries with the lowest GDP	
Belarus	7 620
Cuba	5 259
Medium human development	
Two countries with the highest GDP	
Equatorial Guinea	15 073
Saudi Arabia	13 330
Two countries with the lowest GDP	
Mayanmar	1 027
Congo	970
Low human development	
Two countries with the highest GDP	
Djibouti	2 370
Zimbabwe	2 280
Two countries with the lowest GDP	
Tanzania, U. Rep. of	520
Sierra Leone	470
Arab States	5 038
East Asia-Pacific	4 233
LAC	7 050
South Asia	2 730
Sub-Saharan Africa	1 831
CEE and CIS	6 598
High human development	23 135
Medium human development	4 053
Low human development	1 186
OECD	23 363
High-income OECD	27 135
Developing countries	3 850
Least-developed countries	1 274
World	**7 376**

GDP – Gross domestic product per person per year in 2001 (in 2001 US dollars and corrected for purchasing power parity)

LAC – Latin America-Caribbean; CEE and CIS—Central and Eastern Europe and the Commonwealth of Independent States

Source: UNDP 2003

Global inequality between rich and poor countries is increasing. The gap in GDP between the richest and the poorest countries increased from 30:1 in 1960, to 60:1 in 1990, and to 74:1 in 1997. The gap was only 3:1 in 1820, 7:1 in 1870, and 11:1 in 1913.[68]

The difference between GDP values for industrialised and developing countries is now so wide that it has become insurmountable, because it is many times wider than GDP values in developing countries. The gap is increasing at an alarming rate of between $300 and $500 person per year (see Table 7-12). For instance, the gap between countries with low human development and high-income OECD countries was $15,000 in 1975, but it had increased to $26,200 by 2000.

Table 7-12 The widening gap between GDP values for developing and high-income OECD countries

		Gap in		Increase		
Group	1975	1990	2000	(1)	(2)	(3)
Low human development	15 000	22 100	26 200	11 200	448	75
Sub-Saharan Africa	13 600	19 400	25 600	12 000	480	88
South Asia	14 700	21 200	25 200	10 500	420	71
Arab States	12 700	17 300	23 200	10 500	420	83
East Asia	15 000	21 000	23 100	8 100	324	54
LAC	10 200	16 900	20 500	10 300	412	101
World	10 900	16 900	19 900	9 000	360	83

Groups of countries are arranged in descending order of the gaps in 2000. The gaps are in $US2000 per person per year. They have been corrected for the purchasing power parity—that is, for actual purchasing power in the respective groups of countries. All values have been rounded up to the nearest $100.

GDP – Gross domestic product

Gap – Difference between average GDP for a given group of countries and for high-income OECD countries. For instance, in 2000, the average GDP for high-income OECD countries was $25,600 per person higher than the average GDP in sub-Saharan Africa. The difference in GDP was about 14 times wider than the average GDP in sub-Saharan Africa.

(1) – Increase in the gap between 1975 and 2000. For instance, between 1975 and 2000, the gap increased by $12,000 for sub-Saharan Africa

(2) – An average rate of the increase in the gap per year. For instance, between 1975 and 2000 the gap between sub-Saharan Africa and high-income countries was increasing at an average rate of $480 per person per year

(3) – The percentage increase in the gap from 1975 to 2000. For instance, between 1975 and 2000, the gap increased by 88 per cent—that is, nearly doubled, for sub-Saharan Africa

LAC – Latin America-Caribbean

World – The gap between average GDP for high-income OECD countries and average global GDP. For instance, in 2000, GDP for high-income OECD countries was $19,900 per person higher than global GDP

Source: Calculated using data of UNDP 2000

We have reached a stage in human development when economic growth based on the current unsustainable exploitation of natural resources cannot be greeted with enthusiasm. Our combined ecological footprint is greater than global ecological capacity. In particular, rapid economic growth in Asia points to increasing ecological problems because this region contains the largest proportion of the global population. Even though the region's GDP per person is still relatively low, the combined GDP is high, which means that the combined environmental impact is high—and it is increasing.

Reduced economic inequality between countries and an improved share of global wealth are not only desirable, but also necessary if we want a stable, peaceful, and sustainable future. However, until we repair the damage to the environment and increase ecological capacity, the only way to lessen the harmful impact of global development is for richer countries to reduce their consumption. A mechanism for solving the problem of abject poverty (at least in part) is in place, but it requires our dedicated commitment, support, and collaboration.

The Millennium Development Goals (MDG).[69] The MDG have been designed to help developing countries in their efforts to improve their standard of living. The goals to be achieved by 2015 include:

- Eradicating extreme poverty and hunger
- Providing universal primary education
- Eradicating gender inequality
- Reducing child mortality
- Fighting HIV/AIDS, malaria, and other diseases
- Ensuring environmental stability
- Building global partnership in development

Some progress has been made, but it is not yet certain whether the improvements can be sustained. Yet this little step forward shows that with commitment and cooperation between nations we could improve the living conditions of a greater number of people.

Unfortunately, in most developing countries the progress is sluggish, and in many of them the living conditions are getting worse. For instance, in sub-Saharan Africa only four countries representing 7 per cent of the population in that part of the world made some progress and are likely to achieve the goals by 2015. However, of the 44 countries in this region, 20 are now poorer than in 1990.

The success of the MDG project depends on financial support from OECD countries. The support is in the form of official development assistance (ODA) managed by the Development Assistance Committee (DAC); but, although GDP in OECD countries is increasing, ODA is decreasing. In 2001, ODA fell 52 per cent below the intended target of $116 billion.

Without external financial aid, many developing countries have little chance of improving their living standards. They cannot use military or economic aggression to draw from the resources of other countries. Many developing countries, especially those in the least-developed group, are too weak to use their own resources. They need help to get started.

Rapid urbanisation. We are approaching a time when most of the global population will live in urban areas. The number of people living in urban areas is increasing not only because the global population is increasing, but also because more people are leaving rural areas. The urban population is increasing much faster in developing countries than in industrialised countries.

Up to 1970, the urban population in developing countries was smaller than in industrialised countries (see Table 7-13). Between 1970 and 2000, the urban population in developing countries increased from about 650 million to about 2000 million, whereas in industrialised countries it rose only from about 700 million to 900 million.[70]

Worldwide, between 1950 and 2000 there has been almost a four-fold increase in urban population. About half of the increase is due to the birth rate, and half to an influx from rural areas. Each day, 160,000 people around the world move from rural to urban areas. Globally, the urban population increases at 2.1 per cent per year, and the rural population at only 0.7 per cent per year.

With some exceptions, there is a correlation between poverty levels and urban population growth rates. The poorer the nation the faster its urban development. The average rate of urban population growth in developed countries is 0.4 per cent per year, but in developing countries it is 2.8 per cent per year, and in the least-developed countries it is 4.6 per cent per year.

The fastest regional increase in urban population, 3.8 per cent per year, is in Africa. The rate is 4.7 per cent in Eastern Africa, 4.4 per cent in Middle Africa, 2.7 in Northern Africa, 2.1 in Southern Africa, and 4.3 per cent in Western Africa. In some countries of Africa (Liberia,

Eritrea, Burundi, Somalia, Uganda, Niger, and Sierra Leone) growth in the urban population is close to or more than 6 per cent per year. At this rate, the urban population in these countries will double every 12 years.

Table 7-13 Growth of the urban population, 1950–2050

Year	Industrial countries	Developing countries	Total	(1)	Per cent (2)	(3)
1950	474	276	750	29.3	63.2	36.8
1960	604	413	1 017	33.5	59.4	40.6
1970	711	646	1 357	36.6	52.4	47.6
1980	782	972	1 754	39.3	44.6	55.4
1990	853	1 427	2 280	43.2	37.4	62.6
2000	890	1 976	2 866	47.4	31.1	68.9
2010	911	2 630	3 541	51.6	25.7	74.3
2020	910	3 390	4 300	56.4	21.2	78.8
2030	888	4 255	5 144	62.1	17.3	82.7
2040	846	5 226	6 072	69.0	13.9	86.1
2050	782	6 302	7 084	77.3	11.0	89.0

Urban populations are in millions.

Industrial countries – Urban population in industrial countries

Developing countries – Urban population in developing countries

Total – Global urban population in a given year

(1) – Urban population as a percentage of global population in a given year

(2) – Percentage of urban population in industrial countries

(3) – Percentage of urban population in developing countries

Example: In 1950, the global urban population was 750 million, of which 474 million lived in industrial countries and 276 million in developing countries. In that year, 29.3 per cent of the global population lived in urban areas. About 63.2 per cent of the urban population lived in industrial countries and 36.8 per cent in developing countries.

Source: Data for the urban population in 1950–1999 are from Sheehan 2000. Only the data in steps of 10 years are listed here. The numbers for 2000–2050 are the results of my analysis of the data between 1950 and 1999. The calculated curves reproduced the data points with an accuracy of about 1 per cent. The percentage values of urban populations were calculated using the historical values of global populations for 1950–1990 and my low-level projections for 2000–2050.

In Asia, the urban population increases at 2.7 per cent per year; in Europe, 0.3 per cent; Latin America-Caribbean, 1.9 per cent; Northern America, 1.2 per cent; and Oceania 1.3 per cent. In Australia, the urban population increases at 1.4 per cent per year.

In 1950, 30 per cent of the global population lived in urban districts. By 2000, the proportion had increased to 47 per cent. In North America 77 per cent of the population lives in urban districts. The figures for other regions are: 77 per cent in Oceania, 75 per cent in Latin America-Caribbean, 75 per cent in Europe, and 38 per cent in Africa and in Asia.[71]

Analysis of the data covering 1950 to 2000 shows that by about 2007, half of the global population will live in urban centres.[72] By 2028 about 5000 million people (61 per cent) will live in urban areas—900 million in industrialised countries and 4100 million in developing countries.[73] The total number of people living in urban areas in 2028 will be the same as the number of people living in the world in 1987.

Projections beyond 2028 show that by 2039 there will be 6000 million people in urban districts, the same as the global population at the end of the 20th century. By 2050, the total number of people living in urban districts will be about 7000 million.

The annual growth rates of urban populations are falling in all regions of the world.[74] Even so, the urban population will double in 20 years in Africa and in 27 years in Asia.

The growth of megacities.[75] In 1000 AD, the bustling metropolis of Cordova had a population of 450,000. The second-largest city in the world was Kaifeng in China, with 400,000 inhabitants (see Table A7-2, Appendix A). In 1800, at the onset of the global population explosion, the largest city in the world was Peking, with 1,100,000 people. Close behind was London, Canton, Edo (Tokyo), Constantinople, and Paris.

By 1900, London had become the largest city in the world. Its population exploded to 6.5 million. New York came from nowhere to take second place. The population of Paris increased to 3.3 million. Berlin took fourth place with 2.7 million.

From about 1950, there has been an explosion of megacities—those with a population of 10 million or more. In 1960, there were only two, Tokyo and New York, but by 1975 there were five: Tokyo, New York, Shanghai, Mexico City, and Sao Paulo. By 1990, the number was twelve. The projected number of megacities for 2015 is 26, almost exclusively in developing countries.

Bangkok and Paris may join this group: Bangkok's population is projected to be 9.8 million in 2015; and Paris, 9.7 million. The fastest-growing cities are Mumbai, Lagos, Karachi, Delhi, metro Manila, Jakarta, Istanbul, Hangzhou, Hyderabad, and Lahore.

Worldwide, the number of cities with a population of more than 1 million increased from 326 in 1990 to 411 in 2000. The projected number in 2015 is 622.

Benefits and problems of urbanisation.[76] Life in urban centres is attractive because it offers such benefits as employment and entertainment, as well as access to clean water, sanitation facilities, medical centres, social services, and education (see Box 7-2).

**Box 7-2
Benefits of urbanisation**

Urbanisation offers a number of attractive benefits that are not readily available in rural areas. They include better access to secondary education, lower mortality of children under five years, better access to drinking water, and better access to sanitation.

Education and under-five mortality, 1992–1998

Country	Secondary education		Under-five mortality	
	Urban	Rural	Urban	Rural
Bolivia	72.5	20.8	71.9	134.3
Brazil	68.6	31.3	49.1	79.4
Dominican Rep.	54.3	22.2	54.7	70.0
Nicaragua	58.1	18.0	48.8	64.3
Kenya	49.4	23.1	88.3	108.6
Madagascar	52.6	16.8	127.1	173.8
Mozambique	13.9	1.4	150.4	236.9
Togo	32.4	7.7	101.3	157.4
Bangladesh	42.2	15.0	96.2	130.9
Philippines	82.0	59.6	45.8	62.5
Uzbekistan	99.7	99.6	51.8	56.8
Vietnam	81.3	62.3	30.4	48.3

Secondary education – Percentage of children receiving secondary education
Under-five mortality – Number of deaths per 1000 live births for children aged up to five years

Example: In Bolivia, 72.5 per cent of children in urban districts have access to secondary education, but only 20.8 per cent in rural areas. The under-five mortality is on average 71.9 per 1000 live births in urban centres, but 134.3 per 1000 live births in rural areas.

Box 7-2 (cont.)

Better access to water and sanitation, 1990s
(Per cent)

Country	Access to water		Access to sanitation	
	Urban	Rural	Urban	Rural
MENA	97	72	92	53
Sub-Saharan Africa	77	39	70	35
South Asia	86	78	73	20
East Asia-Pacific	95	58	77	20
LAC	88	42	82	44
World	**90**	**62**	**79**	**25**

Access to water – Percentage of the population with access to safe drinking water

Access to sanitation – Percentage of the population with access to sanitation

MENA – Middle East and North Africa

LAC – Latin America-Caribbean

Source: Harrison and Pearce 2001

In a number of countries, the proportion of children attending secondary education is substantially higher in urban areas than in rural areas. For instance, in the Philippines, 82 per cent of children living in urban areas benefit from secondary education, compared with 60 per cent in rural areas.

Child mortality rates in urban areas are lower than in rural areas. For instance, the rate in the Philippines is 46 per 1000 live births in urban areas and 63 in rural areas.

Worldwide, 90 per cent of the urban population has access to safe drinking water, compared with 62 per cent in rural areas. In the Middle East and North Africa (MENA), the respective figures are 97 per cent and 72 per cent. Access to sanitation is, on average, 79 per cent in urban areas and 25 per cent in rural areas. In MENA, the respective figures are 92 per cent and 53 per cent.

However, these benefits do not compensate for many serious problems, such as slums and squatter accommodation, air pollution, noise pollution, stress, traffic congestion, traffic accidents, and crime. People living in urban centres depend also on an external supply of food, water, energy, and other items.

Worldwide, one-third of the urban population lives in slums (see Table 7-14). The number of people living in slums is higher than the combined population of the richest countries, and comparable to the population of industrialised countries. The highest proportion of urban dwellers living in slums is in sub-Saharan Africa (72 per cent).

Table 7-14 Urbanisation and the propagation of slums

Region	Total population (millions)	Urban population (per cent)	Urban slum population (per cent)	Urban slum population (thousands)
Sub-Saharan Africa	668	34.6	71.9	166 208
South-Central Asia	1 512	30.0	58.0	262 354
East Asia-Pacific	1 376	39.0	36.3	194 323
West Asia	192	64.9	33.1	41 331
LAC	528	75.8	31.9	127 567
North Africa	146	52.0	28.2	21 355
South-East Asia	529	38.3	28.0	56 781
CEE and CIS	411	62.9	9.6	24 831
Developing countries	4 946	40.9	43.0	869 918
Developed countries	1 194	75.5	6.0	54 068
World	**6 140**	**47.7**	**31.6**	**923 986**

Regions are arranged in descending order of the percentage of urban slum population.

LAC – Latin America-Caribbean

CEE and CIS – Central and Eastern Europe and Commonwealth of Independent States

Total population – The population in a given region or group of countries in 2001

Urban population – Percentage of the population living in urban areas

Urban slum population – Urban population living in slums (in per cent and in thousands)

Example: In 2001, 34.6 per cent of the population of sub-Saharan Africa lived in urban areas. In that year, 71.9 per cent of urban population (166,208,000 people) lived in slums.

Source: UNDP 2003. Statistical information is for 2001.

Currently, more than 1 billion people worldwide are exposed to dangerous levels of air pollution. The development of essential services, such as sewage disposal, is often too slow to cope with rapidly increasing urban populations. In many cities, raw sewage is dumped into waterways or allowed to flow through the streets. Garbage collection services are inadequate, and in some places up to 50 per cent of the garbage is dumped around inhabited areas.

Urban living may offer a better chance of employment, but it also adds to the stress of living. Children in urban areas may have more chance of an education, but they are also attracted to street life and addictive drugs. Urban areas also provide a conducive environment for serious crime, including homicide.

People living in urban centres depend heavily, if not entirely, on water flowing from their taps and on the supply of electricity or gas. They depend on buying food in shops and on petrol to fuel the ever-increasing fleet of motor vehicles. They depend on many other goods that have to be brought to them. A drawn-out shortage of goods and services in urban areas can have devastating consequences.

It is also easier to kill people concentrated in urban centres. Big cities are the targets for cruise missiles and for terrorist attacks, but people living in rural areas are more secure.

The concentration of people in small areas increases the probability of epidemics, and the effects of natural disasters are also incomparably greater in the congested areas of large cities than in rural areas. It has been estimated that if the Kobe-type earthquake had occurred in Tokyo, the cost of the damage would have been between $1000 billion and $3000 billion.[77]

Large cities, including megacities, are located mainly in coastal regions and can therefore be affected by rising sea levels caused by global warming and by coastal storms. Health risks associated with higher ambient temperature are also greater in the concrete jungles of large cities.

Urban areas impose a high demand on natural resources. The ecological footprints of urban districts are large—several hundred times larger than the area they cover. The urbanisation process contributes also to environmental degradation through the pollution of waterways, coastal water, and the atmosphere.

Globally, people in urban centres grow only 20 per cent of the food supply. Thus, people in rural regions have the responsibility of growing food not only for themselves, but also for city dwellers. That burden is increasing with the higher levels of urbanisation.

Urban areas contribute substantially to wasting water.[78] City sprawls with their impermeable surfaces and networks of roads do not allow runoff water to recharge groundwater aquifers or to be used by vegetation. Large volumes of clean rainwater flow into gutters to be wasted. City sprawls contribute also to the fragmentation of forests, to the loss of habitat, to the spread of invasive species, and to the loss of wetlands.[79]

Global crime.[80] Global crime, including drug-related activities, is increasing. At the end of the 20th century there were about 200 million drug users in the world, the equivalent of the combined population of Germany, France, and UK. In the last 10 years of the 20th century, the production of opium increased three-fold and coca leaf two-fold. Drug addiction and drug-related crime are spreading to new regions. In 1995, the value of the illegal drug trade reached $400 billion, which was almost the same as the global trade in gas and oil. Crime is the fastest-growing industry in the world.

Organised crime syndicates are growing. They include the Six Triads in China; the Medellin and Cali cartels in Colombia; the Mafia in Italy; the Yakuza in Japan; the Juarez, Tijuana, and Gulf cartels in Mexico; and the Cosa Nostra in the US. Other parts of the world, including Nigeria, Russia, and South Africa, are also experiencing organised crime.

These organisations have international connections. They paralyse legitimate business and threaten law and order. In Russia, for instance, crime organisations allegedly control about 65 per cent of GDP.

Micro-organisms as agents of mass destruction.[81] The leading cause of death in the world is cardiovascular disease (CVD), accounting for 31 per cent of all fatal casualties (see Table 7-15). CVD is a group of about 20 diseases, including heart disease and stroke.

Table 7-15 The leading causes of death, 1998

Cause Worldwide	Per cent	Cause Low-income countries	Per cent
Cardiovascular diseases	31	Infectious diseases	45
Infectious diseases	25	Non-communicable conditions	35
Cancer	13	Injuries	11
Injuries	11	Perinatal	6
Respiratory and digestive	9	Maternal	2
Maternal	5	Nutritional	1
Other	6		
Children 0–4 years		**Ages 0–44**	
Infectious diseases	63	Infectious diseases	48
Perinatal	20	Injuries	19
Non-communicable conditions	8	Non-communicable conditions	18
Injuries	6	Perinatal	10
Nutritional	3	Maternal	3
		Nutritional	2

Source: WHO 1999a

The next most deadly group is made up of infectious diseases, which account for an estimated 25 per cent of all fatal casualties. However, infectious diseases also contribute to the mortality associated with CVD, cancer, and respiratory disease. Consequently, the proportion of deaths caused by infectious diseases is much higher than the estimated 25 per cent. Infectious diseases are claimed to be the main cause of death in low-income countries, and for the young and middle-aged section of the global population. Micro-organisms are therefore silent, stealthy, and effective agents of mass destruction.

Table 7-16 The most deadly infectious diseases, 1998

Cause	Deaths (millions)		
	Under age 5	Over age 5	Total
Acute respiratory infections	1.9	1.6	3.5
AIDS	0.4	1.9	2.3
Diarrhoeal diseases	1.8	0.4	2.2
Tuberculosis	0.1	1.4	1.5
Malaria	0.7	0.4	1.1
Measles	0.9	0.0	0.9
Total	5.8	5.7	11.5

Source: WHO 1999a

Table 7-17 Major outbreaks of infectious diseases, 1970–1999

Disease	Year(s)	Place	Effects
Anthrax	The 1970s	Zambia	10,000 people infected
Rift Valley fever	1977	Egypt	200,000 people infected, 800 deaths
Visceral leishmaniasis	1985–1987	Southern Sudan	100,000 deaths
Typhus	1996–1998	Burundi	100,000 people infected
Meningitis	1996	Africa	187,000 people infected
Dengue fever	1981	Cuba	344,000 infected
Cholera	1991	Latin America	500,000 people infected
Meningitis	1974	Sao Paulo	30,000 people infected
Diphtheria	1995	Russia	50,000 people infected
Hepatitis A	1989	Shanghai	300,000 people infected
Hepatitis C	1989	North America	170,000,000 people infected
Dengue fever	1982	New Delhi	1 in 5 infected

Source: WHO 1999a

Worldwide, infectious diseases kill 13 million people per year, or 1500 per hour. Most infections and deaths occur in developing countries and the main casualties are children. Worldwide, 63 per cent of children under the age of five, mainly in developing countries, die of infectious diseases.

The leading killers are acute respiratory infection (ARI), acquired immune deficiency syndrome (AIDS), diarrhoeal disease, tuberculosis (TB), malaria, and measles (see Table 7-16). These six diseases account for about 90 per cent of fatal casualties caused by infectious diseases. Infectious diseases, which now devastate developing countries, are likely to erupt as a worldwide epidemic.

WHO (1999a) lists 12 large-scale outbreaks of infectious diseases between 1970 and 1999, each affecting thousands of people (see Table 7-17). WHO also claims that in 1998–1999 there were 155 outbreaks of infectious diseases worldwide (see Table A7-3, Appendix A)

Acute respiratory infections (ARI). These infections, which include pneumonia and influenza, kill about 3.5 million people each year, 1.9 million of them being children under the age of five. Respiratory infections kill more children than any other infectious disease.

In many developing countries parents ignore respiratory infections in their children and do not seek professional advice even if it is available. However, the higher the parents' education level the greater the likelihood of seeking advice.

Countries in which less than 50 per cent of children suffering ARI are taken to health-care providers include Chad, Mali, Niger, Togo, Haiti, Benin, Bangladesh, Cameroon, Eritrea, Guatemala, Madagascar, Ivory Coast, Mozambique, the Central African Republic, Bolivia, Malawi, Dominican Republic, and Colombia. The proportion of children referred to health-care providers ranges from 19 per cent (Chad) to 49 per cent (Colombia).[82]

Acquired immune deficiency syndrome (AIDS). Worldwide, HIV infections increased from 0.1 million in 1980 to 58 million in 2000. The number of deaths rose from 0.1 million in 1984 to 22 million in 2000.[83] The number of new HIV infections in 2000 was estimated at 5.3 million and the number of deaths at 3 million, with 92 per cent of them in developing countries.[84]

Each day, 22,000 people are being infected with HIV and 9150 are dying.[85] By the end of 2004, the cumulative number of people killed by AIDS was close to 34 million.

By 2010, the cumulative number of people killed by the disease will be 59 million. An estimated 28,500 people per day will be infected

(10.4 million per year) and the number of deaths will be 12,800 per day (4.7 million per year). If the trend continues, by 2020 the number of people infected will be 39,150 per day (14.3 million per year) and the number of deaths 18,000 per day (6.6 million per year). The cumulative number of people killed by the AIDS virus will be 116 million.

Poverty is a big factor in the spread of AIDS.[86] At the end of 2000, more than 40 million adults and children were living with HIV/AIDS.[87] The largest number was in sub-Saharan Africa, followed by Latin America, and South and South-East Asia (see Table 7-18).[88]

Table 7-18 Adults and children living with HIV/AIDS, 2000

Region	Number
Sub-Saharan Africa	25 300 000
Latin America	7 400 000
South and South-East Asia	5 800 000
North America	920 000
Eastern Europe and Central Asia	700 000
East Asia-Pacific	640 000
Western Europe	540 000
North Africa and Middle East	400 000
Caribbean	390 000
Australia and New Zealand	15 000
World	**42 105 000**

Number – The number of adults and children living with HIV/AIDS in 2000
Source: UNFPA 2001

The proportion of the population aged 15–49 years living with HIV/AIDS in 2001 was 1.32 per cent in developing countries, 0.28 per cent in OECD countries, 3.55 per cent in the least-developed countries, and 9 per cent in sub-Saharan Africa. The global average was 1.2 per cent.[89]

About 83 per cent of deaths caused by HIV infections occur in sub-Saharan Africa. The rest occur in Asia (6 per cent), Latin America-Caribbean (5 per cent), and other regions (6 per cent).[90]

AIDS kills mainly people under the age of 24, the breadwinner generation. Schools and colleges are also understaffed or being closed because teachers are dying of AIDS. In 1999, HIV infections left 860,000 children without teachers in sub-Saharan Africa. The greatest numbers of children without teachers were in South Africa (100,000), Kenya (95,000), Zimbabwe (86,000), Nigeria (85,000), and Uganda (81,000).[91]

AIDS also kills children under the age of five. They account for 16 per cent of all deaths from HIV infection. It is estimated that AIDS is responsible for about 13 million orphans.[92]

The economy of many countries is greatly affected by AIDS.[93] This epidemic is now spreading to India, where it is expected to infect up to 25 million people by 2010.[94]

Diarrhoeal diseases. Diarrhoeal diseases are associated with poor sanitation, poor waste disposal, polluted water, polluted food, and malnutrition. Diarrhoeal diseases kill almost exclusively people in low-income and middle-income countries, and mainly children. Of the 2.2 million deaths reported in 1998, only 7000 (0.3 per cent) were in high-income countries. The number of children aged five and under that were killed by diarrhoeal diseases in that year was 1.8 million. Diarrhoeal diseases compete with ARI as the primary killer of small children.

Tuberculosis (TB).[95] Each year, TB infects 8.4 million people and kills 2–3 million people, almost exclusively in low-income and middle-income countries.

About 37 per cent of all infections (3.1 million per year) are in South-East Asia, 25 per cent (2.1 million) in the Western Pacific,[96] and 19 per cent (1.6 million) in Africa. In other regions the incidence is lower: 8 per cent (670,000 infections) in the Eastern Mediterranean, 5 per cent (420,000) in Europe, and 5 per cent in the Americas.

Most deaths from TB are in South-East Asia (38 per cent of the global total), followed by Africa (28 per cent) and the Western Pacific (19 per cent). About 8 per cent of deaths from TB occur in the Eastern Mediterranean, 4 per cent in the Americas, and 3 per cent in Europe.

The TB bacterium (*Mycobacterium tuberculosis*) and HIV are the deadly twins reinforcing each other in their destructive activities. If the immune system of an infected person is strong, TB bacteria can remain dormant in the body for years, or even for the lifetime of the infected person. The AIDS virus weakens the immune system and makes the person susceptible to TB. The TB bacterium also makes the body less resistant to the HIV infection. Worldwide, about 15 per cent of people infected with HIV die of tuberculosis.

A person infected with TB can infect 20–40 people before being detected, but only 10 per cent of those who are infected will infect other people.

Nearly one-third of the global population is infected with TB. However, the rate is increasing, and WHO estimates that 1 billion more people will be infected by 2020. By then, TB alone will have killed 35 million people.

Inadequate or incomplete treatment of TB leads to the development of a multi-drug resistant tuberculosis (MDR-TB),[97] which is nearly impossible to cure. The emergence of this form of TB is claimed to have been caused by neglect and complacency.[98] The problem with MDR-TB is that we do not have effective drugs or vaccines to combat it, and we do not expect to develop any before 2010. MDR-TB has been described as a time bomb ready to explode. Should it explode before 2010, its effects will be out of our control. TB is a global problem, not one limited to developing countries.

MDR-TB has been detected in 100 countries,[99] or about 60 per cent of all countries of the world. In some, such as those of the former Soviet Union, where prisoners are kept in overcrowded conditions, MDR-TB is spreading rapidly. The number of HIV infections is also increasing in these countries. There is therefore a strong possibility that the combination of HIV and TB will trigger a global explosion of MDR-TB.

Early intervention allows successful treatment of about 95 per cent of drug-susceptible TB cases. Each set of treatments extending over six months costs only $20. However, the cost of treating MDR-TB can be more than $8000. Furthermore, the drugs are also toxic.

A cheap and effective treatment for TB is the directly observed treatment, short-course (DOTS). It consists of direct and early detection of infection and a standard series of treatments extending over 6–8 months. However, DOTS relies on financial support and government cooperation.[100] If these two elements are not consistently applied, the likelihood of an MDR-TB explosion will remain high.

Malaria.[101] The malaria parasites (*Plasmodium*),[102] transmitted by 50–60 species of mosquitoes of the genus *Anopheles*, infect 300–400 million people each year and kill 1.5 million. The disease affects almost exclusively the developing countries, and one region—sub-Saharan Africa—accounts for 90 per cent of deaths. About 70 per cent of the fatalities are among children under the age of five.

The global mortality rate for malaria fell from 48 per 100,000 in 1950 to 16 per 100,000 in 1970, and remained about that level until 1997. In sub-Saharan Africa, the mortality rate fell from 183 per

100,000 in 1950 to 104 in 1970, but rose to 147 in 1990 and 164 in 1997. This increase is attributed, at least partly, to the emergence of the chloroquine-resistant parasite *Plasmodium falciparum.*

Records of malaria infections are incomplete, but the largest number of reported cases in 2000 was in Burundi, where close to 50,000 people were affected, followed by Zambia, São Tome and Principe, Malawi, and Mozambique (see Table 7-19).[103]

Table 7-19 Countries with a large number of cases of malaria in 2000 (per 100,000 people)

Country	Number	Country	Number
Burundi	48 528	Guinea-Bissau	16 454
Zambia	34 274	Ghana	15 348
São Tomé and Principe	31 614	Yemen	15 200
Malawi	27 682	Sudan	13 932
Thailand	21 000	Brazil	13 000
Mozambique	18 108	Côte d'Ivoire	12 162
Gambia	17 376	Benin	11 915
Solomon Islands	16 971	Guinea	11 161

Number – Only the number of cases greater than 10,000 per 100,000 people are listed here.
Source: UNEDP 2002

The future of malaria treatments is uncertain because we do not know how rapidly the parasites will develop a resistance to drugs. We also do not know how rapidly we can develop affordable and effective alternatives. At present, an affordable replacement for chloroquine is pyrimethamine in combination with sulphonamide. However, in various parts of the world, mainly in South-East Asia and South America, the malaria parasite *P. falciparum* is already resistant to sulphonamide.

Measles.[104] Worldwide, measles infections kill about 900,000 small children each year, almost exclusively in developing countries. Of the 888,000 deaths in 1998 reported by WHO (1999b), 882,000 were in low-income and middle-income countries. Most deaths are in South Asia and sub-Saharan Africa.

Infection by the measles virus is often linked with such complications as pneumonia and diarrhoea. Measles can also lead to blindness, deafness, brain damage, lung damage, and stunted growth.

Global immunisation coverage against measles increased from 55 per cent in 1987 to 80 per cent in 1990, and remained constant. The vaccination coverage is only 56 per cent in low-income countries, and 51 per cent in sub-Saharan Africa. The lowest coverage, only 15 per cent, is in the Democratic Republic of Congo. An annual investment of only $600 million could increase the global vaccination coverage to 95 per cent and reduce the child death rate by 70 per cent.

Success and failures in fighting infectious diseases. The number of casualties from infectious diseases is increasing. Between 1970 and 1992, disease fatalities for meningitis increased from 6480 per year to 12,223; for hepatitis B, from 8310 to 26,000; for salmonellosis, from 22,000 to 42,900; for shigellosis, from 13,800 to 23,900; and for syphilis, from 91,000 to 113,000. Between 1980 and 1992, disease fatalities for Lyme disease increased from 100 to 12,669.[105] The average number of cases of dengue fever rose from 908 per year in the 1950s to 514,139 per year in the 1990s.[106]

Table 7-20 Emerging and re-emerging diseases and pathogens, 1975–1999

Year	Disease/Pathogen	Year	Disease/Pathogen
1975	AIDS	1989	Hepatitis C
1976	Legionnaires disease	1988	Hepatitis E
1976	Cryptosporidiosis	1988	Human herpesvirus 6
1976	Ebola haemorrhagic fever	1991	Venezuelan haemorrhagic fever
1977	Hantavirus	1992	Vibrio cholerae O139
1977	Campylobacter jejuni	1994	Brazilian haemorrhagic fever
1979	Creutzfeldt-Jakob disease	1994	Human and equine morbillivirus
1980	Human T-cell lymphotropic virus	1995	Kaposi's sarcoma virus
1980	Hepatitis D	1996	Australian bat lyssavirus
1982	Escherichia coli O157:H7	1997	H5N1 (avian flu)
1982	Lyme borreliosis	1999	Nipah virus
1986	Salmonella enteritidis PT-4		

Source: Pimentel et al 1998; WHO 1999a

However, we also have some success stories. Between 1978 and 1992, the number of deaths from diarrhoeal disease in Mexico was reduced from 250 per 100,000 in 1978 to 50 per 100,000 in 1992. Sex education in Uganda reduced HIV infections from 35 per cent in 1992 to 17 per cent in 1996. The number of deaths from malaria in Vietnam plummeted from 4200 per year in 1990 to 40 in 1997 because of the availability of free treatments. Globally reported cases of measles were reduced from more than 4 million in 1982 to below 1 million in

1996. HIV infections in the Thai army were reduced from 4 per cent in 1993 to 2 per cent in 1997. Worldwide, the number of polio cases was reduced from 35,000 in 1988 to 5000 in 1996.

These few examples show that the fight against infectious diseases is not yet lost, and that with more dedication we could have greater control over outbreaks. The task is not easy, as we have to deal with the increasing resistance of micro-organisms and with emerging or re-emerging diseases (see Table 7-20).[107]

WHO (1999a) claims that more than 30 new diseases emerged in the last 20 years of the 20th century. Between 1996 and 2000, there were 41 outbreaks of emerging or re-emerging diseases in the world, with three of them in Australia (Ross River virus, Hendra virus, and equine morbilli virus).

Costs and priorities. Funds for combating infectious diseases are not readily available. Worldwide, only 5 per cent of GDP is devoted to health care. Only 1.5 per cent of donor assistance for developing countries is assigned to controlling infectious diseases, and only 7 per cent to dealing with other health-related problems and the issues of nutrition and population.

In 1995, the world was spending hundreds of billions of dollars on military activities, but only $15 million on fighting AIDS, TB, and malaria. The number of deaths from armed conflicts between 1945 and 1993 was 26 million, but the number of deaths from AIDS, TB, and malaria was 150 million.

The prudent allocation of funds can reduce the loss of life and economic loss. The cost of prevention and early treatment of infectious diseases are low, but the cost of dealing with epidemics is high. For instance, the cost of dealing with the plague in India in 1995 was $1.7 billion; the cost of fighting the outbreak of cholera in Tanzania in 1998 was $36 billion; and the cost of fighting cholera in Peru in 1991 was $770 million.

Infectious diseases are responsible for a substantial annual collective loss of productive life years, known as disability-adjusted life years (DALYs).[108] For instance, in 1998, acute respiratory infections were responsible for the loss of 83 million life years, diarrhoeal diseases for 73 million, and HIV/AIDS for 71 million (see Table A7-4, Appendix A). The total for all infectious and parasitic diseases was 365 million DALYs.[109]

In 2001, the total for all infectious and parasitic diseases was 359.4

million DALYs. The figure for respiratory infections was 94 million; AIDS, 88.4 million; diarrhoea, 62.5 million; TB, 36 million; malaria, 42.3 million; and measles, 26.5 million.[110]

The spread of infectious diseases. Poor sanitation and hygiene, hunger and malnutrition, lack of satisfactory housing, intensified immigration, climate change, lack of multi-sector coordination, lack of satisfactory health services, and lack of vaccines all contribute to the spread of infectious diseases. Overcrowding in prisons and public transport, and in the shared facilities of large cities, exacerbates the problem. As the frequency of interactions between people increases, so does the likelihood of large-scale epidemics, particularly in developing countries.[111]

Box 7-3
Intensified international travel

The number of people travelling overseas each year is increasing. With this increased traffic, the probability of the spread of infectious diseases is also increasing.

International tourist arrivals, 1950–2050

Year	Arrivals (million)	Year	Arrivals (million)
1950	25	2000	677
1960	69	2010	950
1970	166	2020	1 270
1980	286	2030	1 640
1990	459	2050	2 524

Source: The data for international tourist arrivals between 1950 and 1999 are from Mastny (2000a). Only the data in steps of 10 years are shown here. Tourist arrivals for 2000–2050 are based on my analysis of the data. The data for 1950–1999 have been reproduced by calculations with an average accuracy of 2 per cent.

Percentage increase in tourist arrivals between 1993 and 1997

Region	Increase (per cent)	Region	Increase (per cent)
Americas	32	Middle East	46
Europe	27	South Asia	32
Africa	44	East Asia-Pacific	29

Source: WHO 1999a

Micro-organisms now travel over large distances. International tourism has increased dramatically, from 25 million annual arrivals in 1950 to 677 million only 50 years later. The projected figure for 2050 is 2500 million annual arrivals (see Box 7-3). The number of travellers increased from 69,000 per day in 1950 to 2 million per day in 1998.[112]

Box 7-4
Microbes on the move

With increasing international travel, the probability of the spread of microbes to distant countries is also increasing.

Imported malaria to UK, 1987–1992
(Total of 8353 cases)

Agents	Contribution (Per cent)	Agents	Contribution (Per cent)
Visiting friends and relatives	49	Expatriates	5
Other visitors	19	Immigrants	11
Tourists	16		

Tourists returning from Thailand affected by infectious diseases, 1995

Disease affected	Tourists affected (Per cent)	Disease	Tourists (Per cent)
Diarrhoeal diseases	64	Respiratory infections	8
Fever (causes unknown)	6	Malaria	4
Hepatitis	4	Gonorrhoea	4
Other	10		

Source: WHO 1999a

Visits to developing countries

For every 100,000 travellers returning from developing countries the following numbers were affected by infectious diseases:

Disease	Number of sufferers
Traveller's diarrhoea	30,000–80,000
Malaria	2,500
Acute febrile respiratory tract infection	1,500
Gonorrhoea	200
Typhoid	33

Source: WHO 2000

In 1978, poliovirus was taken from Europe to Canada. Drug-resistant strains of *Pneumococcus* were first detected in Spain, but they found their way all over the world—Korea, Taiwan, Hong Kong, the Philippines, Thailand, Singapore, Malaysia, South Africa, Uruguay, Chile, Argentina, Brazil, Colombia, Mexico, and the US (see also Box 7-4).

International trade also helps the bugs to travel. In 1985, mosquitoes responsible for the spread of such diseases as yellow fever and dengue were carried in a cargo of tyres from Asia to the US and quickly invaded 17 states.

Microbes can travel in ships' ballast water, which is discharged en route or at their destination. For instance, 80 million tonnes of ballast water is brought into the US each year. *Vibrio cholerae* (the cholera bacterium) is routinely found in ballast water.

Drug-resistant strains of bacterial species that develop in industrialised countries due to overuse of drugs can find their way to poorer countries where inadequate health care allows them to multiply. Conversely, exotic bacteria can be transported to industrialised countries where they can cause fatal infections. We share one world, and the likelihood of global pandemics is high and increasing.

Superbugs. Micro-organisms have the ability to adapt to changing environments. This helps them to build resistance to antibiotics. In only 10 years, European strains of *Pneumococcus* increased their resistance to penicillin from 4 per cent to more than 40 per cent.

Worldwide, we produce millions of kilograms of more than 100 types of antibiotic (23 million kilograms per year in the US alone), and are using them to fight even trivial ailments. The more we use them the bigger the problem we create, because we are breeding a greater number of powerful microbes. We "train" them if we do not finish a prescribed series of antibiotic treatments. We kill weak bacteria, and leave the strong ones to reproduce and do even more damage.

Antibiotics that used to be 100 per cent effective in killing bacteria are now much less efficient. Examples are penicillin, streptomycin, chloroquine, mefloquine, lamivudine, chloramphenicol, and cotrimoxazole. Diseases such as gonorrhoea, tuberculosis, malaria, hepatitis B, and typhoid, which could once be treated with antibiotics, are now hard to cure.

From the time the first antibiotic (penicillin) was introduced, its effectiveness began to decline. *Staphylococcus aureus* was the first bacterium that developed a resistance to penicillin, and other bacteria soon followed. We now have to use increasingly stronger forms of drugs. Penicillin was discovered in 1928 and introduced in 1941. By 1946, 14 per cent of hospital strains of *Staphylococcus aureus* developed a resistance. By 1950, this had increased to 56 per cent. Between 1960 and 1970, microbial resistance spread to communities. Between 1980 and 1990, 80 per cent of bacterial strains in communities and 95 per cent in hospitals developed a resistance to penicillin (see Box 7-5).

Box 7-5
The rise and fall of penicillin

Penicillin, a powerful antibiotic in the early 1940s, slowly lost its effectiveness in the treatment of a number of infectious diseases. Its usefulness still depends on the type of disease and the country of application.

Rise and fall of penicillin against *Staphylococcus aureus*

Year	Event
1928	Penicillin discovered
1941	Penicillin introduced
1945	Fleming predicts decreasing power of penicillin
1946	14 per cent of hospital strains of *Staphylococcus aureus* are resistant to penicillin.
1950	59 per cent of hospital strains of *Staphylococcus aureus* are resistant to penicillin.
1960s	Some strains of *Staphylococcus aureus* in communities are resistant to penicillin.
1980s	95 per cent of hospital strains of *Staphylococcus aureus* and 80 per cent of community strains are resistant to penicillin.

Source: Heymann 2000

Percentage of *Pneumococci* strains resistant to penicillin in the 1990s

Country	Resistance	Country	Resistance
South Korea	70	Kenya	25
Hungary	55	Bulgaria	25
Mexico	48	Egypt	22
Hong Kong	31	Saudi Arabia	20
Singapore	25	Israel	15

Source: WHO 2000

Another example of the declining effectiveness of drugs is the lower response to antimalarial drugs: chloroquine, sulphadoxine-pyrimethamine, quinine, and mefloquine (see Table 7-21). In just four years from 1976, the response to chloroquine fell from 30 per cent to only 2 per cent.

Table 7-21 The decreasing response to antimalarial drugs

Drug	Year	Response (per cent)	Drug	Year	Response (per cent)
Chloroquine	1976	30	Quinine	1976	88
	1977	19		1980	81
	1978	13		1984	75
	1979	5		1988	68
	1980	2		1992	61
SP	1976	78	Mefloquine	1976	99
	1977	77		1980	99
	1978	58		1984	99
	1979	36		1988	91
	1980	17		1992	70
	1981	2			

Response – The percentage of successful treatments using a given drug. For instance, the success of treating malaria by chloroquine decreased from 30 per cent in 1976 to only 2 per cent in 1980.

SP – sulphadoxine-pyrimethamine

Source: WHO 2000

Vancomycin has been known as the strongest antibiotic, and it is still used as a last resort in treating infectious diseases. However, the inevitable has happened: we already have new strains of bacteria that are resistant even to vancomycin.

At first, we only had vancomycin-resistant *Enterococci* (VRE). *Enterococci* belong to a group of bacteria made up of 17 species. These are common bacteria that form part of the normal flora in gastro-intestinal and genital tracts. Usually, they do no harm, but sometimes they can cause such conditions as endocarditis, urinary tract infections, and biliary tract infections. About 90 per cent of enterococcal infections are caused by just one species, *E. faecalis*, and 7 per cent by *E. faecium*.

VRE were first identified in Europe in 1986, and since then they have been detected in many other countries. Five species of *Enterococci* have been identified as resistant to vancomycin, but two of them, *E. faecium* and *E. faecalis*, have proved to be outstanding.

You might have heard about the dreaded golden staph or MRSA. The initials stand for methicillin-resistant *Staphylococcus aureus*. Methicillin is an antibiotic from the penicillin family and *Staphylococcus aureus* is a common bacterium, usually present on the skin or in the nose of healthy people. Sometimes it gets into the bloodstream and can cause various infections ranging from minor to fatal, such as pimples, impetigo, boils, carbuncles, endocarditis, arthritis, and osteomyelitis. An example of the increasing resistance of *Staphylococcus aureus* to methicillin is given in Table A7-5, Appendix A.

Staphylococcus aureus is a champion of resistance and, judging by the experience with other drugs, it was expected that the strongest known antibiotic, vancomycin, would eventually be defeated by this bacterium (see Table 7-22). The first justification of this expectation came in 1996 when vancomycin-intermediate *Staphylococcus aureus* (VISA) was identified. About eight years later, strains of this bacterium developed full resistance to vancomycin.[113] These strains are known as vancomycin-resistant *Staphylococcus aureus* (VRSA). If MRSA has been filling us with fear, we now have the more frightening VRSA.

Table 7-22 *Staphylococcus aureus*—a champion of resistance

Antibiotic	Introduced	Defeated
Penicillin	1941	1940s
Streptomycin	1944	mid 1940s
Tetracycline	1948	1950s
Erythromycin	1952	1950s
Methicillin	1959	late 1960s
Gentamicin	1964	mid 1970s
Vancomycin	1958	1997
Linezolid	2000	2001

Defeated – First reports of successful resistance of Staphylococcus aureus to a given antibiotic

Example: Penicillin was introduced in 1941. The first cases of effective resistance of *Staphylococcus aureus* against this antibiotic were reported in the 1940s—that is, soon after its introduction.

Source: Berghuis 2003

Just in time, we developed new drugs: linezolid (Zyvox), daptomycin, and quinupristin/dalfopristin (Synercid). However, that success did not last long, and the first victim was linezolid.

Introduced in April 2000 in the US and hailed as the most powerful antibiotic in existence, linezolid was defeated only one year later by

Staphylococcus aureus.[114] Follow-up research not only confirmed the resistance to linezolid, but also demonstrated an intermediate resistance to quinupristin/dalfopristin.[115]

The emergence of drug-resistant bacteria offers a disturbing prospect for our future. The more we fight them, the stronger they grow, and soon we might have no effective weapons. Who—or what—will win the race? It takes a long time to design, screen, and approve a new drug. A large percentage of designed drugs is routinely rejected. Of 7 million new compounds screened for testing, only one is likely to be approved for use 15 to 20 years later.

It has been suggested that we have 10–20 years to find an effective way of dealing with the growing number of drug-resistant bacteria.[116] If we fail, we might not be able to control the spread of infectious diseases.

The History of Medicine

2000 BC	Here, eat this root.
1000 AD	That root is heathen. Here, say this prayer.
1850 AD	That prayer is superstition. Here, drink this potion.
1920 AD	That potion is snake oil. Here, swallow this pill.
1945 AD	That pill is ineffective. Here, take this penicillin.
1955 AD	Oops … bugs mutated. Here, take this tetracycline.
1960–99	39 more "oops" … Here, take this more powerful antibiotic.
2000 AD	The bugs have won! Here, eat this root.

—Anonymous[117]

Will a return to nature's cures help us to fight diseases, or have we already created an invincible army of bacteria? Have we managed to destroy nature, the forests, and other forms of vegetation, to the extent that it would be hard to find that healing root, leaf, or herb? Can we win the war against the superbugs?

The future. From about 2007, a growing majority of people will be concentrated in urban districts. The urban environment offers many attractions, but it also has many negative features that contribute to social decline.

Micro-organisms are becoming stronger, and our defences against them are falling. We are running out of weapons, and our future is uncertain. Experience shows that the problem with infectious diseases will get worse. The world has shrunk, and micro-organisms travel rapidly and easily over long distances. The likelihood of global pandemics is increasing.

The gap between gross domestic product and the genuine progress indicator will continue to increase. We shall continue to live with the illusion of economic progress as the environmental deficit continues to increase. We live on a rich planet, but are not managing it well.. With prudent management of resources and better distribution of global wealth we could have a secure future.

We have experienced spectacular progress in the past 200 years, and can see almost endless possibilities for technological development. However, the benefits of progress are available to only a small number of countries. The standard of living has improved substantially for a small number of people, but the majority are suffering severe hardship.

Our economic progress is unsustainable because it is based on the destruction of the physical environment and on unequal wealth distribution. The economic gulf between the rich and the poor is already wide, and it continues to increase.

The world is becoming unstable socially and politically. The widening socio-economic gulf between countries, and the continued exploitation of those who are less fortunate, create a perfect breeding ground for terrorism and resentment against richer countries. The world is divided into three groups of countries: the excessively rich, who help themselves to almost all of the planet's resources; the poor, who take very little; and the excessively poor, who take practically nothing.

This division will grow deeper. The rich countries will grow richer. The poor countries will strive to improve their standard of living, but will remain far behind the rich group. The very poor countries have virtually no chance: their living conditions have not improved in the past and are not expected to improve in the future.

In the first half of this century, the number of people living in rich countries will change very little: from 1.18 billion in 2000 to 1.24 billion in 2050. The number of people living in developing countries will increase from 4.9 billion to 7.9 billion in the same period. The dramatically unequal global distribution of wealth will be even more unbalanced.

Income inequality can be tolerated only to a degree. We are already close to a breaking point, and any further increase in hate and despair would have devastating consequences for all of us. Considering the unbalanced distribution of global wealth, it is surprising that the world is still relatively stable.

In its present state, the Earth is not able to support all inhabitants at the level of prosperity of rich countries. Sooner or later, people living in poor and very poor countries will realise that they are fighting a losing battle. Sooner or later, poor and very poor countries will realise that they have no future. What will they do then?

If their people are not devastated by infectious diseases they will be dying of hunger and thirst. Will they calmly die alone, or will they decide to die together? Will they die where they are or will they make a desperate attempt to penetrate the rich enclaves and, in a desperate attempt to free themselves from deprivation and misery, destroy the Earth?

Chapter 8
Conflicts and Increasing Killing Power

'Mankind must put an end to war, or war will put an end to mankind.'

—John Fitzgerald Kennedy (1917–1963), 35th president of the United States

The world is a dangerous place, as many prominent people have reminded us over the years (see the quotes immediately below). However, we have created this danger through overexploitation of nature, neglect, greed; pollution, and socio-economic discrimination. We have added to the danger by inventing and building a range of devastating weapons that have not made us feel more secure. We live in fear of world war, which could cause enormous loss of life and widespread destruction of the material possessions that support us. We are in danger of terminating our existence on this planet or of reducing life to a deplorably primitive state of survival.

But the world is still a dangerous place and freedom always has enemies. It is to you that we still turn to weather the seas, patrol the skies, cross mountains and deserts to preserve our national security, protect our interests, and advance our ideals.[1]

Despite our relative dominance, the world remains a dangerous place. Many challenging conditions exist today and others will emerge over time. Most derive from the volatile mix of factors that have prevented global stability since the end of the Cold War. While these 'threats and challenges' are less significant individually than the global military problem posed by the former Soviet Union, collectively they present a formidable barrier to the emergence of a stable, secure, and prosperous international order. Moreover, the general decline in US and allied defence resources since the Cold War, combined with our more robust global engagement posture, make it increasingly difficult to deal effectively with these diverse global conditions.[2]

The world remains a dangerous place full of authoritarian regimes and criminal interests whose combined influence extends the envelope of human suffering by creating haves and have nots. They foster an environment for extremism and the drive to acquire asymmetric capabilities and weapons of mass destruction. They also fuel an irrepressible human demand for freedom and a greater sharing of the better life. The threats to peace and stability are numerous, complex, oftentimes linked, and sometimes aggravated by natural disaster. The spectrum of likely operations describes a need for land forces in joint, combined, and multinational formations for a variety of missions extending from humanitarian assistance and disaster relief to peacekeeping and peacemaking to major theatre wars, including conflicts involving the potential use of weapons of mass destruction. The army will be responsive and dominant at every point on that spectrum. We will provide to the nation an array of deployable, agile, versatile, lethal, survivable, and sustainable formations, which are affordable and capable of reversing the conditions of human suffering rapidly and resolving conflicts decisively. The army's deployment is the surest sign of America's commitment to accomplishing any mission that occurs on land.[3]

The brutal aggression launched last night against Kuwait illustrates my central thesis: notwithstanding the alteration in the Soviet threat, the world remains a dangerous place with serious threats to important US interests wholly unrelated to the earlier patterns of the US/Soviet relationship.[4]

We were asked to perform a myriad of missions, ranging from peacekeeping to peace enforcement to peacemaking, and in the deserts of South-West Asia, the country asked us to fight and win the last major war of the 20th century. The future looks like more of the same. So, while the threat of a major war has greatly diminished, the world remains a dangerous place, as regional instability, inflamed by ethnic hatred and religious fanaticism, gives rise to a new category of threats. If history has taught us anything, it is that somewhere, at some time, the United States will confront a regional, and eventually, a near-peer competitor, so we must prepare for that inevitability now. Our current force structure is strained and we need to retool to prepare for short-notice operations overseas, in areas with immature infrastructures incapable of accommodating the movement of our heavy forces, or in conditions not suitable for employment of our light forces.[5]

The mission of the army remains unchanged: to fight and win the nation's wars and to support the national security strategy and the national military strategy. While the mission remains unchanged, the world remains a dangerous place with a growing array of potential threats to our national interests.[6]

A close reading of what these people have said reveals that their statements of power and authority are in fact statements of defeat. They acknowledge that the world is a dangerous place, but do not suggest that we could change it to a safe and peaceful place. They show that the only way to be safe is to be strong, vigilant, and ready to fight. It is not the world of peace, but the world of war.

Causing fear, both to prevent and to win wars, is the purpose of our Defence Department. Weapons, organisation, thoughtful strategy and effective tactics will help us; but the courage of the American citizen makes the rest possible ... but I do keep meeting fine and intelligent people who believe modern weapons have abrogated the need for courage.[7]

These statements paraphrase what others have said before—"A strong defence is the surest way to peace"[8] and "To be prepared for war is one of the most effectual means of preserving peace".[9]

The fortress world.[10] Great inequality between nations is expressed not only in levels of prosperity, but also in military budgets. A typical distribution of military budgets is presented in Table 8-1.[11] Almost half of the global military budget is shared by two countries (the US and Japan), and about three-quarters by 12 countries. For almost all countries of the world, military budgets are 100 times lower than the military budget of the US.

In 1995, global military manpower was estimated at 23 million. The manpower of Europe was 6.3 million, the North Atlantic Treaty Organisation (NATO) countries, 4.7 million; and the US, 1.4 million.

The US is a big exporter of arms, many of which go to Third World countries. Between 1986 and 1996, the value of arms exports from the US varied between $22.2 billion and $33.6 billion per year, but the value of arms imports varied between $1.3 billion and $4.2 billion per year. In the same period, the value of arms exports from NATO countries was between $10.5 billion and $18.1 billion per year, and the value of imports was between $7.5 billion and $15.4 billion per year. The biggest Western exporter of arms outside the US is the UK, followed by Russia, France, Sweden, and Germany. The biggest Western (or allied) importers are Japan, Turkey, and Finland.

We can distinguish certain associations between nations that divide the world between those that are militarily strong and well prepared to protect their interests and those who are militarily weak (see Table 8-2). These associations are dictated by similarities in economic development and political orientation rather than by geography. However, it is also important to understand the geographical distribution of military strength.

Table 8-1 Typical global distribution of military budgets, 2000

Country	Billions (2004 US$)	(1)	(2)	Country	Billions (2004 US$)	(1)	(2)
USA	327.8	41.7	41.7	Greece	5.9	0.8	77.7
Japan	49.8	6.3	48.1	Syria	4.8	0.6	78.3
UK	38.8	4.9	53.0	Belgium	3.5	0.4	78.8
France	36.7	4.7	57.7	Poland	3.3	0.4	79.2
Germany	30.6	3.9	61.6	Norway	3.2	0.4	79.6
China (P.R.)	25.0	3.2	64.7	Denmark	2.6	0.3	79.9
Italy	24.4	3.1	67.8	Portugal	2.4	0.3	80.2
South Korea	13.9	1.8	69.6	North Korea	1.5	0.2	80.4
Iran	13.5	1.7	71.3	Libya	1.1	0.1	80.5
Turkey	10.9	1.4	72.7	Czech Republic	1.1	0.1	80.7
Russia	10.1	1.3	74.0	Hungary	0.9	0.1	80.8
Canada	9.0	1.1	75.2	Cuba	0.8	0.1	80.9
Spain	7.6	1.0	76.1	Sudan	0.4	0.05	80.9
Netherlands	6.5	0.8	76.9	Luxembourg	0.1	0.02	81.0

(1) – Percentage of global. (2) – Cumulative percentage of global.

Example: The military budget of the USA in 2000 was $327.8 billion, which corresponded to 41.7 per cent of the global military budget. The combined military budget of the USA and Japan was 48.1 per cent of the global military budget.

Source: Compiled and calculated using SIRI 2003b database. See also CDI 1999a. To facilitate a comparison of military expenditure in various years, I have converted all the values of military budgets discussed in this chapter to the constant currency of 2004 US dollars.

Table 8-2 Fortress world

Country/Group	(1)	(2)	(3)	Region	(1)	(2)	(3)
US	327.8	41.7	62.5	North America	336.8	42.9	42.9
Allies	258.3	32.9	74.6	Western Europe	174.9	22.3	65.1
China	25.0	3.2	77.8	East Asia	120.6	15.4	80.5
Russia	10.1	1.3	79.1	Middle East	73.1	9.3	89.8
Rogues	22.1	2.8	81.9	Rest of the world	80.1	10.2	100.0
Rest of the world	142.3	18.1	100.0	World	785.6	100.0	
World	785.6	100.0					

NATO	(1)	(2)	(4)	G12	(1)	(2)	(5)
USA	327.8	41.7	62.5	US	327.8	41.7	59.9
NATO Europe	176.6	22.5	33.7	G12 Europe	162.3	20.7	29.7
Turkey	10.9	1.4	2.1	Japan	49.8	6.3	9.1
Canada	9.0	1.1	1.7	Australia	7.2	0.9	1.3
Total NATO	**524.3**	**66.7**	**100.0**	**Total G12**	**547.1**	**69.6**	**100.0**

(1) – Military budget in 2000 in billions of 2004 US dollars

(2) – Percentage of the global military budget

(3) – Cumulative percentage of the global military budget

(4) – Percentage of the NATO military budget

(5) – Percentage of the G12 military budget

Allies – NATO Europe, Turkey, Australia, Japan, and South Korea

Rogues – Cuba, Libya, Iran, Sudan, Syria, and North Korea

North America – Canada and the USA

G12 – The group of thirteen industrialised countries: Australia, Belgium, Canada, France, Germany, Italy, Japan, Netherlands, Spain, Sweden, Switzerland, the UK, and the USA

Example: The US military budget in 2000 was 41.7 per cent of the global budget, 56.6 per cent of the NATO budget, and 59.9 per cent of the G12 military budget. The NATO military budget was 73.7 per cent of the global military budget.

Source: Compiled and calculated using SIPRI 2003b database

North America (the US and Canada) has the largest military budget. The next region is western Europe. Together, North America and western Europe have about 65 per cent of the global military budget. The heavily populated East Asian region has only 15 per cent.

The NATO share of the global military budget is 67 per cent. Most of it comes from the US, but NATO Europe also makes a substantial contribution. The G12 industrialised countries have 70 per cent. Here again, the US has the biggest share, and the European G12 members the second biggest.

The combined share of the US and its allies is 75 per cent. Thus most of the world's military strength is concentrated in a small group of politically and economically allied countries. China accounts for 3 per cent of the global military budget, and Russia 1 per cent. The alleged potential enemies of the US, the so-called rogue countries (Cuba, Libya, Iran, Sudan, Syria, and North Korea), total only 3 per cent.

With the rogue countries having a relatively low share of the military budget, the danger of a direct attack by them is small. However, prompted by desperation or foolishness, even the weak might dare to attack the mighty. Clandestine activities are also possible, and they include terrorism.

If we could eliminate or diminish the reasons for an attack, we could also reduce the fear and the sums of money we spend on military activities. If there is no solution to human suffering, deprivation, and poverty, to the glaringly unequal distribution of wealth, and to the death of small children from hunger and disease, then there is also no solution in guns and bombs.

> Every gun that is made, every warship launched, every rocket fired signifies, in the final sense, a theft from those who hunger and are not fed, those who are cold and are not clothed. This world in arms is not spending money alone. It is spending the sweat of its labourers, the genius of its scientists, the hopes of its children. The cost of one modern heavy bomber is this: a modern brick school in more than 30 cities. It is two electric power plants, each serving a town of 60,000 population. It is two fine, fully equipped hospitals. It is some 50 miles of concrete highway. We pay for a single fighter with a half million bushels of wheat. We pay for a single destroyer with new homes that could have housed more than 8000 people. This, I repeat, is the best way of life to be found on the road the world has been taking. This is not a way of life at all, in any true sense. Under the cloud of threatening war, it is humanity hanging from a cross of iron.[12]

The US military budget.[13] The US is an unchallenged military giant. Indeed, it is so strong that it can fight two major wars at the same time.[14] However, with this questionable distinction comes a heavy burden of the share in wars and conflicts (see Table 8-3). Between 1917 and 2001, the number of Americans killed in action was more than 600,000. More than 1 million were wounded in action, and more than 90,000 were missing in action. The financial burden of wars during that period is estimated at $7000 billion. The US has spent more than $1000 billion on wars during the hypothetical global peace—that is, after the two world wars. During this time, 100,000 Americans have been killed in action, 260,000 wounded in action, and more than 10,000 missing in action.

Table 8-3 The US burden of wars, 1917–2001

War	Years	KIA	WIA	MIA	Total	Cost
World War I	1917–1918	116 708	204 002	4 452	325 162	590 949
World War II	1941–1945	407 316	671 846	78 751	1 157 916	4 818 217
Korean War	1950–1953	36 570	103 284	8 177	148 031	410 133
Vietnam War	1961–1975	58 198	153 363	1 973	213 534	879 404
Lebanon	1982–1984	263	169		432	79
Grenada	1983–1985	19	119		138	95
Panama	1989–1990	23	324		347	203
Iraq	1991	383	467	2	852	9 041
Southwest Asia	1991–2001	26	NA	NA	NA	10 530
Somalia	1992–1995	43	175		218	2 553
Haiti	1992–1995	4	3		7	1 915
Rwanda	1992–1995	0	0	0	0	668
Kosovo War	1999	2	0		2	2 446
Former Yugoslavia	1992–2001	9	0		9	21 379
World wars total		**524 024**	**875 848**	**83 203**	**1 483 075**	**5 409 166**
After world wars total		**95 540**	**257 904**	**10 152**	**363 596**	**1 338 445**
Total (1917–2001)		**619 564**	**1 133 752**	**93 355**	**1 846 671**	**6 747 612**

KIA – Killed in action; WIA – Wounded in action; MIA – Missing in action; NA – Not available

Cost – Cost in millions of 2004 US dollars. I have converted the values listed by Berry et al 2003 to the constant currency of 2004 US dollars

Source: Based on Berry et al 2003

About a half of the US's discretionary budget is devoted to military needs (see Table 8-4). Education gets only 7 per cent, agriculture 3 per cent, space exploration 2 per cent, and environmental protection 2 per cent. The annual budget for the National Aeronautics and Space Administration is about the same as the cost of three aircraft carriers.[15]

A review of US discretionary budgets dating back to 1981 shows that the share of military expenditure was consistently about 50 per cent of the total discretionary budget. Projections extending to 2010 show that the share will steadily increase (see Table 8-5).

Table 8-4 Typical US discretionary budget, 2004
(billions of 2004 US dollars)

Agency	2004	(1)
Agriculture	19.5	2.5
Commerce	5.4	0.7
Defence	379.9	48.6
Education	53.1	6.8
Energy	23.4	3.0
Health and Human Services	66.2	8.5
Homeland Security	26.7	3.4
Housing and Urban Development	31.3	4.0
Interior	10.6	1.4
State and International Assistance Programs	26.8	3.4
Justice	17.7	2.3
Labour	13.7	1.8
Treasury	11.4	1.5
Veterans Affairs	28.1	3.6
Corps of Engineers	4.0	0.5
Environmental Protection Agency	7.6	1.0
Environmental Protection Operating Program	4.3	0.5
National Aeronautics and Space Administration	15.5	2.0
National Science Foundation	45.5	0.7
Small Business Administration	0.8	0.1
Other Agencies	19.2	2.5
Total discretionary budget	**782.2**	**100.0**
Total budget	2,229.4	

(1) – Percentage of discretionary budget

Discretionary budget – About one-third of the US budget that must be requested each year and has to be approved by Congress. The part of the budget that is spent automatically by the government is known as the mandatory budget.

Total budget – Includes the deficit of $307 billion in 2004.

Example: The discretionary budget for agriculture in 2004 is only $19.5 billion (2.5 per cent of the total discretionary budget), but for defence it is $379.9 billion (48.6 per cent of the total discretionary budget).

Source: Based on EOPUS 2003a. See also EOPUS 2003b for discretionary budgets between 1976 and 2008.

Note: Figures presented in EOPUS 2003a do not add up to the total presented in that document. I have therefore cross-checked the figures with EOPUS 2003b, where they tally correctly, and have corrected the relevant entries.

**Table 8-5 The burden of US military spending, 1981–2010
(Military expenditure as a percentage of discretionary budgets)**

Year	Per cent	Year	Per cent	Year	Per cent
1981	51.8	1991	58.5	2001	46.2
1982	59.7	1992	53.9	2002	46.9
1983	61.5	1993	50.2	2003	48.7
1984	61.1	1994	48.9	2004	48.6
1985	62.9	1995	50.4	2005	49.2
1986	64.4	1996	50.7	2006	49.8
1987	62.8	1997	49.7	2007	50.5
1988	62.7	1998	49.1	2008	51.1
1989	61.9	1999	47.2	2009	51.6
1990	59.2	2000	49.2	2010	52.1

Per cent – US discretionary military budget as a percentage of the discretionary budget in a given year

Years 1981–2002 – Calculated using actual values of discretionary budgets as listed in EOPUS (2003b)

Years 2003–2008 – Calculated using estimated values of discretionary budgets listed in EOPUS (2003b)

Years 2009 and 2010 – Extrapolated values

Source: Based on EOPUS 2003b

Records of US military budgets, extending back to 1941 (see Table 8-6), show that its lowest military spending was in 1941, but that it increased rapidly to $839.1 billion in 1945. From about 1952, the annual military budget varied between $270 billion and $440 billion. Average annual military spending between 1952 and 2003 was $346.1 billion, which corresponded to $948.2 million per day or $39.5 million per hour.

US military spending is increasing and is projected to reach $439.0 billion per year by 2008. The extrapolated value for 2010 is $457.7 billion ($1254 million per day or $52.3 million per hour). Its total military expenditure during the Cold War period, 1947–1989, was $13,857.4 billion. Average annual military spending during that period was $322.3 billion. Average annual military spending after the Cold War (1990–2010) is projected to be $371.6 billion. Even with the Cold War over, the world has not become a safer place.[16]

Meanwhile, the US national deficit in 2002 was $158 billion. It increased to $304 billion in 2003 and $307 billion in 2004. The cost of war in Iraq has been estimated at $8 billion per month and the cost of occupation at $1.5 billion to $5 billion per month.[17]

Table 8-6 US military budgets, 1941–2010 (billions of 2004 US dollars)

1941	68.3	1951	217.9	1961	321.1	1971	361.3	1981	303.7	1991	363.2	2001	326.2
1942	230.1	1952	412.7	1962	336.5	1972	331.0	1982	330.6	1992	383.4	2002	359.4
1943	562.7	1953	435.8	1963	329.7	1973	299.0	1983	357.2	1993	368.1	2003	384.1
1944	736.2	1954	399.8	1964	335.5	1974	286.8	1984	362.1	1994	349.1	2004	392.3
1945	839.1	1955	334.2	1965	309.8	1975	281.9	1985	387.0	1995	330.2	2005	405.8
1946	428.3	1956	310.7	1966	342.3	1976	271.2	1986	414.2	1996	311.3	2006	413.4
1947	116.4	1957	316.7	1967	406.3	1977	269.6	1987	422.2	1997	310.1	2007	419.7
1948	89.6	1958	312.9	1968	444.0	1978	270.1	1988	426.9	1998	304.4	2008	439.0
1949	127.8	1959	313.5	1969	422.8	1979	277.2	1989	433.0	1999	305.1	2009	448.4
1950	133.4	1960	319.7	1970	396.4	1980	287.3	1990	415.4	2000	317.1	2010	457.7

Total per decade

3,331.8	3,373.9	3,644.2	2,935.2	3,852.2	3,342.0	4,046.1

Cold War total (1947–1989): $13,857.4 billion

Years 1941–2002 – Actual budget values as listed in EOPUS (20003b)

Years 2003–2008 – Estimated values as listed in EOPUS (2003b)

Years 2009 and 2010 – Extrapolated values

Source: Based on EOPUS 2003b. I have converted the values listed in this source to the constant currency of 2004 US dollars.

Global trends in military spending.[18] Global military expenditure decreased from $1600 billion in 1985 to $756.2 billion in 1997. This dramatic decrease corresponded with the end of the Cold War, but the resultant reduction in military spending occurred mainly in Europe and the former Soviet Union. Europe reduced its military spending by 71 per cent, the former Soviet Union by 81 per cent, and the Warsaw Pact countries by 79 per cent. However, US military expenditure fell only 23 per cent in the same period.

Between 1993 and 2002 there was only a small increase in global military expenditure (see Table 8-7). The largest percentage increase during that period of time was in South Asia. However, we should notice that the combined military budget of this region is small when compared with that of the US.

Regional military expenditure decreased in some years during this period and increased again later. So, if we take the change between the minimum values and the values in 2002, we can see a substantial increase for all regions. The global increase was 14 per cent between the minimum in 1998 and the value in 2002.

Table 8-7 Regional trends in military expenditure, 1993–2002 (Billions of 2004 US dollars)

Region	1993	1994	1995	1996	1997	1998	1999	2000	2001	2002	(1)	Change (2)	(3)
Africa	**8.0**	**8.4**	**7.8**	**7.5**	**7.7**	**8.3**	**9.1**	**9.6**	**9.7**	**10.4**	**1.2**	**29.7**	**39.1**
North	2.7	3.2	2.9	3.0	3.3	3.4	3.6	3.9			0.5		
Sub-Saharan	5.4	5.2	4.9	4.5	4.5	4.8	5.5	5.7			0.7		
Americas	**418.3**	**396.6**	**377.0**	**356.4**	**356.4**	**348.8**	**349.9**	**361.8**	**367.3**	**399.8**	**46.1**	**-4.4**	**14.6**
North	396.6	373.8	352.0	332.5	330.3	323.8	324.9	336.8	340.1	373.8	42.9	-5.8	15.4
Central	3.0	3.7	3.3	3.4	3.5	3.5	3.7	3.7	3.8	3.6	0.5	17.9	17.9
South	19.1	18.9	21.7	19.9	22.7	21.8	21.3	21.2	23.4	22.9	2.7	19.9	21.3
Asia-Pacific	**130.4**	**131.5**	**133.6**	**139.1**	**139.1**	**138.0**	**140.2**	**145.6**	**152.1**	**159.7**	**18.1**	**22.5**	**22.5**
Central Asia	0.0	0.4	0.4	0.4	0.5		0.5	0.5			0.1		
East Asia	108.4	109.7	111.9	116.3	116.3	114.1	115.2	120.6	126.0	132.6	15.4	22.2	22.2
South Asia	13.0	13.0	13.7	13.9	14.6	14.7	15.9	16.5	17.6	18.8	2.1	44.2	44.2
Oceania	8.4	8.4	8.0	8.0	8.0	8.4	8.1	7.9	8.0	8.0	1.0	-3.9	1.4
Europe	**213.0**	**208.6**	**193.4**	**192.3**	**192.3**	**190.1**	**192.3**	**195.6**	**196.7**	**196.7**	**24.9**	**-7.7**	**3.4**
CEE	27.8	28.1	21.8	20.4	21.3	18.4	19.3	20.5	21.8	23.3	2.6	-16.4	26.6
Western	185.8	180.4	171.7	171.7	170.6	171.7	172.8	174.9	174.9	173.8	22.3	-6.4	1.9
Middle East	**58.1**	**58.8**	**55.3**	**56.2**	**61.4**	**66.0**	**65.2**	**73.1**	**80.2**		**9.3**		
World	**827.9**	**804.0**	**768.2**	**750.8**	**756.2**	**749.7**	**756.2**	**785.6**	**805.1**	**851.8**	**100.0**	**2.9**	**13.6**

North America – Canada and the USA.

CEE – Central and Eastern Europe.

(1) – Percentage of the global figure in 2000.

(2) – Percentage change between 1993 and 2002. Negative values indicate a decrease.

(3) – Percentage change between minimum value and the value in 2002. The minimum values are in italic.

Example: The military budget in Africa in 2000 was $9.6 billion, which represented only 1.2 per cent of the global military budget in that year. The annual military budget in Africa increased from $8.0 billion in 1993 to $10.4 billion in 2002, which represented a 29.7 increase. The minimum value of military expenditure in Africa was $7.5 billion in 1996. The increase between this minimum value and the value in 2002 is 39.1 per cent.

Source: Based on SIPRI 2003b. I have converted the values listed there to the constant currency of 2004 US dollars.

The world at war. The world has not been at peace since World War II. Every year there are continuing religious, ideological, political, ethnic, or territorial conflicts in various parts of the world. Examples of the continuing conflicts in 2001 are listed in Table 8-8. In addition, we have a series of suspended, dormant, and unresolved conflicts ready to explode.[19]

Table 8-8 Examples of continuing conflicts, 2001

Onset	Parties involved in conflict	Causes of conflict
1948	India v. Pakistan	Ethnic and religious
1949	China v. Tibet	Autonomy and religious
1961	Iran v. Kurds	Independence
1963	Indonesia v. Irian Jaya separatists	Ethnic and economic
1969	Philippines v. New People's Army	Ideological
1969	India v. Aceh separatists	Autonomy and religious
1975	Israel v. Hamas and Hazballah	Religious and territory
1975	Angola v. UNITA	Ethnic and economic
1977	Indonesia v. Christians and Muslims in Maluccan Islands	Religious and territory
1978	Colombia v. National Liberation Army	Drug trade and ideology
1978	Colombia v. FARC	Drug trade and ideology
1978	Sri Lanka v. Tamil Eelam	Religious
1981	Peru v. Sendero Luminoso	Drug trade and ideology
1982	India v. Assam insurgents	Independence
1983	Sudan v. Sudanese People's Liberation Army	Ethnic and religious
1984	Philippines v. Moro Islamic Liberation Front	Religious
1986	Uganda v. Lord's Army	Power
1988	Burundi: Tutsi v. Hutu	Ethnic
1989	India v. Jammu and Kashmir Liberation Front	Ethnic and religious
1990	Colombia v. Autodefensas Unidas de Colombia	Ideology
1990	Rwanda: Tutsi v. Hutu	Ethnic
1991	Algeria v. Armed Islamic Group	Religious
1994	Russia v. Chechnya	Independence
1996	China v. Uighur	Independence
1997	Tajikistan v. Islamic militants	Religious and drugs
1997	Uzbekistan v. Islamic Movement of Uzbekistan and drug gangs	Religious and drugs
1997	DRC v. Rwanda, Uganda and indigenous rebels	Ethnic
1998	Ethiopia v. Eritrea	Territory
1999	Philippines v. Abu Sayyaf	Religious
1999	Kyrgyzstan v. Islamic militants and drug gangs	Religious and drugs
2000	Guinea v. rebels in Sierra Leone and Liberia	Power

DRC – Democratic Republic of Congo.

FARC – Revolutionary Armed Forces of Colombia.

UNITA – National Union for Total Independence of Angola.

Source: Berry et al 2003

The Heidelberg Institute for International Conflict Research maintains a detailed database[20] of world conflicts back to 1945.[21] Conflicts are defined using 28 parameters. The types can be then grouped in a number of ways depending on the combination of selected parameters. For instance, conflicts can be labelled as violent or non-violent, and as wars or crises of different intensity (see Table 8-9).

Table 8-9 The relationship between different types of conflicts in the KOSIMO database

Non-violent conflict		Violent conflict		
Low intensity		Medium intensity	High intensity	
Latent	Active	Crisis	Severe crisis	War

Source: HIIK 2004

The number of conflicts is steadily increasing. The striking feature is that they increased rapidly in the five years after 1945. As soon as World War II was over, people quickly engaged in other forms of conflict. After this short period of rapid increase, the increase continued at a slower rate.

The number of conflicts fluctuates between consecutive years, so to describe the increase it is better to look at trends (see Table 8-10). The trend values for the total number of conflicts increased from 100 per year in 1945 to 200 per year in 2003.

The number of conflicts increased at 17 per decade, or at the rate of about six new conflicts every three years. The annual number of high-intensity conflicts, including wars, increased from 10 in 1945, to 24 in 1975, and to 35 in 2003.

An interesting feature of high-intensity conflicts is that they tend to take place within countries. The annual number of high-intensity intrastate conflicts increased from seven in 1945 to 33 in 2003. A large number of conflicts are about resources, and their frequency is increasing as people try to establish a place on the overcrowded planet.

Table 8-10 Trends in the global number of conflicts, 1945–2003

Type of conflict	Trend values		
	1945	1975	2003
High intensity	10	24	35
High intensity intrastate	7	20	33
High intensity interstate	3	4	2
Medium Intensity	7	24	43
Low intensity	83	102	122
Total	**100**	**150**	**200**

Trend values – The global number of conflicts per year. The number of conflicts fluctuates considerably between years. It is therefore more meaningful to give the trend values rather than the actual number of conflicts in a given year.

Example: The global number of high-intensity conflicts increased from 10 per year in 1945 to 35 per year.

Source: Based on HIIK 2004

With limited resources and a rapidly growing population in many regions of the world, we can hardly expect everybody to be comfortable and happy. However, with better cooperation and better care for the environment we could make living conditions more bearable for a larger number of people. All our energy and skills should be directed to improving life, improving the share of resources, finding a way of using wisely what we have, and working together for a better future. Conflicts drain our energy: they destroy human, financial, and physical resources; demolish what we have built and developed; and harm the environment.

The number of conflict-related activities used in a positive way (to resolve conflict) or a negative way (to incite them) is staggering. They include bilateral diplomacy (for example, visits, negotiations, notes of protests, and breaking treaties); multilateral diplomacy (conferences, arbitration, and boycotting conferences); information and propaganda (to find solutions or spread misinformation and increase conflict); economic instruments (trade agreements, embargoes, blackmailing, and closure of borders); military instruments (peacekeeping forces, mobilisation, manoeuvres, terrorist attacks, and war); secret agencies and services; informal subversive instruments (supporting selected factions), alliances (forming or leaving them); regional or universal integration or isolation; and internal instruments (negotiations, demonstrations, bribery, censorship, police actions, martial law, expulsion of citizens, resistance, and coups d'état).[22]

This is where our time, energy, financial resources, and attention are diverted. We are busy and we are working hard, but getting nowhere. The world is increasingly restless and dangerous.

The burden of armed conflicts.[23] Violent conflicts are now more intense than in the past and the main casualties are civilians. About 50 per cent of civilian casualties are children. Since 1990, close to 220,000 people have died in wars and conflicts between states, and 3,600,000 have died in internal wars and conflicts. The heaviest burden of death was in sub-Saharan Africa where more than 1.5 million people were killed in the last 10 years of the 20th century.

In many poor countries, such as Burma, Colombia, Liberia, Peru, Sierra Leone, Sudan, and Uganda, children are actively involved in armed conflict. Their age varies between seven and 16. It is estimated that children serve as soldiers in more than 50 countries. It is also estimated that 300,000 children are serving as soldiers in 30 countries. For instance, in Rwanda the number of child soldiers is estimated at

14,000–18,000, some being seven years old. In Colombia, the number is more than 15,000, some being eight. In Pakistan, Peru, Sierra Leone, and Colombia, thousands of children as young as eight serve as soldiers.[24]

About 90 per cent of all wars and armed conflicts since 1945 have taken place in developing countries.[25] A typical regional distribution of conflicts according to their intensity is presented in Table 8-11. Most of the high-intensity conflicts occur in poor countries. Of 35 such conflicts in 2003, 12 were in Africa and 11 in Asia. Of the 14 wars in that year, seven were in Africa.[26]

Table 8-11 Typical distribution of conflicts by region and intensity, 2003

Intensity	Europe	Africa	Americas	Asia	MENA	Total
High intensity	1	12	5	11	6	**35**
Wars	0	7	3	1	3	**14**
Severe crisis	1	5	2	10	3	**21**
Medium intensity	7	12	4	16	6	**45**
Low intensity	30	29	18	40	21	**138**
Total	**38**	**53**	**27**	**67**	**33**	**218**

MENA – Middle East and North Africa

Example: In 2003, there was only one high-intensity conflict in Europe, but 12 in Africa.

Source: HIIK 2004

Conflicts destroy the fragile economy of developing countries, often beyond the point of recovery. For instance, the total cost of the 16-year war in Mozambique was several times higher than the country's GDP before the war.

Wars and armed conflicts are also responsible for a large number of refugees and displaced people. Worldwide, the number of refugees rose from 2.7 million in 1972 to 11.7 million in 1999. The total number of displaced people living in foreign countries rose from 84 million in 1975 to 150 million at the end of the 20th century.[27] The total number of refugees, internally displaced, and returning refugees who were of concern to the United Nations High Commissioner for Refugees (UNHCR) in the year 2000 has been estimated at 21.5 million.[28]

Poor countries have to dig deep into their nearly empty coffers to support their military activities. They do this often by sacrificing such essential services as health and education.

In the late 1990s, the average expenditure in countries with high human development—such as Norway, Sweden, Canada, Belgium,

Australia, the US, and Japan—was 2 per cent of GDP for military purposes, 7 per cent for health, and 5–8 per cent for education. In countries with medium and low human development, the percentage of GDP devoted to military activities was in general either comparable to or higher than the percentage devoted to health or education. For instance, in Ethiopia, 9.4 per cent of GDP was set aside for miliary expenditure, but only 1.2 per cent for health and 4 per cent for education. In some countries the burden of military expenditure was exceptionally high: in Eritrea it was 22.9 per cent of GDP and in Angola 21.2 per cent.[29]

Deo confidimus.[30] We still do, but it seems that our priorities are shifting into other channels.[31] In eastern and central Europe, people put more trust in armed forces than in the church. In Latin America, people still put a high degree of trust in God as represented by the church, but right next to God and the church are the armed forces. In all these regions, the least-trusted institutions are political parties.

Can we reduce the number of conflicts?[32] It has been shown that there is a correlation between GDP and a country's level of political stability. Countries with a low GDP are more likely to be politically unstable than countries with a high GDP.

The probability of a change from a democratically elected government to an authoritarian system is high for countries with a very low GDP. The highest probability of change from a democratically elected government to an authoritarian system is for countries with a GDP up to $1000 per person per year, but the probability decreases gradually to zero for countries with a GDP between $7000 and $8000 per person per year. Examples of countries with a GDP lower than $1000 per person per year include, Ethiopia, Eritrea, the Democratic Republic of Congo, Mozambique, Rwanda, Zambia, and Niger.

The probability of survival for a democratically elected government in low-income countries is low. These countries will be restless, will fight, and will take out their anger against ethnic groups and against the government. The most likely way of controlling them will be through an authoritarian system, which will result in internal oppression and a potential threat to international stability.

On the other hand, countries with a high GDP of $5000 per person per year are more likely to change from an authoritarian to a democratic system of government and become stable. The threshold for the transition to a high level of stability seems to be a GDP of

$7000 to $8,000 per person per year, which corresponds to 25–30 per cent of the average GDP of rich countries.

Most countries with a GDP lower than $5000 per person are also characterised by a negative democracy score.[33] The democracy score varies between negative 2 and positive 2. The lower the score, the less democratic the country. For instance, China (with a GDP of $4000 per person), has a democracy score of negative 1.2, and the UK (with a GDP of $24,000 per person) has a score of 1.3. In general, the lower the GDP the lower the democracy score.

Countries with a reasonably high GDP are more likely to be democratic and stable. They do not have to be as rich as the most affluent countries, such the US or Australia. However, if their GDP is no worse than about four times less than the average for rich countries, they will be likely to accept their relatively low standard of living and remain politically stable.

Consequently, if we want to reduce the number of armed conflicts in the world and make the world a safer place we should help poor countries to improve their economy. People in general want to live in peace. If they are not pushed beyond their limit of endurance, if they are not forced into a desperate and hopeless situation, they will live in peace.

It may be against our accepted doctrines or customs, but the answer to local unrest and global insecurity is not survival of the fittest. It is not in our military power and fortifications. It is not in a selfish and ruthless accumulation of wealth, but in a better share of the planet's resources. We do not have to be equal, but we also do not have to be so unequal. Maybe by now being so closely linked and so dependent on one another we might realise that our self-centred way of living does more harm than good for all of us.

If all this sounds utopian, perhaps we are still too primitive. Perhaps we have allowed ourselves to be indoctrinated by the idea that a higher level of civilisation means only a much higher level of fighting skills and more advanced warfare technologies.

If only we could live in peace, if only we could cooperate, if only we could care more for those who suffer. If only we could care for our children's future. If only we could stop fighting and bickering. If only we could stop wasting money, time, and human resources on useless or damaging activities, we could turn this planet into a hospitable and pleasant place to live.

With proper planning and collaboration we could gradually improve living conditions for many people of the world. If we could gradually increase the number of countries with a GDP of no less than one-quarter of the average GDP of affluent countries, we could reduce gradually the number of conflicts. We could also reduce the danger of overt or covert aggressive activities. We could gradually reduce the sums of money we spend on military activities, and use it to strengthen the process of healing.

Increasing killing capacity.[34] Our capacity to kill and to destroy grew rapidly during the 20th century. We are now able to kill more people more easily than at any time in human history. We can also cause other forms of damage more efficiently and on a much larger scale than ever before.

The destructive capacity of our military forces can be measured using the so-called Weapons Lethality Index (WLI: see Table 8-12). For a sword, the WLI is 1, for a modern battle tank it is 160,000, and for a strategic nuclear missile with a 25-megaton warhead it is 10,500,000,000. If you can kill one person at a time with a sword you can kill more than 10 billion people with a nuclear warhead. If we could have all the people of the world assembled in one place, we could kill them all with just one bomb. From a sword to weapons of mass destruction in a relatively short time is indeed breathtaking progress towards self-annihilation.

Table 8-12 Weapons Lethality Index

Weapon	Lethality Index
Sword	1
Early musket	3
World War I/II rifle	38
Modern assault rifle	210
World War I tank	3 400
World War I fighter/bomber	11 500
World War II tank	110 000
World War II fighter/bomber	150 000
Modern battle tank	160 000
Conventional fighter/bomber	480 000
Short-range nuclear missile (20-kiloton warhead)	41 500 000
Nuclear-armed fighter/bomber (350-kiloton warhead)	310 000 000
Strategic nuclear missile (25-megaton warhead)	10 500 000 000

Example: An early musket is about three times more efficient killing instrument than a sword.

Source: Zuckerman 1996

Nuclear warheads. It is estimated that between 1945 and 2000 the five members of the original nuclear club constructed 128,060 nuclear warheads (see Table 8-13).[35] The race between the US and the Soviet Union was prompted by fear of aggression. Its aim was to maintain equilibrium through the threat of mutually assured destruction (MAD).

Table 8-13 Number of nuclear warheads constructed between 1945 and 2000

Country	Warheads constructed	Per cent of total
USA	70 000	54.7
FSU	55 000	42.9
France	1 260	1.0
UK	1 200	0.9
China	600	0.5
Total	**128 060**	

Source: Norris and Arkin 2000

The global nuclear arsenal increased from two in 1945 to 20,368 in 1960, and to a maximum of 69,478 warheads in 1986. From that year on, the arsenal decreased to reach 31,535 in 2000 (see Table 8-14). Russia was holding 20,000 nuclear warheads in that year and the US 10,500. The remaining three members (the UK, France, and China) accounted for 1035 warheads.[36] The number of warheads in India, Pakistan, and Israel is unknown, but is estimated to be 150–270.[37]

Table 8-14 Global nuclear stockpiles, 1945–2000
(Number of nuclear warheads)

Year	USA	Russia	UK	France	China	Total
1945	2	0	0	0	0	2
1950	298	5	0	0	0	303
1955	2,280	200	10	0	0	2,490
1960	18,638	1,700	30	0	0	20,368
1965	32,400	6,300	310	32	5	39,047
1970	26,600	12,700	280	36	75	39,691
1975	28,100	23,500	350	188	185	52,323
1980	24,300	36,300	350	250	280	61,480
1985	23,500	44,000	300	359	426	68,585
1990	21,000	38,000	300	504	432	60,236
1995	14,000	28,000	300	500	400	43,200
2000	10,500	20,000	185	450	400	31 535

Source: Norris and Arkin 2000

The latest estimate of the number of strategic warheads in the five countries of the original nuclear club is 14,968 (see Table 8-15). They are mainly in the US and Russia. In addition, the world has a number of tactical nuclear weapons and so-called hedge weapons, which can be used in case of emergency (see Table 8-16).[38] The total number of these weapons is just over 8000. The estimated total of nuclear warheads is at least 23,000. We do not need all of them to destroy ourselves.

Table 8-15 Current number of nuclear strategic systems and warheads

Country	Delivery system	Number of systems	Number of warheads
USA	ICBM	550	2 325
	SLBM	432	3 616
	Bombers	92	1 578
Total USA		1 074	7 519
Russia	ICBM	756	3 800
	SLBM	348	2 272
	Bombers	69	788
Total Russia		1 173	6 860
France	SLBM	64	384
UK	SLBM	48	185
China	ICBM	20	20
Total: ICBMs		1 326	6 145
Total: SLBMs		892	6 457
Total: bombers		161	2 366
Grand total		**2 379**	**14 968**

ICBM – Intercontinental ballistic missile (land based)
SLBM – Submarine-launched ballistic missile (sea based)

Source: Based on Berry et al 2003

Table 8-16 Estimated number of tactical and hedge nuclear weapons

Country	Number
USA	3 300
Russia	4 000
France	80
UK	0
China	390
Israel	200
India	60
Pakistan	30
North Korea	2
Total	**8 062**

Source: Berry et al 2003

The combined global explosive power of known nuclear warheads is estimated at 5000 megatons, which corresponds to 250,000 bombs of the type dropped on Hiroshima. We have enough weapons-grade plutonium in the world to build at least 85,000 more nuclear warheads.[39] The explosive power needed for mutually assured destruction was estimated at only 400 megatons.[40]

Reduction of the nuclear arsenal began gradually in the US from about 1965 and more rapidly in the Soviet Union from 1986. The end of the Cold War prompted much of the reduction that followed later. With this threat removed, there was no need for the MAD race. Many weapons in the old stockpiles became obsolete.

Keeping the nuclear machinery in a state of readiness is costly. In 1993, the US Department of Energy introduced the so-called stockpile stewardship program. The annual budget for this program was then $3.1 billion, but it increased to $4.5 billion in 2000. The budget was increased to $5.6 billion in 2002 and to $6.1 billion in 2003.[41] Meanwhile, an annual investment of $1.5 billion could combat the worldwide spread of TB, $1 billion could combat malaria, and $9.2 billion could combat HIV/AIDS.[42]

Nuclear weapons are too powerful to be used in combat. They are therefore useless, but are claimed to serve as a deterrent against world war. However, they can also serve to ignite world war. We now have clandestine operators for whom the accepted rules of engagement do not apply and who do not care about self-destruction—let alone about mutually assured destruction. This invisible enemy is undeterred by the threat of nuclear weapons and, in turn, would not hesitate to use them.

Conventional nuclear technology, including delivery systems, is complex and not practicable for small operators. However, simple nuclear devices can be constructed and used to devastating effect, or for blackmailing. With suicide bombers so readily available, the delivery of simple nuclear bombs is no longer an insurmountable problem. We can create the most sophisticated and expensive anti-ballistic missile systems ready to repel attacks from the sky, but such systems are useless against terrorism.

However, even sophisticated nuclear technology is no longer a secret shared by a few countries. A deranged dictator with access to the technology could unleash a nuclear holocaust.

Chemical weapons.[43] Weapons of mass destruction include chemical and biological weapons. Chemical weapons employ various synthetic agents produced specifically for military purposes or adapted from a wide range of commercial or industrial agents. They include: choking agents, which irritate the eyes and respiratory tract; blood agents, which rob the tissues of oxygen; nerve agents, which cause convulsions and death by respiratory paralysis; and vesicants, which burn the skin, eyes, and respiratory tract.

Chemical weapons can be manufactured easily in almost any country in the world using simple equipment. They can be delivered by various methods, ranging from ballistic or cruise missiles to artillery shells and landmines.

Chemical agents can also be delivered using much simpler methods, as demonstrated by Aum Shinrikyo cult members. They took sarin nerve agent onto Tokyo underground trains in plastic bags, which they pierced just before leaving the trains. The number of casualties was low—12 dead and 5498 injured —but the panic was great.

Biological weapons.[44] Biological agents can be used to cause infectious diseases or poisoning. Biological weapons are made up of micro-organisms (bacteria, histoplasma,[45] rickettsia,[46] viruses, and fungi) or toxins produced by plants or animals.

Examples of viral diseases caused by biological weapons include (see Appendix E): chikungunya fever, dengue fever, eastern equine encephalitis, tick-borne encephalitis, Venezuelan equine encephalitis, hepatitis A, hepatitis B, influenza, yellow fever, and smallpox. The incubation time for viral agents ranges between one and 150 days, the infection rate is high, and the mortality rate varies from very low (1 per cent) to high (60 per cent).

Examples of bacterial diseases caused by biological weapons include: anthrax, brucellosis, cholera, glanders, melioidosis, tularemia, typhoid fever, and dysentery. The incubation time for bacterial agents varies from one day to 20 days, the infection rate varies from low to high, but the mortality is usually high or extremely high.

The rickettsial group includes Q-fever, mooseri, prowazeki, psittacosis, and tsutsugamushi. The incubation time for rickettsial agents ranges from three to 15 days, the infection rate is high, and the mortality ranges from low to very high.

Examples of fungal diseases are coccidioidomycosis and histoplasmosis. Their symptoms become apparent in one to 18 days. Finally, examples in the toxin group are botulinum toxin, ricin, and mycotoxins. The symptoms appear in a few hours for some and a few days for others.

If we consider the production, maintenance, and delivery of weapons of mass destruction, the cost is 1000 times lower for biological weapons than for nuclear weapons, and 500 times lower than for chemical weapons. A few kilograms of a biological agent can cause as much damage as hundreds, or even thousands, of kilograms of a chemical agent.

Biological agents can be delivered in various ways: in capsules carried by ballistic missiles or sprayed from cruise missiles; by fighter bombers, guided aircraft, light aircraft, helicopters; and by many other methods. Their versatility is great. They can kill people directly by infection or poisoning, or indirectly by contaminating water, crops, and food. They can be used to destroy vegetation, terrestrial wildlife, fish, and livestock.

Biological agents are tasteless, odourless, and invisible when dispersed. Production is relatively easy and cheap, as is delivery. The action of such agents is usually delayed, so they can be used in clandestine operations. They are singularly dangerous and can cause large-scale devastation.

No country is strong enough or clever enough to defend itself against the variety of weapons we have invented, or against numerous other forms of aggression. Borders do not protect us, and we can no longer feel safe even if neighbouring countries are peaceful and cooperative. An attack can come suddenly from a distant place, or an attacker can come unnoticed from without or within. Even a single person can inflict serious damage.

Light weapons.[47] Light weapons and small arms are also described as weapons of mass destruction. Armed conflict and the production of light weapons go hand in hand: one supports the other.

It is estimated that the use of light weapons is responsible for 90 per cent of conflict casualties. About 90 per cent of casualties are civilians, and 80 per cent are women and children.[48] Each year, 550,000 people are killed by light weapons and small arms.[49] Millions of people are displaced because of armed conflicts in which light weapons have been used. The use of light weapons and small arms disrupts economic, social, and political development. It also supports and encourages criminal activities, which are becoming more frequent and more lethal.

Light weapons are readily available to anyone, in large or small quantities, through legal, covert, and illegal channels. It is estimated that since 1947, 60–70 million AK rifles have been produced by Russia, China, Bulgaria, Egypt, Iraq, Poland, Romania, and North Korea. About 10 million Uzi rifles have been produced by Israel, 8 million M-16s by the US, 7 million G-3s by Germany, and 5–7 million FAL assault rifles by Belgium.[50]

Light weapons constitute a profitable market and there are hundreds of manufacturing facilities around the world, in industrialised and developing countries. Millions of light weapons are sold each year by manufacturing companies; millions are also sold or donated by military establishments from surplus supplies.

For instance, between 1950 and 1975, the US donated over 2 million M-1 rifles to Turkey, South Korea, France, South Vietnam, Greece, Iran, Pakistan, Norway, Denmark, Israel, Venezuela, and Indonesia. When East and West Germany were reunited, the German government donated 304,000 Kalashnikov rifles with a good supply of ammunition, and 5000 RPG-7 grenade launchers, to Turkey.

Between 1980 and 1999, the US government sold 33,275 M16 rifles, 3120 40-mm grenade launchers, and 267,000 hand grenades to El Salvador; 38,000 M16 rifles, 60,000 hand grenades, 120,000 81-mm high-explosive mortar rounds, 2944 anti-personnel mines, and 4000 anti-tank mines to Lebanon; 4800 M-16 rifles to Somalia; 347,588 hand grenades, 185,000 anti-personnel mines, and 40,000 anti-tank mines to Thailand; and 1000 M-16 rifles to Zaire.

In addition, 158,000 M16A1 rifles were sold to Bosnia, Israel, and the Philippines; 124,815 M14 rifles to the Baltic countries and Taiwan; 26,780 pistols to the Philippines, Morocco, Chile, and Bahrain; 1740 machine guns to Morocco and Bosnia; and 10,570 grenade launchers to Bahrain, Egypt, Greece, Israel, and Morocco. Between 1989 and 1993, about 100 million pistols, revolvers, and rifles were sold through US-approved commercial channels to Argentina, Brazil, Colombia, Costa Rica, El Salvador, Guatemala, Mexico, and Peru.

The main European exporter of light weapons is the UK, accounting for $5200 million of exports in 1995, or 16.3 per cent of the global market. Other weapon-exporting countries are France ($2200 million of exports in 1995), Germany ($1200 million), Sweden ($310 million), the Netherlands ($230 million), Italy ($150 million), Belgium ($130 million), Spain ($80 million), and Denmark ($20 million). The total value of light arms' exports in 1995 was $32 billion.

At the end of the Cold War, huge quantities of surplus light arms were distributed all over the world. Many of them ended up on the black market. Light weapons are often left behind when the fighting is over, and are acquired by anyone with the money to pay for them. After the Vietnam war, the US left behind 1.8 million small arms, including 800,000 M16 rifles, 850,000 other rifles, and 90,000 45-calibre pistols. It also left 300,000 small arms in Cambodia.

Light weapons can also be stolen from military establishments. The illegal trade is increasing and it is often associated with the international drug trade.

Light weapons are attractive and convenient killing instruments because they are widely available, durable (even 20-year-old weapons can be still used), and inexpensive to buy and operate. They are portable and lethal.

Landmines.[51] Landmines have been described as weapons of mass destruction in slow motion. Anti-personnel mines can be divided into four groups according to the injuries they inflict:

(1) Relatively small mines, with a diameter of about 9 centimetres and containing 40 grams of explosive material. They amputate a foot or leg when someone steps on them.

(2) Slightly larger mines, with a diameter of about 11 centimetres and between 150 and 240 grams of explosive material. They cause more traumatic amputation of legs.

(3) Butterfly mines, which are attractive to children because they look like toys. Children can play with them for hours, but when the wings are moved in a certain way the mine explodes, amputates an arm, and may cause partial or complete blindness.

(4) Fragmentation mines, triggered by an attached wire. They explode at a height of 50 to 100 centimetres, and usually cause death.

Landmines can be distributed by ground personnel or helicopters. They are usually scattered in areas used by civilians, and the aim is to deprive them of access to water and food. People are injured or killed when they draw water, gather wood or food, cultivate crops, or tend livestock. Children are maimed or killed when they play. People who already suffer from water and food shortages are deprived of their basic means of survival.

Strada (2002) reports that of the 1950 landmine victims he has treated, about 93 per cent were civilians and 29 per cent of them were children. It is estimated that landmines kill or maim 15,000 to

20,000 people each year, mainly civilians. Carreño (2001) estimates the number of civilians killed or maimed each year by landmines at 26,000. She also claims that 120 million landmines are scattered around the world.

Landmines are cheap, treacherous, and deadly. They cost $3 to $15 each, and they can be made mainly of plastic to make them difficult to detect.

About 23 million landmines have been scattered in Egypt, 16 million in Iran, 15 million in Angola, 10 million in Afghanistan, 10 million in Cambodia, 10 million in China, 10 million in Iraq—and millions in Bosnia-Herzegovina, Croatia, Mozambique, Eritrea, Somalia, Sudan, Ukraine, Ethiopia, and Jordan. In addition, ICBL (2002) has estimated that there are 230 million anti-personnel landmines stockpiled in 94 countries.

According to ICBL, 143 countries have agreed to ban the production and distribution of landmines. The list of countries that are still producing anti-personnel landmines includes Cuba, the US, Russia, Egypt, Iran, Burma, China, India, North Korea, South Korea, Pakistan, Singapore, and Vietnam.

Urban populations as an easy target. As the urban population grows, it is becoming easier to kill or incapacitate a great number of people. If nuclear technology is out of reach for terrorists or fanatics, biological or chemical weapons can easily fill the bill.

An electromagnetic bomb (E-bomb)[52] is also a simple and cheap weapon of mass destruction that can be used against urban centres. The idea behind this device is to produce powerful bursts of electromagnetic radiation, which would paralyse computer systems and destroy stored information. When used on cities, E-bombs would not kill people or blow up buildings, but they would create chaos. E-bombs can be also used to paralyse sophisticated military technology, making it useless. They can be made for a few hundred dollars each.

Terrorism. Terrorism is a new factor that contributes much to making the world a dangerous place. People are afraid to travel overseas or even move around in their own countries. People living in large cities can no longer relax and go on with their daily activities; they live with increasing fear of an open or insidious attack, any time and anywhere.

One factor that makes terrorism so dangerous is that individuals are ready to commit suicide for their causes. People who live in fear of terrorism see it as a hideous and cowardly activity, but terrorists see it differently. They have been taught to hate. They must strongly believe that what they are doing is right.

We have declared war against terrorism, but this is a new kind of warfare. In this battle, the old methods do not make sense, as we are fighting an elusive enemy.[53] It is worse than fighting a shadow, because we cannot identify the source. But even if we think we can, the source moves to a new location, or we do not have enough evidence to justify taking action.

The new type of warfare requires a new strategy. We can try to identify the leaders, and remove or incapacitate them. However, we should also study the excuses terrorists use to justify their activities, then try to eliminate the excuses or at least lessen their effectiveness. Leaders can be replaced, so going to the root of the problem would be more effective. The use of force, and the use of boastful and arrogant language, only plays into the hands of terrorists and increases their support.

What is it that makes young people wire up explosives to their bodies and blow themselves up? What language, what excuses, and what arguments do terrorists use to justify the killing of innocent people, and how are they then able to increase their support? What language and what arguments do teachers and parents use to indoctrinate their children?

Terrorists claim to be freedom fighters—freedom from the interference, oppression, and exploitation by rich nations. If we use force against terrorists, many of the allegedly oppressed will be hurt or killed, and terrorists will prove to be right. We might be able to win a battle, but not the war, and there is a danger of alienating the possible supporters and increasing the support for terrorists.

If terrorists are accusing us of being the bullies of the world, it makes little sense to use arrogant and boastful language. It also makes little sense to use our superior force against a much weaker opponent, contrary to the view of most world leaders. Doing so just proves that we are the bullies.

We all share the same speck of dust in the universe, and we have to find a place for everyone on it. We cannot solve problems by fighting. There is no justifiable excuse for terrorism, but without delving into the roots of terrorism we cannot hope to find a solution. Now is the time to do so while developing countries are still cooperating with us. Terrorism might well be a wake-up call for us, and our future might depend on how we respond to it.

Supply sources for terrorists. We have created a wide range of resources that terrorists can use to support their activities. The richest and the most tempting of them are in the countries of the former Soviet Union and the United States.

In the arms race, prompted by the fear of aggression, we have created not only a wide range of conventional weapons, but also a huge arsenal of weapons of mass destruction – nuclear, chemical and biological. Many of them are unaccounted for, particularly in countries of the former Soviet Union. Many of them can be bought from embittered individuals or disgruntled groups—or they can be stolen. Weapons-grade plutonium and uranium is also available to anyone who can afford to pay. It is estimated that there are 130 terrorist groups in the world, so there is no shortage of customers.[54]

Sophisticated nuclear warheads may be out of reach for terrorists, but they can build a dirty bomb. Such a device would contain non-fissile radioactive material packed with conventional explosives. Its main purpose would be to contaminate large areas of land. Experts might argue that contamination from such explosions would be low, but the public would be hard to convince. A dirty bomb would create panic and economic disruption. If terrorists aim to destroy the economies of rich countries they have many ways of doing so.

It seems that the anthrax spores that created so much panic in the US were not produced in one of the rogue countries, but in the US, and probably in one of its military establishments.[55] They were identified as the so-called Ames strain isolated in the US Department of Agriculture in Ames, Iowa, in 1980 and used by the US to produce anthrax weapons.

In October 2001, a watchdog organisation called Project on Government Oversight (POGO) prepared a report in which it discussed the security risk of 10 nuclear weapons facilities in the US where 1000 tonnes of weapons-grade plutonium is stored. To show how vulnerable to terrorism these facilities are, members of POGO carted away about 100 kilograms of nuclear material from one installation without being noticed.[56] POGO has prepared a new report in which it discusses the vulnerability of nuclear reactors in the United States.[57]

Terrorism has no place in civilised society; however, it may well be telling us that our old style of living based largely on greed, selfishness, and aggression is working against us.

The issues we face are too complex to discuss fully here. We live in a dangerous and deeply divided world. The time has come for a radical change in the way we think and act. This applies to those who live in rich countries and in developing countries. The question is not whether we should change our style of living and our attitude to one another, but whether we should do it willingly or wait until we are forced. However, if we wait, will we be able to survive?

The future. The best future imaginable is one based on the Sustainability First principles.[58] The worst is the Fortress World[59] or what UNEP (2002) labels as the Security First development.

The Fortress World future is characterised by global apartheid with increasing global inequality, suffering, and conflicts. In the Fortress World future, the rich and strong will rule the world, with a total neglect of the poor. The world will belong to the 20/20 club: rich countries accounting for about 20 per cent of all nations, and the rich 20 per cent of society in poorer countries.

It will be a time of global decay, of rapidly increasing inequality and poverty for most people; a time when global law (whatever that will mean) and order will be enforced by military authorities. This will be a time of global authoritarian governance.

The Fortress World is already here. It is still mainly a structure, but we are only one small step from converting it to an active and functioning organisation. Even now, the functional characteristics of the Fortress World are present. Inequalities between countries are increasing. Rich nations show little or no interest in helping developing countries to develop their economies and improve their standard of living. Rich countries keep together and guard their interests.

The structure of the Fortress World can be seen in the global distribution of military budgets. Military strength is concentrated in a small number of rich countries. The most powerful nation is the US, but there are also alliances such NATO, the European Union, and the G-12 group of industrialised countries.

Because of their geographical positions, the US and Canada could easily form a fortress, as could NATO Europe. Japan and Australia are geographically isolated. Japan is the second-strongest military power in the world, but Australia is militarily weak.

The Fortress World is not the future we want. It would be a tragic and brutal future, with rich nations confined to their little islands of uneasy affluence, surrounded by an ocean of human misery.

If the continuing trends of divisions and unrest continue, the number of conflicts will increase from 100 per year in 1945 to 280 per year in 2050. The number of high-intensity conflicts, including wars, will increase from 10 to 55. The world will grow more restless and more insecure. Force will be needed to maintain a degree of peace. The threat of terrorism will increase.

We find ourselves in a vicious circle, and there seems to be no way out. History has taught that fighting makes no sense, yet to be defenceless is foolish. So we have to devote large sums of money to maintaining a readiness for war, to improving our means of defence or attack, to inventing new and more powerful weapons, and to perfecting methods of destruction.

Could we live without fighting? Fighting seems to be natural, but we are supposed to be the most intelligent species, guided by reason not instinct. Animals fight, but they do not have the capacity to destroy the Earth—only humans can do so. The MAD race demonstrated dramatically a particular human characteristic: we feel secure if we are strong enough to destroy the enemy. The contest of strength and the process of producing new and more efficient weapons will continue. Tribal, ethnic, and religious conflict will increase, and tensions between nations will grow. The likelihood of global military confrontation is high and rising.

We may think the world is a dangerous place now, but it will become even more dangerous. Wars of the future will involve much more sophisticated and efficient ways of killing and destroying.[60] For instance, hundreds or thousands of small robotic aeroplanes will be used to electronically illuminate battlefields, to scout enemy territory, or to deploy mines and sensors. Mines will be much more sophisticated and much more deadly; now they just sit in the ground until a victim happens along. Future mines will be active—that is, equipped with detection devices to scan wide areas of land or sky, and launch projectiles at tanks, helicopters, or other objects identified as hostile. Active mines have already been produced, so the future is right on our doorstep.

Future wars will also make use of so-called arsenal ships. The idea behind such ships is to equip them with huge quantities of cruise missiles and other deadly projectiles, but with only a small crew. These ships will be partly submerged, and will be able to do a lot of damage in a short time.

Future wars will also include space, which is no longer an empty place, but is full of communication and surveillance satellites. No strategist worth his salt would be so foolish as to overlook them. There is a lot of important and useful orbiting equipment to destroy or incapacitate, and plans are already being made to do so.

Anti-satellite weapons do not have to be sophisticated. Launching a cluster of junk in the path of a satellite could do the trick. This cheap and effective method could even be within reach of terrorist groups. More sophisticated ways of destroying a satellite would be by using high-power microwaves (HPM) or ultra-wide band (UWB) devices. Such electromagnetic devices would incapacitate a satellite without breaking it up.

We shall no doubt have many other new and exciting combat ideas. Many technological developments and scientific discoveries that we could use to improve our lives will be considered as war applications.

"I know not with what weapons World War III will be fought, but World War IV will be fought with sticks and stones."[61]

A better world. We are now facing hugely complex environmental problems which are difficult to solve. However, we are also facing complex social problems, and problems of mind and culture. Without solving them, our future will remain insecure. If by magic we all could be transported to a new pristine world, would we not soon turn it into the world we left behind?

We live on a rich and beautiful planet. It is a miracle of nature, a small oasis of life in the vast expanse of space—the only such planet in our solar system or the near galactic neighbourhood. This is a priceless and irreplaceable jewel we should guard and protect with the utmost care.

Chapter 9
In a Nutshell

'There is no more difficult art to acquire than the art of observation, and for some men it is quite as difficult to record an observation in brief and plain language.'

—Sir William Osler (1849–1919), Canadian-born physician and author, *Aphorisms from His Bedside Teachings and Writings*

-1-
The population explosion

For countless generations the global population was small and rising exceedingly slowly. However, from the early 1800s it exploded, reaching 6 billion at the end of the 20th century. The main thrust occurred in the last 70 years of that century, when 4 billion people were added to the global community.

The population explosion came suddenly. It took the world by surprise, and we are still not sure how to react or how it will end, except that it will leave a great deal of damage behind. We are not even sure of surviving it.

We are probably coming to the end of the population explosion, but even when it is over its effects will linger. The most optimistic estimate is that we will add 3 billion people to the existing global population, reaching a total of 9.4 billion about 2060. From that year on, there should be a slow decrease.

The population in a handful of richer countries is almost stable, and will remain about 1.2 billion. The increase will be in developing and poor countries. The ratio of poor to rich, now 4:1, will be 7:1 in a few decades.

The population explosion puts enormous stress on the environment, and for the first time in human history we are reaching and sometimes exceeding the ecological limits of our planet. We cannot control the population explosion, but we could reduce its impact by living within our global means and by caring for the environment.

-2-
Diminishing land resources

The Earth is a small planet and it can hold only a relatively small number of people. Only a small fraction of the earth's surface is land, and only 68 per cent of that land is useful. The total useful land area is only 8.8 billion hectares. Adding the small productive areas of water gives a total of 11.1 billion hectares of biologically productive surface area, or only 22 per cent of the total surface area of our planet. This small percentage has to support all our consumption and demands.

At the current average level of global consumption, our planet can support only about 4.8 billion people, which means that even now we have about 1.4 billion too many. The surplus is constantly increasing. Consumption per person is increasing, but our biologically productive surface area is not. Even if we assume that the average consumption per person will remain constant, by 2020 there will be a surplus of 3 billion people on the planet. By 2050, the surplus will be 4.5 billion.

Industrialised countries need an average of 6 hectares of biologically productive surface area per person [ha/p] to support their current levels of consumption. However, the global average is now only 1.8 ha/p. By 2020, the figure will be 1.5 ha/p, and by 2050 it will be 1.2 ha/p. By then, the biologically productive surface area per person will be about one-fifth of what is required to support the current consumption of industrialised countries. By then, also, consumption per person will have increased, so the shortage of biologically productive surface area will be even more severe.

For a satisfactory diet we need about 0.5 ha/p of arable land. Today we have only 0.24 ha/p, and by 2020 we shall have 0.20 ha/p. If we include the loss of arable land we shall find that the global average will be around 0.17 ha/p.

Global land resources per person are diminishing, mainly because the population is increasing. However, productive land is also being lost through urbanisation, water and wind erosion assisted by incompetent management, overgrazing, waterlogging, salinisation, acidification, overuse of agricultural chemicals, heavy-metal contamination, organic pollutants, oil spills, and other human activities. Increasing consumption is supported by ruthless and thoughtless overexploitation of natural resources, systematic deforestation, environmental damage, and large-scale loss of species.

By using intensified farming methods we still produce enough food for everybody, yet millions of people die of hunger each year because food distribution is inadequate. We have huge genetic resources and, in principle, with consolidated effort and planning, we could improve the production of food and make it available where it is needed. However, if we allow current trends to continue, the number of people dying of hunger is likely to increase.

-3-
Diminishing water resources

Only about 2.53 per cent of global water is in the form of fresh water, but even a much smaller percentage is available for human consumption. About 68.7 per cent of global fresh water is in glaciers and permanent snow cover, and 30.06 per cent is groundwater. Together they make up 98.76 per cent of global freshwater reserves. They are known as non-renewable freshwater resources.

Only about 0.3 per cent of global fresh water is in lakes, rivers, marshes, and wetlands. They are known as renewable resources because much of their water is renewed each year. However, we cannot empty lakes and rivers each year, so the water we can use has to come from precipitation over land areas. This water is known as runoff water.

Much runoff water is lost in floods, but by building dams we can recover some of it and boost the availability of water. In the final analysis, even though our planet has huge water resources, only a minute fraction of it is available. Fresh water is a precious commodity.

For countless generations we had more than enough fresh water to satisfy all our needs. Now we are running out of water. In the past 100 years, global water withdrawals have increased nearly seven-fold. By 2050, global withdrawals of fresh water will be more than 11 times greater than in 1900. The two regions most affected by water shortage are Asia and Africa.

The number of countries and people experiencing water shortage is increasing. At the end of the 20th century a large part of the global population (about 2 billion people) had very low or catastrophically low water availability.[1] About 1.1 billion people did not have access to clean water, and 2.6 billion people (about half the global population) did not have access to sanitation facilities.

Between 1950 and 2000, water availability declined by 40 per cent in industrialised countries and 70 per cent in developing countries. By 2030, the availability of renewable water in developing countries will decline further, to 80–90 per cent below the 1950 level. In industrialised countries, it will remain at nearly the same level as at the end of the 20th century.

In the best case, the number of people living with severe water stress will increase from the current 2.4 billion to 3.2 billion in 2030. However, in the most likely case of business as usual, the number of people living with severe water stress will rise to 4.5 billion.

In some countries, including India, China, and the US, fresh-water consumption is supplemented by excessive withdrawal of groundwater. In many countries, the rate of annual withdrawal of groundwater is higher than the rate of recharge. These countries live on an increasing groundwater deficit, the largest being in India, China, and the United States.

To boost the supply of fresh water, thousands of dams—small, large, and super large—have been constructed all over the world, and many more are added each year. They solve some problems, but create social, political, and ecological problems. They contribute to global warming, destroy productive land areas, disturb the habitat of species, contribute to deforestation, and displace people, which adds to the problem of ecological refugees. They also interfere with the sharing of water, thus aggravating international conflicts. Tensions over the sharing of fresh-water resources are increasing.

-4-
The destruction of the atmosphere

We are destroying the stratospheric ozone layer and increasing our exposure to lethal ultraviolet radiation. We are also polluting the lower regions of the atmosphere with greenhouse gases, and bringing about climate change.

In addition, we are polluting the atmosphere with a wide variety of harmful chemicals that have a direct, harmful effect on our health. As the urban population increases, mainly in developing countries, the people in large cities suffer most.

The process of global warming is faster than previously thought. In the 20th century, the mean global temperature increased by 0.6°C, and it is projected to increase by up to 6°C by the end of this century.

We are dumping about 6.6 billion tonnes of carbon into the atmosphere every year, and by 2020 the figure will be about 10 billion tonnes. We could reduce carbon emissions and achieve a level of 7 billion tonnes per year by 2030. However, with business as usual, emissions will rise to 15 billion tonnes in that year.

Carbon emissions from developing countries are increasing rapidly: by 2020 they will meet and surpass those of industrialised countries. However, carbon emissions per person will rise faster in industrialised countries.

In the past 150 years, we have dumped 280 billion tonnes of carbon into the atmosphere—the first 140 billion tonnes in 125 years and the next 140 billion tonnes in the following 25 years. The third 140 billion tonnes will be dumped in only another 15 years.

The concentration of carbon dioxide in the atmosphere is increasing rapidly. The concentration was constant in pre-industrial times, but it rose 32 per cent towards the end of the 20th century. The concentrations of other greenhouse gases are also increasing.

Independent analyses show that by 2050 carbon dioxide in the atmosphere will be 60–100 per cent greater than in pre-industrial times. By the end of this century, the concentration will be at least twice as high, and it could reach 3.5 times.

Independent analyses show that to stabilise the carbon dioxide concentration in the atmosphere the industrialised countries would have to reduce their emissions by an amount five to nine times the reduction required by the Kyoto Protocol. Current trends indicate that this is not going to take place. Instead, carbon emissions by developed and developing countries will continue to increase.

Climate change is expected to have harmful effects in many areas. It will affect our health, agriculture, water availability, biodiversity, and weather. Millions of people living in coastal areas and on islands will be affected by rising sea levels and coastal storms. The rapidly increasing economic cost of weather-related losses could mean global bankruptcy in this century.

-5-
The approaching energy crisis

Fossil fuels support nearly all our energy consumption. Our other energy sources, including nuclear and hydro, contribute only 15 per cent. In principle, we have a wide variety of alternative sources of energy: not only hydroelectricity and nuclear fission, but also solar, wind, marine, and geothermal. We also have a potentially inexhaustible source in nuclear fusion. However, the development of these sources is slow, and they are not expected to play a big role in the near future, at least not before 2020.

In the past 100 years, the global consumption of energy has increased nearly 14 times. The consumption of fossil fuels increased more than 23 times. The greatest rate of increase was after 1950. Between 2000 and 2020, global consumption of energy will increase by 60 per cent. Industrialised countries consume twice as much energy as developing countries, and consumption per person is more than eight times greater. Energy consumption in developing countries is increasing faster than in industrialised countries, and by 2050 it will be the same as for industrialised countries. However, consumption per person will continue to be higher in industrialised countries.

The contribution of fossil fuels to global energy consumption is projected to increase from the current 85 per cent to 87 per cent in 2020. Between 2000 and 2020, global consumption of fossil fuels is projected to increase by 60 per cent. This will continue to create serious environmental and health problems. The problem of global warming will remain unresolved.

Our main source of fossil fuel energy is crude oil. Currently, it contributes 47 per cent to the consumption of fossil-fuel energy. The remaining 53 per cent is shared nearly equally by natural gas and coal. The contribution of crude oil to total energy consumption is 40 per cent. Between 2000 and 2020, the consumption will rise by 55 per cent. The consumption of natural gas will increase by 87 per cent and of coal by 41 per cent.

Nearly half the estimated ultimately recoverable reserves of crude oil have been exhausted, which means we are coming to a peak of global crude oil production. There is a strong likelihood that global production will slow down, and perhaps start to decrease, before 2020. However, demand for this energy source will continue to increase. The widening gap between supply and demand will deepen the global energy crisis.

Crude oil reserves are shared mainly by a small group of countries. The big consumers of crude oil are small producers, and the small consumers are big producers. The dependence of energy-hungry OECD countries on OPEC is heavy and increasing. OPEC can easily control the global supply of crude oil. However, even with collaboration between OECD countries and OPEC the increasing shortage of crude oil is likely to boost the price and seriously affect many aspects of our life.

A chain is as strong as its weakest link. Civilisation's weakest link is transport, because it relies almost entirely on crude oil. Unless we take immediate steps to develop new technology, such as fuel cells or hybrid cars, the world could be brought to a standstill.

-6-
Social decline

The world is deeply divided between haves and have-nots. Only about 11 per cent of countries, containing 16 per cent of the global population, are rich. Their share of global income is 59 per cent. Poor countries, including China and India, account for 76 per cent of the global population, but only 29 per cent of global income. Between these extremes there are a few middle-income countries accounting for 8 per cent of the population and 12 per cent of global wealth. The gulf between poor and rich countries is wide and increasing.

The range in GDP is from $1300 per person per year in the least-developed countries and $27,000 per person per year in high-income OECD countries. The lowest GDP is $500 per person, and the highest is $54,000. The average GDP for developing countries is $4000 per person and for OECD countries $23,000.[2]

The gap in GDP between rich and poor countries is excessive. On average, the gap between developing and developed countries is about six times average GDP for developing countries, and is increasing by $400 per year.

The highest levels of income inequalities within countries are found in sub-Saharan Africa, Latin America-Caribbean, and the East Asia-Pacific region.

Absolute poverty is defined as an income of $1 per person per year, corrected for local purchasing power. About 1.2 billion people are in this category—roughly the same as the combined population of developed countries. The largest number of people in absolute poverty is in South Asia and sub-Saharan Africa.

A more accurate way of measuring poverty is to include factors such as illiteracy, access to clean water, lifespan, the percentage of underweight children, and long-term unemployment. In about one-third of countries, 20–60 per cent of the population lives in poverty. Even in high-income countries, poverty can be relatively high. The highest level among rich countries is 16 per cent, in the United States.

Overcrowding in polluted cities is increasing. Close to 1 billion people, mainly in developing countries, live in slums. Cities depend critically on the external supply of goods and services, and they are also easy targets for terrorists or weapons of mass destruction. Their other problems include crime, corruption, and susceptibility to infectious diseases.

Infectious conditions such as tuberculosis, malaria, AIDS, pneumonia, diarrhoeal diseases, and measles kill millions of people each year. The overuse of antibiotics has contributed to the development of drug-resistant strains of micro-organisms that have defeated even the strongest antibiotics, such as vancomycin and linezolid. We are racing against time to develop a defence against the superbugs.

For the first time in human history, social decline and social inequalities are not only local, but also global. The unequal distribution of wealth is approaching breaking point. Deprivation increases immigration pressures, breeds resentment against richer countries, and encourages terrorism. We live in constant danger of a global confrontation on a scale never experienced before.

-7-
Conflicts and increased killing power

Between 1945 and 2003, the number of high-intensity conflicts increased from 10 per year to 35 per year. The number of all conflicts increased from 100 per year to 200 per year. The largest number of violent conflicts after 1945 has been in the Middle East and sub-Saharan Africa.

In 2002, the world was spending about $850 billion per year on military activities,[3] often at the expense of such basic needs as health and education. From about 1998, the global military budget has been increasing by $25 billion per year.

The world is divided markedly according to military budgets. The US and its allies account for about three-quarters of the global total. The largest budget is in the US, accounting for 40 per cent of all military expenditure. The G-12 industrialised countries (of which the US is a member) have 70 per cent of the total.

The global arsenal of weapons is increasing. We have amassed more than enough nuclear warheads to destroy civilisation. Small arms and light weapons are also weapons of mass destruction: they have killed an estimated 30 million people since 1945. Landmines have been described as weapons of mass destruction in slow motion.

Terrorism is a new threat to global security, a factor that can easily unsettle the fragile political balance between nations. Conventional rules of combat do not apply to this invisible and elusive enemy. Terrorists use oppression and deprivation as an excuse for their activities, and we have created a perfect environment for their operations.

The world is a dangerous place. As the division between rich and the poor increases, the degree of tension rises. We have the capacity to destroy the Earth in one final display of power.

The worst possible future is the Fortress World, in which the rich will rule with complete disregard for the poor and military strength will enforce law and order. It will be a future of uneasy peace, social decay, and deep human misery for most of the population. Fortress World has already been constructed, and we are only one step away from converting it to a form of global governance.

What should we do? The critical global problems and issues are now so closely connected that it would be difficult, or even impossible, to untangle them and solve them one at a time. However, we can still identify certain areas that require focused attention:

1. We have to eliminate gross inequality between countries. No country of the world should have an income per person that is less than about 25 per cent of the figure for rich countries.
2. We must start moving away from fossil fuels. We should work harder to develop all alternative energy sources, but in particular we should put more effort into developing nuclear fusion technology.
3. Globalisation is a threat to the free market and to political, economic, and social stability. It is also a threat to sustainable development. We must solve this problem soon, before it becomes uncontrollable.
4. The patenting of life forms is getting out of hand and should be resolved urgently. It has a crippling effect on research and development. It also poses a serious threat to sustainable development.
5. We have to reduce consumption and pollution. The burden rests mainly with industrialised countries, in particular with a handful of countries whose consumption is excessive. Our planet is under great stress, which has to be reduced if we want to survive.
6. We have to invest generously in the conservation, restoration, and protection of the environment. This is the best and the most profitable investment we can ever make. It is a new type of investment that might not create personal profit or a quick return. It is an investment for a whole country, region, or even for the whole world. It is also an investment for future generations. The environmental changes now taking place are so enormous that they will affect every one of us.

A chance to choose. Time is running out, but we might still be able to choose and use danger as an opportunity. The future does not have to be based on business as usual; it could be based on sustainable development. However, this will require a radical change of direction and a radical change to our customary ways of living. It will require closer collaboration between nations, more equitable sharing of the planet's resources, a big cut in profit for individuals and corporations, better support of research and development, and eventually a substantial reduction of money, time, and effort devoted to military activities.

This change of direction does not apply just to industrialised countries; developing countries have to play their part. We must all learn to live together, and we must learn fast. However, developed countries need to make the greatest commitment—and extend a helping hand to the rest of the world.

Chapter 10

Landmarks of Progress

'Technological progress has merely provided us with more efficient
means for going backwards.'

—Aldous Huxley (1894–1963), British writer

Critical global events do not comply with our reckoning of time, which
we like to slice into such units as centuries, decades, years, and months.
We like to be precise about when certain events took place or when they
will occur. We cannot be so precise with critical global events.

An explosion is a sudden event that can be timed precisely but
how to time the population explosion? It began about 200 years ago
and is continuing, although it seems to be slowing. In the time frame
of the Earth's history, it has all happened so suddenly. However, in the
time frame of the average human life this event is too long and it is
difficult to locate it precisely in time.

Trends fluctuate, and it is futile to take one point and claim that it
marks an important event. We have to analyse trends over a sufficiently
long period; only then might we be able to identify distinct events.

All the dates presented here are as accurate as the data and the
mathematical analysis can make them. There is nothing profoundly
mystical about them. Not all of them mark global events, but they are
all associated with critical global trends and events. It is also by no
means a complete list.

Many events that shape our future are continuing, such as armed
conflicts, social degradation, urbanisation, air pollution, degradation
of land, and diminishing water resources. For some of them—for
instance, water—we can still distinguish certain outstanding features
against a background of change. For others, such differentiation is
problematic or impossible.

The following list is aimed at helping people to understand the
approximate time relation between critical events that are likely to
influence our future. Briefly, my aim is to assess which ecological
limits we have reached, or can expect to reach in the future.

1800: the onset of critical global trends[1]

The growth rate of the global population changed from slow to fast over a relatively short period in the early 1800s, a period that can be taken as the starting point for the population explosion. The population explosion is the primary driving force behind critical global trends and events, and thus its onset can be also taken as marking the beginning of what we now recognise as critical global trends.

We could try to be more precise in determining this starting point by defining it, for instance, as corresponding to 10 per cent of what is now the expected maximum. My low-level projections show a maximum of 9.4 billion people in 2064, which agrees well with the latest projections from other sources. On this basis, the starting point of critical global trends can be identified as the moment when the global population reached 940 million people—that is, about 1800.

The early 1800s mark not only the starting point of the population explosion, but also the beginning of the end of a specific chapter in human history. Human interactions have always been characterised by violence, selfishness, greed, and aggression. We have been destroying, plundering, killing, enslaving, exploiting, burning, ruining, slaughtering, stealing, cheating, and robbing for countless generations. More recently, we have perfected our destructive power to the extent that we can destroy not only ourselves, but also much of life on Earth. What we have been doing on a local scale we can now do on a global scale. We are already destroying the environment on a global scale, and we have the power to do even greater damage.

If we are going to survive this century and enter the next, our attitude to one another and to nature will have to change. There seems to be no future in our customary style of living and behaviour. The early 1800s therefore can be considered not only as marking the starting point of critical global trends, but also as the beginning of an end to our ruthless dealings with one another and with nature.

1857: the limit of lavish global consumption is reached[2]

The biologically productive surface area of the Earth is 11 billion hectares. Whereas consumption per person in the past was lower, the lavish consumption of countries such as the US, Australia, New Zealand, Finland, and Norway currently corresponds to a footprint of about 9 hectares per person [ha/p]. At these current levels, the Earth can support a maximum of 1.2 billion people, which is approximately the present population of industrialised countries only—or the global population in 1857. To put it another way, at this level of consumption we now have a surplus of 5 billion people, and we would have to discover four Earth-like planets to accommodate them all. By 2025 we would need six additional planets, and by 2050 we would need seven.

1920: the limit of moderately high consumption is reached[3]

We could define moderately high consumption by the industrialised countries' average consumption, which requires about 6 ha/p to support. At this level, the Earth can support about 1.9 billion people—which corresponds to the global population in 1920. For this level of consumption, we have a surplus of about 4.3 billion people, and would have to find at least two more planets to accommodate them.

1930: the global availability of water declines to 'high'[4]

The thresholds of water availability have been defined as very high, high, adequate, low, very low, and catastrophically low (see Appendix B). For countless generations, the global availability of water has been very high—that is, more than 20,000 cubic metres per person per year [$m^3/p/y$]. The current global average is 6850 $m^3/p/y$. Without knowing it, we have already taken two steps down from very high to high, and then to adequate water availability.

The first step down occurred when average global water availability decreased to below the level of 20,000 $m^3/p/y$. Using the average global annual volume of runoff water, we can calculate that this event must have taken place about the time when the global population reached 2 billion, or about 1930.

1958: US crude-oil discoveries peak

A plot of crude oil discoveries in the lower 48 states of the US shows an erratic time-dependent behaviour. In some years, the number of discoveries was high; in other years it was low. However, the data show a clear trend of an increase between around 1900 and 1950, followed by a clear gradual decrease after around 1950. An analysis of the data shows that crude oil discoveries in the lower 48 states reached a maximum in 1958.[5]

US crude oil discoveries can be described by a typical bell-shaped Gaussian distribution. At their peak, the average rate of discoveries was around 3 billion barrels per year [Gb/y]. If the trend of the fall continues, the discoveries around 2010 will be about the same as around 1900—that is, only around 0.3 Gb/y per year.

1960: the global limit of an adequate diet is reached[6]

We need 0.5 hectares of arable land per person to support the typical diet of people in industrialised countries. The global total of arable land is 1.5 billion hectares, offering an adequate diet for only about 3 billion people, which corresponds to the population in 1960. We can now feed only about half the world's population at the standard expected in industrialised countries.

1962: global crude oil discoveries peak

According to records of the US Geological Survey (USGS), global crude oil discoveries reached a maximum in 1962.[7] The most productive period was between 1955 and 1965, when the discovery rate was 40 billion barrels per year [Gb/y]. By 1990, the rate had fallen to 7 Gb/y.

The distribution of global crude oil discoveries can be fitted using Gaussian distribution, which shows that by 2010 the annual rate is likely to be one-twentieth of the maximum, or 2 billion barrels. The rate of discoveries will represent a move back in time by about 100 years, but the demand for crude oil will continue to move forward.

1968: global natural gas discoveries peak

The rate of global discoveries of natural gas was increasing slowly between 1915 and 1950, from 0.4 trillion cubic metres per year [Tm³/y] in 1915 to 1.5 Tm³/y in 1950.

The most intensive period of discoveries was between 1965 and 1972. A fitted curve gives maximum discoveries of 9 Tm³/y in 1968. By 1990, the rate had dropped to 2 Tm³/y.[8]

1970: undeveloped fisheries disappear[9]

World fisheries are in various stages of exploitation, described as undeveloped (when they are not yet used for a large-scale commercial fishing); developing (when catches are still increasing); fully developed (when the maximum sustainable yield has been reached); and declining (when the yield is decreasing).

In 1951, about 55 per cent of fisheries were still undeveloped; by 1970, there were no undeveloped fisheries left in the world. A rapid decline occurred in just two years, between 1968 and 1970, when the proportion of undeveloped fisheries plummeted from 30 per cent to zero.

1975: US crude oil production peaks

Maximum production does not coincide with maximum discoveries: the time lag depends on the abundance of discovered fields and on demand for the product.

Crude oil production in the US is divided into activity in the lower 48 states and in Alaska, and these regions reached a maximum at a different time. Production in the lower 48 states reached a maximum in 1970, and by 2000 it had dropped to about 50 per cent of that peak. Production in Alaska was slow between 1960 and 1976, but then increased rapidly to reach a maximum in 1988. Alaskan production helped to boost the overall level for the United States.[10]

The combined production shows two maxima: one of 4 Gb/y in 1970 and one of 3.8 Gb/y in 1985. Since 1985 there has been a steady decline in production in the United States. In 2001, total production reached 2.8 Gb/y.[11] A fitted curve gives a maximum production of 3.8 Gb/y in 1975.

The decline in the production of crude oil is accompanied by a slow, but steady decline of crude oil reserves in the United States: they decreased from 34 billion barrels in 1977 to 24 billion barrels in 2001.[12] The US therefore has no hope of satisfying domestic demand from its own oil fields, and has to look for supply elsewhere.

1976: global crude oil production per person peaks

Using BP's data[13] for global crude oil production between 1965 and 2001, we can calculate that production per person per year reached a maximum of 5.5 barrels about 1976. It fell to 4.4 barrels per person per year in 1983 and has remained nearly constant from that year on.

By keeping production per person constant, we are still coping with the global demand for crude oil. However, we are in danger of overusing our global crude oil reserves.

Oil shocks in the 1970s distorted the Gaussian distribution of global production and shifted the maximum to a later year. We still have reserves we have not used earlier, and thus we have a good chance of keeping production per person constant for a little longer.

1985: the limit of global ecological capacity is reached[14]

Global consumption of all resources is increasing steadily not only because global population is increasing, but also because the demand per person is increasing. This means that the global ecological footprint is growing. However, global ecological capacity remains roughly constant. Maybe it is even decreasing because of the continuing damage to the environment.

About 1960 average global consumption could be still supported by only about 50 per cent of global ecological capacity. However, about 1985, for the first time in human history, average global consumption matched global ecological capacity. Since that year we have been living beyond our means, supporting consumption through an increasing ecological deficit.

In 1985 the global population was about 4.9 billion. If we had kept population constant and if we had not increased consumption, there would have been just enough resources to hold the status quo. The poor would have stayed poor and the rich would not have grown richer. However, global consumption continued to increase, and by the end of the 20th century it was about 25 per cent higher than global capacity and therefore unsustainable.

1977: global fish production per person peaks

The fishing catch is determined by three factors: the abundance of fish, the effort invested in catching them, and the catchability factor, which depends on the number of fish in a given place.

Records of global fishing show that the catch per person per year from natural fisheries—that is, excluding aquaculture—is decreasing.[15] The global catch reached a peak of 16.4 kilograms per person per year [kg/p/y] in 1971 and remained nearly constant at 15.3 kg/p/y until 1989. However, from 1989, the catch of fish from natural fisheries declined to reach 9.3 kg/p/y in 1998. When fitted, the global catch shows a maximum of 15.7 kg/p/y in 1977.

Fish numbers are decreasing everywhere, and we are witnessing a developing global fish crisis. The diet and income of many people depend heavily on a sufficient supply of fish. These people are suffering increasing hardship.

1977: the global availability of water declines to 'adequate'

Following the same procedure as for the first step down in the average availability of water globally, we can calculate that in 1977 we took the second step down. In that year, we moved from high to adequate availability. At that point, only three steps separated us from catastrophically low water availability.

Globally, fresh-water availability is still adequate. However, water is not distributed evenly according to population, and billions of people live in countries where the availability is very low or catastrophically low.

1983: the global grain-harvested area peaks

The global area under cultivation for the harvesting of grains increased from 600 million hectares in 1950 to 730 million hectares in 1980, then went into a slow but steady decline. Towards the end of the 20th century, the global grain-harvested area reached 670 million hectares.[16] When fitted, the global grain-harvested area shows a maximum in 1983. The calculated maximum will have to be reassessed using future data.

1984: the global production of cereals peaks

According to the World Bank, cereals include wheat, rice, maize, barley, oats, rye, millet, sorghum, buckwheat, and mixed grains.[17] Grain and cereals mean the same for Dyson,[18] but the data presented by Brown[19] for the production of grain are different from the data presented by Dyson for the production of cereals.[20]

The data presented by Dyson show that global production of cereals is steadily increasing. It rose from about 1.2 million tonnes per hectare in 1950 to about 3 million tonnes per hectare towards the end of the 20th century.

Dyson also presents a plot of production per person. Here the story is different. Global production rose from about 270 kilograms per person per year [kg/p/y] in 1951 to about 380 kg/p/y in 1984. From that year, production fell to reach about 340 kg/p/y by the end of the century. A five-year moving average shows a clear maximum in 1984.

Dyson also discusses cereal production per person per year in selected regions. Production in sub-Saharan Africa reached a maximum of 170 kg/p/y in the early 1960s and fell to 150 kg/p/y by the end of the 20th century. In the Middle East, production steadily fell from 300 kg/p/y in 1950 to 250 kg/p/y towards the end of the century. Production in Europe and the former Soviet Union peaked at 620 kg/p/y in 1988 and dropped to 500 kg/p/y at the end of the century. Production in South, East, and South-East Asia is still increasing.

The production of cereals depends heavily on irrigation. Globally, 30 per cent of production is by irrigation. In South Asia (excluding India), 80 per cent of cereal is produced by irrigation, and in China 70 per cent.[21] The increasing shortage of water will cause a decline in the production of cereals.

If we use the data listed by Brown, global grain production increased from 250 kg/p/y in 1950 to 340 g/p/y in the 1980s.[22] It then decreased to 310 kg/p/y by the end of the 20th century. A fit to the data gives a maximum of 322 kg/p/y in 1987.

1985: the gap between crude oil supply and demand in the United States widens

The gap between the domestic supply of crude oil in the US and demand is increasing. In 1970 the gap was 1.1 billion barrels per year [Gb/y]. It widened to 1.9 Gb/y in 1977 and narrowed to 1.4 Gb/y in 1985. However, from that year on the gap has been steadily increasing, and by the end of 2000 it had reached 3.5 Gb/y.[23]

According to EIA projections, the gap will increase to 4–6 Gb/y in 2010 and 5–7 Gb/y in 2020.[24] These estimates seem to be optimistic, as they are based on almost constant production of crude oil in the US. As noted earlier, production has been falling steadily since 1985. However, even if we assume constant production, 1985 marks the beginning of an increasing dependence on imports.

Crude oil imports in the US increased from 1.8 billion barrels per year [Gb/y] in the early 1980s to 4.2 Gb/y in 2000—that is, from 53 per cent below domestic production to 50 per cent above. The EIA projections show that imports will be 5.5 Gb/y in 2010, 7 Gb/y in 2020, and 7.7 Gb/y in 2025.[25]

About 46 per cent of all imports in 2000 were from OPEC, whose contribution is projected to be 50 per cent by 2025.

1985: North American crude oil production levels off[26]

The production of crude oil in North America increased from 3.8 Gb/y in 1965 to 5 Gb/y in 1973. It decreased to 4.5 Gb/y in 1976, increased to 5.6 Gb/y in 1985, and decreased again to 5.1 Gb/y in 1990. Since then, production has been almost constant at 5.1–5.2 Gb/y. The 1985 figure was the highest recorded in North America and is not likely to be reached again.

Production in North America has three components: the dominant US component and the two smaller components of Canada and Mexico. In 2001 the contribution of the US was 55 per cent, Mexico 25 per cent, and Canada 20 per cent. Future production in North America will depend on the interplay of these components.

The relative contribution of the US is decreasing, and the relative contribution of Canada and Mexico is increasing. It is therefore likely that the production of crude oil in North America will not decrease in the near future.

1987: crude oil production in the former Soviet Union peaks [27]

The production of crude oil in the former Soviet Union peaked at 4.6 billion barrels per year [Gb/y] in 1987. It dropped to 2.6 Gb/y in 1996, but rose again to 3.2 Gb/y in 2001.

However, the consumption of crude oil in this region reached a well-defined plateau of 3.1 Gb/y between 1980 and 1990. From 1990, consumption fell rapidly to 1.4 Gb/y in 1996, and it continued to fall, but more slowly, to reach 1.2 Gb/y in 2001. There was also no dramatic change in proved reserves in this region—63 billion barrels in 1981, 57 billion barrels in 1991, and 65 billion barrels in 2001.

It is clear therefore that greater production after 1996 was not prompted by increased domestic consumption or a substantial increase in reserves, but by a need for cash. This seems to be a good example of forced production. Unless new and rich reserves are discovered, forced production can only lead to a more rapid decline.

1989: the world's fisheries collapse

The continuing depletion of natural fisheries is reflected not only in the declining catch per person, but also in the total catch. It has been pointed out that reports of the global catch give inflated data that do not show clearly the continuing depletion of fisheries.[28] The corrected data show that the catch peaked at 90 million tonnes per year in 1991 then remained close to or below this level. By 2000, the figure was 87 million tonnes per year.

It has been also pointed out that if we exclude the highly variable Peruvian anchoveta, the catch reached a maximum of 82 million tonnes in 1989 and decreased from that year on. By 2000, it had dropped to 75 million tonnes per year.

The global collapse of fisheries has been signalled by a series of depletions for individual fish species in various parts of the world. For example, the production peak for the Pacific herring was reached in 1964, and by 1992 it had dropped 71 per cent below that maximum. For the Atlantic herring the peak was in 1966; by 1992 it had dropped 63 per cent (see Table 10-1).[29] The once plentiful Atlantic cod has been so severely depleted that the Committee on the Status of Endangered Wildlife in Canada has recently listed it as an endangered species.[30]

Table 10-1 The collapse of world fisheries

Species	Maximum catch Year	[Mt/y]	Catch in 1992 [Mt/y]	Decline [Mt/y]	Per cent
Pacific herring	1964	0.7	0.2	0.5	71.4
Atlantic herring	1966	4.1	1.5	2.6	63.4
Atlantic cod	1968	3.9	1.2	2.7	69.2
South African pilchard	1968	1.7	0.1	1.6	94.1
Haddock	1969	1.0	0.2	0.8	80.0
Peruvian anchovy	1970	13.1	5.5	7.6	58.0
Polarcod	1972	0.4	0.0	0.3	94.3
Cape hake	1972	1.1	0.2	0.9	81.8
Silver hake	1973	0.4	0.1	0.4	88.4
Greater yellow croaker	1974	0.2	0.0	0.2	80.0
Atlantic redfish	1976	0.7	0.3	0.4	57.1
Cape horse mackerel	1977	0.7	0.4	0.3	42.9
Chub mackerel	1978	3.4	0.9	2.5	73.5
Blue whiting	1980	1.1	0.5	0.6	54.5
South American pilchard	1985	6.5	3.1	3.4	52.3
Alaska pollock	1986	6.8	0.5	6.3	92.6
North Pacific hake	1987	0.3	0.1	0.2	80.0
Japanese pilchard	1988	5.4	2.5	2.9	53.7
Total		**51.5**	**21.8**	**29.7**	**57.7**

Decline – The decline between the maximum catch in a given year and the catch in 1992

Mt/y – megatonne (million tonnes) per year

Example: The maximum production of Pacific herring was 0.7 million tonnes per year in 1964. By 1992, the production had fallen to 0.2 million tonnes per year—that is, by 0.5 million tonnes per year. The annual production of Pacific herring in 1992 was 71.4 per cent lower than in 1964.

Source: Allen 2000

In search of fish, richer countries are looking for help from poorer countries, which are allowing unsustainable fishing in their territories. Stocks worth billions of dollars are being sold for much-needed petty cash. For instance, Argentina sacrificed a long-term $5 billion profit for a quick cash payment of only $500 million. In Senegal, 60 per cent of export revenue comes from such devastating transactions.[31]

In 1951, about 95 per cent of fisheries were still available, and fish harvests were increasing everywhere. However, as noted earlier, by 1970 there were no undeveloped fisheries left anywhere. By that time also, 10 per cent of global fisheries were in decline. By 1991, only 45 per cent of fisheries were developing, 25 per cent were fully developed, and 30 per cent were in decline.[32] The decline in world fisheries is alarmingly rapid.

This overexploitation is yet another example of unsustainable development. We are drawing too much and too quickly from our dwindling resources. We may be solving some of our current problems, but we are creating greater problems for the future.

1989: the global reserves-to-production ratio for crude oil peaks

The global reserves to production ratio (R/P)[33] for crude oil reached a maximum in 1989. At the production level of that year we had enough oil in the world to last 45 years. The R/P ratio had fallen to 40 years by 2001.[34]

As mentioned in Chapter 6, crude oil reserves are still increasing. However, the R/P ratio is decreasing, which means that production of crude oil is increasing faster than reserves. For about 15 years we have been extracting more crude oil than we have been discovering.

1997: Australian crude oil production levels off[35]

Between 1965 and 1985 the production of crude oil in Australia increased steadily and relatively quickly to 0.24 billion barrels per year [Gb/y] at the end of that period. From then until the end of 1999, crude oil production remained at 0.22 Gb/y. In 2000 production was pushed to 0.3 Gb/y, but in 2001 it dropped to 0.27 Gb/y. A fitted curve shows a maximum reached in 1997.

Proved crude oil reserves in Australia were 1.7 billion barrels in 1981, 1.5 billion barrels in 1991, and 3.5 billion barrels in 2001. With the current proved reserves and current production, Australia has enough crude oil for 14 years. The rapid rise in production between 1965 and 1985, the prolonged plateau between 1985 and 1999, the low R/P ratio, and the fitted curve seem to indicate that Australia has passed or is close to the maximum of its production capabilities.

2002: global crude oil production levels off

The global production of crude oil increased from 11.6 Gb/y in 1965 to 24.1 Gb/y in 1979. This rapid increase was followed by a brief period of decline to 20.6 Gb/y in 1983. From then, production rose to reach 27.2 Gb/y by 2001.[36]

A fitted curve to all data points shows a maximum in 2002. The calculated maximum is 40 years after the maximum of global crude oil discoveries. Under normal circumstances, the expected time lag between the peak of global discoveries and the peak of global production should be about 35 years.[37] However, the 1970s' oil shocks distorted the global production trend, pushing it down and shifting the expected maximum forward.

2007: global urbanisation reaches halfway mark

As pointed out in Chapter 7, the world will be at the halfway mark of urbanisation by 2007. From then on, a growing majority of the population will live in cities and towns. The UNFPA projections are for 2005. By 2047, only a quarter of the population will live in rural areas.

2010: global crude oil production reaches its absolute maximum

The earlier estimate of a peak in crude oil production about 2002 was based on direct analysis of production data without considering estimated reserves. More accurate predictions can be made by including the estimated ultimately recoverable (EUR) reserves in the analysis. The estimates of these reserves vary over a wide range of values, but most of them are in a relatively narrow band between 1600 billion barrels [Gb] and 2600 Gb, with the most probable value at 2200 Gb.

In this analysis, I have used the standard Hubbert model[38] and a wide range of values for the estimated ultimately recoverable reserves. The results are summarised in Table 10-2, where they are also compared with the results of earlier analyses by MacKenzie (2000) and Bartlett (2000). For the most probable level of ultimately recoverable reserves of 2200 billion barrels, the peak of global crude oil production is predicted for 2010.

We can also calculate maximum production using another approach—when half the original reserves are used up. As mentioned in Chapter 6, by 1999 we had extracted 857 billion barrels [Gb] of oil from the reserves, and by the end of 2002 we had extracted another 80 Gb. Thus for global reserves of 2200 Gb we were only 163 Gb away from using up half by the end of 2002. If production continues at the rate of the past few years, the absolute maximum will be reached about 2008.

This date is in reasonably good agreement with the predicted maximum of 2010 based on modelled calculations, and is in close agreement with earlier estimates by other authors.[39] After the absolute maximum is reached, production will go into an irreversible decline.

Table 10-2 Projected maxima of global production of crude oil

EUR	Author		MacKenzie		Bartlett	
[Gb]	Year	[Gb/y]	Year	[Gb/y]	Year	[Gb/y]
1 800	2004	26.5	2007	27		
2 000	2008	28.5	2011	28	2004	26.5
2 200	2010	29.3	2013	30		
2 400	2013	29.9	2016	31		
2 600	2016	31.4	2019	32		
2 800	2018	31.9				
3 000	2020	33.3			2019	33.2
3 500	2028	35.8				
4 000	2033	38.0			2030	39.5

EUR – Estimated ultimately recoverable crude oil reserves

Author – My projections

MacKenzie The projections of MacKenzie (2000)

Bartlett – The projections of Bartlett (2000)

Gb – gigabarrel (billion barrels).

Year – Years of the projected maxima of global production of crude oil

Gb/y – Gigabarrels per year. The projected maximum of global production of crude oil for a given level of EUR reserves.

Example: If we assume that global ultimately recoverable reserves of crude oil are 2200 billion barrels, then the calculated maximum of global production is 29.3 billion barrels per year in 2010. The calculations of MacKenzie 2000 give a maximum production of 30 billion barrels per year in 2013.

We have been producing crude oil for about 100 years, so we might expect to be producing it for another 100 years after reaching the peak. However, the situation in the future will be different. Up till now, demand for crude oil has been increasing on the rising side of the production curve. In the future, demand will continue to increase, but it will be on the falling side of the production curve. After about 2010, therefore, the gap between demand and supply will be increasing.

Assuming reserves of 2200 Gb, and using the fitted curve with the peak in 2010, we can calculate that crude oil production will be reduced to 27.7 Gb/y by 2020. Thus, it will be only 2 per cent higher than average annual production for 2000–2002. However, global demand for crude oil will then be 59 per cent higher.[40] There will be at least twice as many motor vehicles in the world.

By 2030, expected global production will fall to 23.4 Gb/y, and by 2050 it will be 12 Gb/y—or 60 per cent below the peak in 2010. By 2075, production will be 90 per cent below the peak and by the end of this century 99 per cent below.

Global production of crude oil will probably take us to 2020, but life will become more difficult. With increasing demand, which will not be matched by increasing production, the price of crude oil and the cost of living are likely to increase as we approach 2020. After that, the supply of crude oil will decline rapidly.

It is hard to tell which production level we should use to mark the end of the crude oil era. Considering the increasing demand, it might be assumed that the end will come when production is 50 per cent below the peak, or not long afterward. For the figure of 2200 billion barrels of reserves, that will occur in 2047. For the less likely estimate of 3000 billion barrels, it will be in 2068.

The end of the crude oil era will occur in the second half of this century or even earlier. After we reach the production peak, the end might be dictated by the price of crude oil rather than by the remaining reserves.

2012: developing fisheries disappear

Between 1951 and 1970, the number of developing fisheries was nearly constant because declining fisheries were quickly replaced by undeveloped fisheries. In 1970, the proportion of developing fisheries was 55 per cent; it remained constant until 1982, when it started to decline.[41]

Projecting the 1951–64 trend shows that by 2012 there will be, most probably, no developing fisheries left in the world. We might still have some left until 2020, but this appears to be a less-likely projection.

2015: OPEC's dominance in crude oil production begins

OPEC's contribution to global crude oil production increased from 45 per cent in 1965 to 53 per cent in 1973, followed by a gradual decline to 29 per cent in 1985. The contribution rose to 41 per cent in 1993, and has remained at that level.

Considering the gross imbalance of crude oil reserves between OPEC and the rest of the world, and our proximity to the absolute maximum of global production, we can expect that OPEC's contribution will reach 50 per cent in 2008 and will continue to increase. Indeed, it has been projected that the irreversible 50 per cent "crossover" will occur in 2008.[42] However, close examination of the relevant trends shows that the 50 per cent crossover is not likely to occur earlier than in 2015.

2018: arable land reaches its limit[43]

We need a minimum of 0.2 hectares per person [ha/p] to support a simple, mainly vegetarian, diet. With the arable land available, we can feed 7.5 billion people on such a diet. This will be the global population in 2018.

However, if we consider the loss of arable land by degradation, we shall find that this limit might occur earlier, around 2012. Unless we increase the production and distribution of food, or agree to live on a very simple diet, the number of people dying of hunger is likely to increase.

2020: will superbugs rule?[44]

The overuse of antibiotics contributes to the development of drug-resistant micro-organisms, and the fight against them is gradually becoming less effective.

We have strains of bacteria that are resistant to vancomycin which, until 2000, was the most powerful antibiotic. The latest replacement, linezolid (Zyvox), was defeated only a year after its introduction, and we have evidence that another new drug, quinupristin/dalfopristin (Synercid), is likely to suffer the same fate soon.

As bacterial resistance increases, the range of useful antibiotics is decreasing. If Dr Heymann's warning is to be taken seriously, we have 15–20 years, from the beginning of the current century, to find a way of dealing with drug–resistant bacteria.[45] If we fail, the year 2020 might mark our defeat.

2030: global production of natural gas peaks[46]

Natural gas production is increasing. It increased from 1023 billion cubic metres per year in 1970 to 2464 billion cubic metres per year in 2001.[47] Calculations carried out by Walsh (2000) show that conventional production is likely to reach a maximum of 2803 billion cubic metres per year in 2030.

2034: the global availability of water declines to 'low'

The world is expected to take the third step down in the average availability of water in 2034. By then, it will have declined from adequate to low. Water availability will still be adequate for the minority living in richer countries where population growth is low. However, for the majority, water availability will be close to very low or catastrophically low.

2045: weather-related losses cause global bankruptcy[48]

If the current trends in weather-related economic losses and global income continue, the world is likely to experience bankruptcy in or before the year 2045. In that year, annual weather-related economic losses are projected to match annually generated global wealth. This projection will have to be reassessed using future data for global weather-related economic losses.

Epilogue

We have entered a unique century, in which questions about our survival will be answered and our future decided. This century will mark the conclusion of the first-ever global population explosion, with all its damaging and ominous consequences. For the first time in human history, we are approaching and crossing the ecological limits of our planet. Never before has the survival of the human race been so threatened. Never before has there been a convergence of so many critical global trends.

The onset of these critical global trends and events can be traced to about 200 years ago, but now they are in the final stages of their development. We probably have no more than 50 years until the conclusion of these trends. The time is short. With luck and determination we might still enjoy a soft landing, but we would have to work fast and hard for this outcome.

We know what to do to change the undesirable course of events. We know, for instance, about the danger of the build-up of greenhouse gases in the atmosphere—a danger so great that it alone can cause global bankruptcy. Shall we work together on solving this problem or shall we ignore it?

We have many ideas about cleaner energy sources. We know about the futility of wars. We know about excessive inequality in the standards of living between people and countries. We know about hunger and extreme levels of human deprivation. We know about diminishing water and land resources. We know that with better global cooperation, and better use of financial and intellectual resources, we could solve many of these problems. However, to be successful we would also have to change radically our perception of our role in the world.

Albert Einstein expressed this dilemma well:

A human being is a part of the whole, called by us 'Universe', a part limited in time and space. He experiences himself, his thoughts and feelings as something separated from the rest, a kind of optical delusion of his consciousness. This delusion is a kind of prison for us, restricting us to our personal desires and to affection for a few persons nearest to us. Our task must be to free ourselves from this prison by widening our circle of compassion to embrace all living creatures and the whole of nature in its beauty. Nobody is able to achieve this completely, but the striving for such achievement is in itself a part of the liberation and a foundation for inner security.[49]

Shall we follow a constructive course, or continue in our drive to satisfy short-sighted and short-term aims and ambitions? Shall we continue with business as usual, or work together towards a sustainable future?

Appendix A
Selected tables and boxes

Key: Table A2-1 means Table 1 in Appendix A for Chapter 2.

Table A2-1 Projections of global population by region

Region	2002 Millions	2002 Per cent	2025 Millions	2025 Per cent	2050 Millions	2050 Per cent
Asia	3 766	60.6	4 741	60.3	5 297	58.2
China	1 281	20.6	1 455	18.5	1 394	15.3
Excluding China	2 485	40.0	3 287	41.8	3 904	42.9
Western	197	3.2	298	3.8	404	4.4
South Central	1 521	24.5	2 047	26.0	2 474	27.2
Southeast	536	8.6	706	9.0	811	8.9
Eastern	1 512	24.3	1 690	21.5	1 608	17.7
East. excl. China	231	3.7	235	3.0	214	2.4
Africa	840	13.5	1 281	16.3	1 845	20.3
Sub-Saharan	693	11.2	1 081	13.8	1 606	17.6
Northern	180	2.9	249	3.2	302	3.3
Western	247	4.0	403	5.1	605	6.6
Eastern	260	4.2	396	5.0	572	6.3
Middle	102	1.6	191	2.4	324	3.6
Southern	50	0.8	42	0.5	41	0.5
Europe	728	11.7	718	9.1	651	7.2
Northern	96	1.5	103	1.3	103	1.1
Western	184	3.0	187	2.4	178	2.0
Eastern	301	4.8	279	3.6	231	2.5
Southern	147	2.4	149	1.9	139	1.5
L. America-Carib.	531	5.1	697	4.9	815	4.9
Central	140	2.3	188	2.4	225	2.5
Caribbean	37	0.6	45	0.6	50	0.5
Southern	354	5.7	463	5.9	540	5.9
North America	319	5.1	382	4.9	450	4.9
Oceania	32	0.5	40	0.5	46	0.5
Australia	20	0.3	23	0.3	25	0.3
World	6 215	100.0	7 859	100.0	9 104	100.0

Per cent – Percentage of global population in a given year
North America – The United States and Canada

Source: Based on the data of PRB 2002

Table A3-1 Examples of ecological footprints and ecological capacity (hectares per person)

Country	Ecological Footprint	Ecological Capacity	Ecological Deficit/ Surplus
Australia	7.6	14.6	7.0
Austria	4.7	2.8	-2.0
Bangladesh	0.5	0.3	-0.2
Belgium & Luxemburg	6.7	1.1	-5.6
Brazil	2.4	6.0	3.6
Canada	8.8	14.2	5.4
China	1.5	1.0	-0.5
Colombia	1.3	2.5	1.2
Congo	0.9	9.0	8.1
Donmark	6.6	3.2	-3.3
Egypt	1.5	0.8	-0.7
Ethiopia	0.8	0.5	-0.3
Finland	8.4	8.6	0.2
Gabon	2.1	28.7	26.6
Germany	4.7	1.7	-3.0
India	0.8	0.7	-0.1
Indonesia	1.1	1.8	0.7
Italy	3.8	1.2	-2.7
Japan	4.8	0.7	-4.1
Kuwait	7.7	0.4	-7.4
Netherlands	4.8	0.8	-4.0
New Zealand	8.7	23.0	14.3
Norway	7.9	5.9	-2.0
Papua New Guinea	1.4	14.0	12.6
Peru	1.2	5.3	4.2
Switzerland	4.1	1.8	-2.3
United Arab Emirates	10.1	1.3	-8.9
United Kingdom	5.3	1.6	-3.7
United States of America	9.7	5.3	-4.4
World	**2.3**	**1.9**	**-0.4**

Example: The average consumption level in Australia corresponds to an ecological footprint of 7.6 hectares per person while the ecological capacity is 14.6 hectares per person. Australia therefore has an ecological surplus of 7 hectares per person, which can be used to increase consumption or population, or both. Austria has a deficit of 2 hectares per person.

Source: Wackernagel et al 2002

Box A3-1
The human threat to the 25 biodiversity hotspots
(summary for 1995)

A small number of only 25 biodiversity hotspots have been identified as giving a much-needed shelter for a great number of species. The combined area of these shelters is now small and is decreasing. Human intrusion into these areas is increasing and the species living there are threatened with extinction.

• Human population density in the 25 biodiversity hotspots is usually higher than the average global density.

• The annual population growth rates in the 25 biodiversity hotspots are usually higher than the average global growth rate.

• The areas covered by natural vegetation in the 25 biodiversity hotspots are decreasing.

Overview of the 25 biodiversity hotspots

	Original	Current
• Combined area of the 25 hotspots[a]	17.5	2.1
• Combined human population[b]		1.1
	Hotspots	**Global**
• Average population density[c]	73	46
• Average population growth[d]	1.8	1.3

Per cent – The per cent of the remaining original area
[a] – Million square kilometres; [b] – Billions; [c] – Number of people per square kilometre; [d] – Per cent per year

List of the 25 hotspots of biological diversity

	(1)	(2)	(3)	(4)
Tropical Andes	40	2.8	21 567	6.3
Mesoamerica	56	2.2	6 159	12.0
Caribbean	136	1.2	7 779	15.6
Atlantic Forest Region	79	1.7	6 546	2.7
Chóco-Darlén-Western-Ecuador	44	3.2	2 668	6.3
Brazilian Cerrado	7	2.4	4 517	1.2
Central Chile	29	1.4	1 666	3.1
California Floristic Province	108	1.2	2 196	9.7
Madagascar and Indian Ocean Islands	26	2.7	10 475	1.9
Eastern Arc Mts & Coastal Forests	50	2.2	1 521	16.9
Guinean Forests of West Africa	104	2.7	47 475	1.6
Cape Floristic Province	42	2.0	5 735	19.0

Box A3-1 (cont.)

Succulent Karoo	3	1.9	1 985	2.1
Mediterranean Basin	111	1.3	13 235	1.8
Caucasus	76	-0.3	16 059	2.8
Sundaland	121	2.1	15 701	5.6
Wallacea	54	1.9	2 029	5.9
Philippines	198	2.1	6 350	1.3
Indo-Burma	98	1.5	7 528	7.9
Mountains of South-Central China	25	1.5	3 678	2.1
Western Ghats and Sri Lanka	341	1.4	2 535	10.4
Southwest Australia	13	1.7	4 431	10.8
New Caledonia	8	2.1	2 635	2.8
New Zealand	11	1.0	2 001	19.2
Polynesia/Micronesia	58	1.3	3 557	10.7

(1) – Human population density (the number of people per square kilometre)

(2) – Population growth rate (per cent per year)

(3) – Total number of endemic species (plants, birds, mammals, reptiles, and amphibians)

(4) – Per cent of original areas remaining in natural vegetation

Source: Cincotta and Engelman 2000

Table A3-2 Global distribution of forests (natural and plantations) by region

Region	Land area	Forest Area		
	[Mha]	[Mha]	(1)	(2)
Africa	2 978	650	22	17
Asia	3 085	548	18	14
Europe	2 260	1 039	46	27
North and Central America	2 137	549	26	14
Oceania	849	198	23	5
South America	1 755	886	51	23
World	**13 064**	**3 869**	**30**	**100**

(1) – Percentage of the land in a given geographical region

(2) – Percentage of global forest area

Mha – megahectare (million hectares)

Example: The land area in Africa is 2978 million hectares, of which 650 million hectares—that is, 22 per cent, is covered by forests. Africa's share in global cover of forests is 17 per cent.

Source: FAO 2000b. In this report, Europe includes Belarus, Republic of Moldova, Russian Federation, and Ukraine.

Table A3-3 How rich are the forests of the world?

Region	Forest area [Mha]	Volume			Biomass		
		Density [m³/ha]	Total [Gm³]	Per cent	Density [t/ha]	Total [Gt]	Per cent
Africa	650	72	46	12	109	71	17
Asia	548	63	35	9	82	45	11
Europe	1 039	112	116	30	59	61	14
North and Central America	546	123	67	17	95	52	12
Oceania	198	55	11	3	64	13	3
South America	886	125	111	29	203	180	43
World	**3 869**	**100**	**386**	**100**	**109**	**422**	**100**

Per cent – Percentage of global volume or mass, respectively
Mha – megahectare (million hectares)
m³/ha – cubic metre per hectare
Gm³ – giga cubic metre (billion cubic metres)
t/ha – tonne per hectare
Gt – gigatonne (billion tonnes)

Source: FAO 2000b. In this report, Europe includes Belarus, Republic of Moldova, Russian Federation, and Ukraine.

Table A3-4 Regional use of fertilisers, 1997

Region	Amount	Per cent
Asia	71.73	52
Europe	24.71	18
North America	23.33	17
LAC	10.98	8
Africa	4.12	3
Oceania	2.75	2
World	**137.25**	**100**

Amount – The amount of fertilisers used in 1997 (in million tonnes)
Per cent – Percentage of global use of fertilisers
LAC – Latin America-Caribbean

Source: Harrison and Pearce 2001

Table A3-5 Regional distribution of the intensity of the use of commercial fertilisers

Region	Intensity
East Asia	265
Europe	156
North America	98
South Asia	87
Latin America	61
WANA	58
Oceania	46
FSU	18
Sub-Saharan Africa	12

Intensity – The average number of kilograms of commercial fertilisers per hectare of cropland
WANA – West Asia and North America; FSU – Former Soviet Unio

Source: UNDP et al 2000b

Table A3-6 Changes in the use of fertilisers between 1968 and 1998

Country	1968 [kg/ha]	1998 [kg/ha]	Change [kg/ha]	Yield gain (per cent)
Switzerland	346	749	403	76
Ireland	258	520	262	58
Netherlands	622	494	-128	78
S. Korea	206	458	252	91
Egypt	115	337	222	87
UK	243	330	87	93
Japan	365	290	-75	11
Israel	113	277	164	23
China	26	259	233	153
Denmark	204	170	-34	51
Italy	76	158	82	101
USA	77	110	33	77
India	11	99	88	114
Kenya	8	28	20	21
PNG	2	22	20	108
World	**43**	**91**	**48**	**77**

Countries are arranged in the decreasing order of the intensity of the use of fertilisers in 1998.
Change – The change in the use of fertilisers between 1968 and 1998. A decrease is indicated by negative numbers.
Yield gain – Percentage increase in the yield of crops between 1968 and 1998
kg/ha – kilogram per hectare
Example: Between 1968 and 1998, the use of fertilisers in the Netherlands decreased by 128 kilograms per hectare, but the yield of crops increased by 78 per cent.

Source: Harrison and Pearce 2001

Box A4-1
Salinity scale

	Salinity [mg/l]
Fresh	< 500
Marginal	500 – 1,500
Brackish	1,500 – 5,000
Saline	5,000 – 50,000
Hypersaline	> 50,000

The salinity of seawater is about 35,000 mg/l.

mg/l – milligram per litre.

< – Less than.

> – More than.

Example: The salinity of fresh water is less than 500 milligrams of salt per one litre of water.

Source: WRC 2000

Box A4-2
Water-related diseases

- **Diarrhoeal diseases** cause suffering to about 1–1.5 billion people per year worldwide, and kill about 2–3 million each year, of which about 70 per cent are children. They affect mainly people living in low- and middle-income countries. Of the 2,219,000 of deaths reported by WHO in 1999, 2,212,000 were in the low- and middle-income countries.

- **Infection** with intestinal helminths (worms such as parasitic roundworms and flatworms). These infections bring suffering to about 1.5 billion people and kill about 100,000 people each year.

- **Malaria** infects 300–400 million people per year and kills about 1.5 million.

- **Schistosomiasis** is a parasitic worm disease caused by *schistosomes*. It causes chronic urinary tract problems, gastrointestinal problems, diseases of liver, and bladder cancer. It infects about 200 million people each year and kills about 20,000.

- **Trachoma** is an infection of mucous membrane of the eyelids by *Chlamydia trachomatis* bacterium. It causes suffering to about 150 million people per year. Each year, about 6 million people are made blind or virtually disabled by trachoma mainly in Africa and Asia.

- **Bancroftian filariasis** is an infection with a tread-like worm, *Wuchereria bancrofti*, which attacks the lymphatic system causing lymphangitis, and extensive obstruction, which may lead to elephantiasis. It infects about 73 million people each year.

Box A4-2 (cont.)

- **Onchocerciasis** is also known as river blindness and is caused by infection with parasitic worms of the genus *Onchocerca* transmitted by blackflies. It infects about 18 million people per year and kills about 40,000.

- **Dengue fever** is also known as breakbone fever or bone-crusher disease. It is caused by a dengue virus transmitted by mosquitoes. It causes suffering to about 1.8 million people each year and kills about 20,000.

- **Trypanosomiasis** is a parasitic disease caused by protozoa of the genus *Trypanosoma*. It infects about 280,000 people per year and kills about 130,000.

- **Poliomyelitis** is the well-known acute viral disease also known as infantile paralysis, acute anterior paralysis, Heine-Medin disease, and Medin's disease. It is caused usually by poliovirus, but sometimes also by coxsackievirus or echovirus. It affects about 100,000 people each year.

- **Dracunculiasis**, also known as a guinea worm disease, is an infection by the guinea worm *Dracunculus*, and is caused by drinking impure water. Dracunculiasis affects about 100,000 people in rural areas each year.

The listed diseases are attributed to the lack of clean water and to poor personal and domestic hygiene, which are also related to an inadequate availability of water, as well as to poor or non-existent water management (poor water storage, inadequate drainage, and the contamination of water with human, animal, or chemical wastes).

Source: Gleick et al 2001; WHO 1999a, 1999b, 2002

Table A4-1 Interstate water-related conflicts

Shared water resources	Countries involved in conflict	Causes of conflict
Amu Dar'ya and Syr Dar'ya	Uzbekistan, Kyrgyzstan, and Kazakhstan	Water scarcity and control
Danube	Hungry and Slovakia	The control (diversion) of water
Ganges	India and Bangladesh	Water scarcity and diversion
Great Lakes	USA and Canada	Water pollution
Han	North and South Korea	The control of water
Indus	India and Pakistan	Water scarcity
Lake Chad	Nigeria, Chad, and Cameroon	Water sharing
Lake Victoria	Uganda, Kenya, and Tanzania	Water sharing
Lauca	Bolivia and Chile	The control of water
Mekong	China, Burma, Thailand, Laos, Cambodia, and Vietnam	Water scarcity and control
Orange	Botswana, Lesotho, Namibia, and South Africa	Water scarcity and diversion
Parana	Argentina, Brazil, and Paraguay	The control of water
Rhine	France, Germany, the Netherlands	Water pollution
Rio Grande	USA and Mexico	Water pollution
Saharan Aquifer	Libya, Egypt, and Sudan	Water scarcity
Salwan and Nu Jing	Burma and China	Water scarcity
Senegal	Senegal and Mauritania	Water scarcity
Szamos	Hungry and Romania	Water pollution
Okavango	Namibia, Botswana, and Angola	Water scarcity
Zambezi	Zambia, Zimbabwe, Mozambique, and South Africa	Water scarcity and diversion

Source: Björklund et al 2000

Box A5-1
The Earth's atmosphere

The Earth's atmosphere is made up mainly of nitrogen (78.08 per cent), oxygen (20.94 per cent), and argon (0.933 per cent). The atmosphere can be divided into a number of layers, according to the distribution of the atmospheric temperature.

- **The troposphere** is the lowest layer of the atmosphere, extending from ground level to about 11 kilometres. The temperature in the troposphere decreases with the increasing altitude at an average rate of 6.5°C per kilometre, to reach about minus 55°C at the altitude of 11 kilometres. This layer contains about 75 per cent of the total mass of the Earth's atmosphere and about 99 per cent of water vapour. It is the layer where all weather events take place.

- **The stratosphere** is the next layer just above the troposphere. It extends from about 11 kilometres to about 50 kilometres. The temperature in the stratosphere increases from minus 55°C to 0°O. The density of the air in the stratosphere is very low, but it is still high enough to allow for hot air balloons to fly. The stratosphere is not turbulent and it is therefore used for air navigation.

  The stratosphere is relatively rich in ozone, which absorbs ultraviolet radiation and makes the stratosphere warm. The ozone layer, which extends from about 10 kilometres to about 50 kilometres above the surface of the Earth, would be only about 3 mm thick if it were brought down to where we live. It is a fragile veil that protects us from the harmful effects of ultraviolet radiation, but we are now destroying it mainly through the accumulation of chlorofluorocarbons (CFCs), which we have been producing and using over a number of years.

  The highest concentration of ozone is between around 20 and 25 kilometres. The stratosphere contains around 90 per cent of all the atmospheric ozone. The remaining 10 per cent is in the troposphere. Much of it is in the form of the harmful urban smog.

- **The mesosphere** follows the stratos[here, and extends from 50 to 90 kilometres. The temperature in the mesosphere decreases from 0°C to minus 100°C. The air in the mesosphere is so thin that electrons can roam freely for a long time before they meet a positive ion and are captured.

- **The thermosphere** follows the mesosphere. It has no clearly defined upper boundary, but fades away into vacuum space. The temperature in the thermosphere increases from minus 100°C at the altitude of 90 kilometres to between 300°C and 1,700°C or even higher at around 500 kilometres. The thermosphere is the place where protons and electrons coming from the Sun excite the air molecules and produce the spectacular displays of aurora borealis (northern lights) and aurora australis (southern lights).

We also can distinguish additional layers at the boundaries of the troposphere, stratosphere, and mesosphere, and within the thermosphere.

- **The tropopause** is the lower layer of the stratosphere, extending from around 11 to 20 kilometres. This layer has a constant temperature of minus 55°C.

- **The stratopause** is the upper layer of the stratosphere, extending from around 40 to 50 kilometresand is characterised by a constant temperature of around 0°C.
- **The mesopause** is the upper layer of the mesosphere, extending from 80 to 90 kilometres. It has a constant temperature of around minus 100°C.
- **The ionosphere** is the lower layer of the thermosphere, extending from around 90 to 400 kilometres.
- **The exosphere** is the upper layer of the thermosphere, extending from around 400 kilometres into vacuum space.

Summary

Layer	Altitude [km]	Temperature [°C]
Troposphere	0–11	From 15 to minus 55
Tropopause	11– 20	minus 55
Stratosphere	11–50	From minus 55 to 0
Stratopause	40–50	0
Mesosphere	50–90	From 0 to minus 100
Mesopause	80–90	minus 100
Thermosphere	90–>700	From minus 100 to >300
Ionosphere	90–400	
Exosphere	400–>700	

km – kilometre

°C – degree of Celsius

> – more than

Source: Lugens and Tarbuck 2000

Table A5-1 Examples of carbon emissions per person, 2000

Country/Region	Emission [t/p/y]	Country/Region	Emission [t/p/y]
Qatar	18.2	Argentina	1.0
USA	5.6	Latin America	0.7
Australia	4.7	China	0.6
Canada	4.4	India	0.3
Russia	3.1	Sub-Saharan Africa	0.2
Germany	2.8	Chad	0.0
Japan	2.5	Industrial countries	3.4
South Africa	2.5	Developing countries	0.5
EU	2.4	**World**	**1.1**

t/p/y – tonne per person per year

EU – European Union

Example: In 2000, carbon emissions in Qatar were 18.2 tonnes per person per year

Source: Aslam 2002; EIA 2000; Mitchell and Hulme 2000

Box A5-2
Examples of extreme weather events

1953 Floods and storms in the Netherlands and Britain caused 1,932 deaths and $3 billion damage.

1954 Floods and storms in China caused 40,000 deaths.

1959 A hurricane in Japan resulted in 5,100 deaths and $600 million damage.

1965 A hurricane in the USA caused 75 deaths and $1.42 billion damage.

1970 A hurricane and floods in Bangladesh caused 300,000 deaths and $63 million damage.

1976 Gales in Europe caused 82 deaths and $1.3 billion damage.

1988 A hurricane in the USA and Central America caused 355 deaths and $3 billion worth of damage.

1990 Winter storms in Western Europe resulted in 230 deaths and $14.8 billion of damage.

1991 Floods in Bangladesh caused 139,000 deaths and $3 billion worth of damage. In the same year, a hurricane in Japan resulted in 62 deaths and $6 billion worth of damage.

1992 A hurricane in the USA caused 62 deaths and $30 billion of damage. In 1998, floods in China resulted in 3,650 deaths and $30 billion damage. In the same year, hurricane in Central America caused 92,000 deaths and $5.5 billion damage.

1999

- In October, a cyclone in Orissa in India, with a wind velocity of up to 300 kilometres per hour, was described as a 'super cyclone'. It killed about 15,000 people, affected 20 million, and caused $2.5 billion worth of damage.

- Venezuela's rainstorm in December caused floods and landslides resulting in over 30,000 deaths and overall damage of around $15 billion.

- In December, winter storms over France, Germany, Spain, and Switzerland were described as 'one in a hundred year' events. They caused 130 deaths and $9.6 billion damage.

- A hurricane threat in the Bahamas was responsible for the largest evacuation in US history, and its total cost amounted to $4.1 billion.

2000

- In March, floods in Argentina left 120,000 people homeless, and floods in Mozambique were described as the worst in living memory.

- In July, there were floods in the Philippines.

- In August and towards the end of the year, there were the worst floods in 230 years in England; floods and mudslides in northern Italy; drought in eastern Africa and central Asia; floods in Thailand, Vietnam, and Cambodia; the worst storms in 100 years in Japan; and floods in New South Wales, Australia.

2001 This pattern of extreme weather conditions continued throughout the year. For instance:

- Central England, Denmark, and Germany experienced the warmest October on record, with temperatures up to 4°C above average in Germany.

- Southern and central Siberia experienced an exceptionally severe winter with temperatures down to minus 60°C. Extremely cold conditions were also experienced in northern India.

Many areas of the world experienced either strong precipitation and flooding or exceptionally weak precipitation and drought.

- The Northern Pacific experienced an abnormal number of typhoons and tropical storms.

- Exceptionally heavy rainfalls were experienced in at least 19 districts of Java.

- The worst flooding on record occurred in the Mekong Delta region.

- Heavy rainfalls and flooding were also experienced in many other parts of the world: Mozambique, Zambia, Chad, Algeria, Argentina, Uruguay, France, Hungry, and Poland.

By contrast, devastating droughts were experienced in central and southern Asia, notably in Iran, Afghanistan, and Pakistan, as well as in other parts of the world, in Kenya, Brazil, northern China, the Korean peninsula, the western United States, and Canada. In central and southern Asia, the drought spell began in 1998 and continued throughout 2001. In the western USA, the precipitation between November 2000 and February 2001 was the lowest on record.

2002 The year 2002 was marked by the now-usual progression of extreme weather events and weather-related disasters, with droughts, heavy rainfalls, tropical storms, and floods.

Source: ENS 2002b, Munich Re 2001, 2002; NCDC 2003; Walker 2000

Table A6-1 The flow of energy at the source (per cent)

Source	Outlets					
	Transport	Industry	Residential and commercial	Generation of electricity	Non-energy use	Loss
Crude oil	44–60	12–20	14	10	6–14	6
Coal	1–2	32–75	5–15	30–49	2	
Natural gas	5	28–44	28	24	6	14
Hydro/Nuclear				100		
Other primary		32	68			
Electricity	2	42–47	40			6–13

Example: Wordwide, 44–60 per cent of crude oil is used to support transport.

Source: Davis 1990; EIA 2002a; Harris 2001; IEA 2000, 2002

Table A6-2 The flow of energy at outlets (per cent)

Outlets	Source					
	Crude oil	Coal	Natural gas	Hydro/ nuclear	Electricity	Other primary
Transportation	96	4				
Industry	17	30	19		17	17
Residential and Commercial	18	14	18		14	36
Generation of Electricity	7–10	35–39	15–18	17–21 (h) 16–20 (n)		
Non-energy	73	9	18			

(h) – Hydro energy; (n) – Nuclear energy

Example: Worldwide, 96 per cent of transport is supported by crude oil.

Source: Davis 1990; EIA 2002a; Harris 2001; IEA 2000, 2002

Table A6-3 Installed electrical capacity by region and globally in 2000 (gigawatts)

Region	Fossil fuels	Nuclear power	Hydroelectric power	Total
North America	660.5	109.8	175.5	962.8
Asia-Pacific	650.7	68.6	159.7	883.4
Western Europe	352.8	128.5	142.1	633.4
EE/FSU	297.9	47.8	80.3	426.0
CSA	64.5	1.7	112.1	181.4
Middle East	94.4	0.0	4.1	98.5
Africa	72.8	1.8	19.9	94.6
World	**2,193.6**	**358.3**	**693.7**	**3,280.1**

Regions are arranged in the descending order of their total installed electrical capacities.

SCA – South and Central America

EE/FSU – Eastern Europe and former Soviet Union.

Example: The total installed capacity for the genration of electricity in North America in 2000 was 962.8 gigawatts (billion watts), of which 109.8 gigawatts was in the form of nuclear power.

Source: EIA 2002c

Table A6-4 Installed electrical capacity by countries, 2000 (gigawatts)

Country	Capacity	Per cent[a]	Per cent[b]
USA	812.7	24.8	24.8
China	293.7	9.0	33.8
Japan	229.2	7.0	40.8
Russia	202.8	6.2	47.0
Canada	110.8	3.4	50.4
France	110.5	3.4	53.8
Germany	108.8	3.3	57.1
India	108.1	3.3	60.4
UK	72.4	2.2	62.6
Brazil	68.8	2.1	64.7
Italy	66.8	2.0	66.7
Ukraine	53.9	1.6	68.3
South Korea	50.0	1.5	69.8
Spain	46.3	1.4	71.2

[a] – Percentage of global installed electrical capacity

[b] – Cumulative percentage of global installed electrical capacity

Example: The installed capacity for the generation of electricity in China in 2000 was 293.7 gigawatts (billion watts), which corresponded to 9 per cent of global installed capacity. The combined capacity of China and the USA for the generation of electricity in 2000 was 33.8 per cent of global installed capacity.

Source: Based on EIA 2002c

Table A6-5 Projections of nuclear generating capacity, 2000–2020 (gigawatts)

Country/Region	2000	Per cent	2005	2010	2015	2020	Change
USA	97.5	27.9	97.7	94.3	88.8	88.0	-9.7
France	63.2	18.1	62.9	62.9	62.9	64.4	1.9
Japan	43.5	12.4	44.3	47.8	50.8	53.4	22.8
Russia	19.8	5.7	21.7	21.3	20.3	14.8	-25.3
South Korea	13.0	3.7	15.9	16.3	19.4	22.1	70.0
UK	12.5	3.6	11.4	9.8	8.1	4.8	-61.6
Ukraine	11.2	3.2	11.2	11.2	11.2	11.2	0.0
China	2.2	0.6	6.6	9.6	11.6	16.6	654.5
Industrialised	322.6	92.2	326.5	319.9	309.9	297.1	-7.9
Developing	27.4	7.8	35.9	43.3	50.6	62.3	127.4
World	**350.0**	**100.0**	**362.4**	**363.2**	**360.5**	**359.4**	**2.7**
Gap	295.2		290.6	276.6	259.3	234.8	20.5

Per cent – Percentage of global nuclear generating capacity in 2000. For instance, the installed nuclear capacity in the USA in 2000 was 97.5 gigawatts (billion watts), which corresponded to 27.9 per cent of global installed nuclear capacity.

Change – Percentage change in the nuclear installed capacity between 2000 and 2020. Increase is given by positive numbers, and decrease by negative numbers. For instance, the installed nuclear capacity in the USA will decrease by 9.7 per cent between 2000 and 2020, but in France it will increase by 1.9 per cent.

Gap – The gap between the total installed nuclear capacities in Industrialised and developing countries. For instance, the gap is projected to decrease from 295.2 gigawatts in 2000 to 234.8 gigawatts in 2020—thta is, by 20.5 per cent. Polynomial analysis shows that the gap will close in around 2046.

Source: Based on EIA 2002a

Table A6-6 Ultimate potentials for the production of hydroelectricity, by region and globally (terawatt-hours per year)

Region	Theoretical	Technical	Economic	(1)	(2)
Central Asia	6 511	2 159	1 302	18.7	18.7
Sub-Saharan Africa	3 583	1 992	1 288	18.5	37.2
Latin America-Caribbean	7 533	2 868	1 199	17.2	54.4
Europe	3 489	2 038	937	13.5	67.9
North America	5 817	1 509	912	13.1	80.9
Former Soviet Union	3 258	1 235	770	11.1	92.0
Pacific OECD	1 134	211	184	2.6	94.6
Pacific Asia	5 520	814	142	2.0	96.7
Middle East and North Africa	304	171	128	1.8	98.5
South Asia	3 635	948	103	1.5	100.0
World	**40 784**	**13 945**	**6 965**	**100.0**	

Regions are arranged in descending order of their economic potential. The ultimate practical potential can be considered as being given by the economic potential.

(1) – Percentage of global economic potential. For instance, the estimated economic potential for the production of hydroelectricity in Central Asia is 1,302 terawatt-hours (trillion watt-hours) per year, which corresponds to 18.7 per cent of global economic potential.

(2) – Cumulative percentage of economic potentials. For instance, the economic potential of Central Asia and sub-Saharan Africa is 37.2 per cent of global economic potential.

Pacific OECD – Australia, Japan, New Zealand, and the Republic of Korea.

Source: Based on Rogner 2000

Table A7-1 The depths of poverty (in per cent)

Country	Income $2	Income $1	Country	Income $2	Income $1
Peru	41.1	15.5	Lesotho	65.7	43.1
Egypt	43.9	3.1	Senegal	67.8	26.3
Honduras	44.4	23.8	Mauritania	68.7	28.6
El Salvador	45.0	21.4	Lao People's Dem. Rep.	73.2	26.3
Yemen	45.2	15.7	Sierra Leone	74.5	57.0
Sri Lanka	45.4	6.6	Malawi	76.1	41.7
Philippines	46.4	14.6	Mozambique	78.4	37.9
China	47.3	16.1	Ghana	78.5	44.8
Paraguay	49.3	19.5	India	79.9	34.7
Côte d'Ivoire	49.4	12.3	Nepal	82.5	37.7
Mongolia	50.0	13.9	Bangladesh	82.8	36.0
Botswana	50.1	23.5	Gambia	82.9	59.3
Ecuador	52.3	20.2	Madagascar	83.3	49.1
Indonesia	55.4	7.2	Central African Rep.	84.0	66.6
Namibia	55.8	34.9	Rwanda	84.6	35.7
Kenya	58.6	23.0	Niger	85.3	61.4
Tanzania, U. Rep. Of	59.7	19.9	Burkina Faso	85.8	61.2
Viet Nam	63.7	17.7	Burundi	89.2	58.4
Zambia	63.7	63.7	Nigeria	90.8	70.2
Zimbabwe	64.2	36.0	Nicaragua	94.5	82.3
Cameroon	64.4	33.4	Uganda	96.4	82.2
Pakistan	65.6	13.4	Ethiopia	98.4	81.9

Countries are arranged in the increasing order of the percentage of the population earning less than $2 per person per day (PPP). Only countries with more than 40 per cent of the population in this category are listed.

Income $2 – Percentage of the population earning less than $2 PPP

Income $1 – For comparison, I also give the percentage of the population earning less than $1 PPP in the relevant countries. For instance in Peru, 41.1 per cent of the population earns less than $2 PPP, and 15.5 per cent earns less than $1 PPP.

PPP – Corrected for purchasing power parity—that is, expressed in terms of the real purchasing power in a given country.

Source: UNDP 2003. See also UNDP 2002. Statistical data are for years between 1990 and 2001.

Table A7-2 The explosion of urbanisation, 1000–2015

1000–1900

Year	City	Population (Million)	Year	City	Population (Million)
1000	Cordova	0.45	1800	Peking	1.10
	Kaifeng	0.40		London	0.86
	Constantinople	0.30		Canton	0.80
	Angkor	0.20		Edo	0.69
	Kyoto	0.18		Constantinople	0.57
	Cairo	0.14		Paris	0.55
	Baghdad	0.13	1900	London	6.5
	Nishapur	0.13		New York	4.2
	Hasa	0.11		Paris	3.3
	Anhilwara	0.10		Berlin	2.7

1950–2015
Explosion of mega cities
(Cities with the population of 10 million or more)

Year	City	Population	Year	City	Population
1950	Tokyo	12.3	2015	Tokyo	28.9
				Mumbai	26.2
1970	Tokyo	16.5		Lagos	24.6
	New York	16.2		Sao Paulo	20.3
	Shanghai	11.2		Dhaka	19.5
				Karachi	19.4
1990	Tokyo	25.1		Mexico City	19.2
	New York	16.1		Shanghai	18.0
	Mexico City	15.1		New York	17.6
	Sao Paulo	15.1		Calcutta	17.3
	Shanghai	13.3		Delhi	16.9
	Mumbai	12.2		Beijing	15.6
	Los Angeles	11.5		Metro Manila	14.7
	Buenos Aires	11.1		Cairo	14.4
	Calcutta	10.9		Los Angeles	14.2
	Beijing	10.8		Jakarta	13.9
	Osaka	10.5		Buenos Aires	13.9
	Seoul	10.5		Tianjin	13.5
				Seoul	13.0
				Istanbul	12.3
				Rio de Janeiro	11.9
				Hangzhou	11.4
				Osaka	10.6
				Hyderabat	10.5
				Teheran	10.3
				Lahore	10.0

Source: Harrison and Pearce 2001; Munich Re 2001; UNFPA 1999, 2001

Table A7-3 Reported outbreaks of infectious diseases, 1998–1999

Reported outbreaks, 1998–1999

Disease	Number of reported outbreaks
Cholera	45
Meningococcal disease	29
Enteric infections	20
Acute haemorrhagic fevers	15
Acute respiratory infections	8
Anthrax	7
Yellow fever	6
Plague	5
Other	20
Total	**155**

Source: WHO 1999a

Table A7-4 Burden of infectious diseases, 1998 (all ages)

Cause	DALYs (Million years)
Acute respiratory infections	83
Diarrhoeal diseases	73
AIDS	71
Malaria	39
Measles	30
Tuberculosis	28
Sexually transmitted diseases	17
Pertussis (whooping cough)	13
Tropical diseases	11
Total	**365**

DALYs – disability-adjusted life years – the number of healthy life years lost collectively due to a disease. In 1998 worldwide, 365 million life years were lost due to infectious diseases.

Source: WHO 1999b

Table A7-5 The increasing resistance to antibiotics

Resistance of *Staphylococcus* to methicillin in UK		Resistance of *Salmonella* in Germany	
Year	Resistance	Year	Resistance
1989	2.1	1990	0.9
1990	2.5	1991	0.9
1991	2.1	1992	1.5
1992	2.7	1994	8.1
1993	4.1	1995	8.2
1994	7.1	1996	16.6
1995	13.7		
1996	20.1		
1997	32.2		

Resistance – In per cent

Source: Heymann 2000; WHO 2000

Appendix B
Scales of water availability

The scale of water availability as used by Shiklomanov.[1] This is a simple and straightforward scale. It uses the volume of runoff water and divides it by the number of people. For instance, the average global volume of fresh water is 42,800 cubic kilometres per year [km^3/y], which for 6 billion people corresponds to 7133 cubic metres per person per year [$m^3/p/y$]. Using this procedure, we can construct the following scale of water availability:

- *Very high* – the amount of runoff water is more than 20,000 $m^3/p/y$. (Examples: Brazil, Canada, Chile, Fiji, New Zealand, Norway, and Russia.)
- *High* – runoff water between 10,000 and 20,000 $m^3/p/y$. (Examples: Australia, Austria, Croatia, Indonesia, and Ireland.)
- *Adequate* – runoff water between 5000 and 10,000 $m^3/p/y$. (Examples: Chad, Iraq, Switzerland, and the US.)
- *Low* – runoff water between 2000 and 5,000 $m^3/p/y$. (Examples: China, Germany, India, Italy, Mexico, Pakistan, Philippines, Sudan, Turkey, and Britain.)
- *Very low* – runoff water between 1000 and 2000 $m^3/p/y$. (Examples: Belgium, Egypt, Iran, Peru, Poland, and South Korea.)
- *Catastrophically low* – runoff water below 1000 $m^3/p/y$. (Examples: Barbados, Israel, Kuwait, Rwanda, Singapore, Somalia, and Yemen.)

Water stress/scarcity index (WSSI).[2] This is a confusing scale because of its name and abbreviation. I use here the original name, but have changed the original abbreviation WSI to WSSI to distinguish it from another scale called the water scarcity index (see below), for which the abbreviation WSI is also used.

The WSSI scale is constructed using Falkenmark's idea of the flow unit,[3] which is defined as one million cubic metres of runoff water

---

[1] Shiklomanov 2000; see also Shiklomanov and Rodda 2003.
[2] Ohlsson 1989
[3] Falkenmark 1989

per year. The scale is constructed according to the number of people per flow unit but expressed as the number of people per 10,000 cubic metres of runoff water per year. For instance, with 600 people per flow unit we have six people per 10,000 cubic metres of runoff water, and therefore a WSSI value of 6. Using this procedure, the following scale is constructed for water availability:

- *Relative sufficiency* – WSSI between 0 and 5
- *Water stress* – WSSI between 6 and 10
- *Water scarcity* – WSSI between 11 and 20
- *Absolute water scarcity* – WSSI greater than 20

A related scale based on Falkenmark's idea of the flow unit.[4] This scale is based on the inverse calculations of the WSSI scale.

Water availability is considered to be adequate if there are 600 or fewer people per flow unit. This gives a threshold for adequate water availability of 1667 cubic metres of water per person per year [m^3/p/y]. This number can be rounded to 1700 m^3/p/y, and the following scale of water availability can be constructed:

- *Well-watered conditions* – the amount of runoff water is more than 10,000 m^3/p/y
- *Relative sufficiency* – runoff water between 1700 and 10,000 m^3/p/y. With this much water, a country or district may suffer only occasional shortages.
- *Water stress* – runoff water between 1000 and 1700 m^3/p/y
- *Water scarcity* – runoff water between 500 and 1000 m^3/p/y
- *Absolute water scarcity* – runoff water below 500 m^3/p/y

We can see immediately that there is a simple relation between this scale and the WSSI scale. To calculate the number of cubic metres of water per person per year, divide 10,000 by the WSSI figure. For instance, if the WSSI is 10, water availability is 1000 cubic metres per person per year, the borderline between water stress and water scarcity.

Water scarcity index (WSI).[5] This index belongs to a slightly different category. WSI, which is expressed in terms of the percentage runoff water withdrawal, does not describe the physical availability of

---

4   Falkenmark 1989
5   WMO and UNESCO 1997

water, but rather the degree of stress a country experiences depending on its water withdrawals.

A country may have a relatively high availability of water, but it may still suffer high water stress if withdrawal is high. Conversely, a country may have relatively low water availability, but may not suffer water stress if withdrawal is low. Examples in the first category could be found in industrialised countries where demand for water per person is high. Examples in the second category could be found in developing countries where demand for water is low.

For instance, US water availability in the 1990s was high, but withdrawal was between 20 per cent and 40 per cent, giving the country medium to high water stress. In Kenya and Somalia, water availability was catastrophically low, but withdrawal was less than 10 per cent, resulting in low water stress.[6]

The WSI is helpful, but it should be treated with care because it may be misleading. We should remember that the WSI is used to measure water stress and not the physical availability of water. Just because people of certain countries lead a poor life and use very little water it does not mean that water availability in these countries is high or even adequate. Likewise, if people in rich countries use more water than they need and suffer water shortage it does not mean that water availability is low. Such countries may suffer a high degree of water stress because their demand is unnecessarily high. Their problem is created not so much by the physical environment, but by attitudes and behaviour. However, there are also countries that suffer low water availability and high water stress.

The scale for the water scarcity index is as follows:

- *Low water stress* – withdrawal is less than 10 per cent of runoff water
- *Moderate water stress* – withdrawal is 10 per cent to 20 per cent
- *Medium to high water stress* – withdrawal is 20 per cent to 40 per cent
- *High water stress* – withdrawal is more than 40 per cent

In 1995, medium to high water stress was experienced in the US, South Africa, Morocco, Algeria, Spain, Italy, Germany, Poland,

---

[6] Engelman and LeRoy 1993, Appendix II; WMO and UNESCO 1997

Syria, Israel, Jordan, United Arab Emirates, Oman, Kazakhstan, Turkmenistan, India, and Burma.

Countries with high water stress included Libya, Egypt, Iraq, Kuwait, Saudi Arabia, Yemen, Iran, Uzbekistan, Afghanistan, and Pakistan.

Appendix C
How to read large numbers

We have two systems of large numbers: the elegant and logical British system and the American system.[7] These terms are labels that are useful for identification, but otherwise meaningless. They have nothing to do with the origin of the two systems.

The British system was devised by a French physician and mathematician, M. Chuquet, in the 1480s. For no apparent reason the American system was introduced and adopted in France in the 17th century. A little later, other countries, including the US, accepted it. In 1948, France corrected its mistake and now uses the original system. The popularity of the American system is increasing.

The British system is elegant and logical because the names of large numbers have a straightforward meaning. This does not apply to the American system.

We know that a million is the figure 1 followed by six zeros: 1,000,000. The number and name are the same in both systems. However, when it comes to larger numbers, we have created unnecessary confusion.

In the British system, a billion is a number with two lots of six zeros (1,000,000,000,000). The prefix "bi" in the word billion means two. This prefix is also used in words such as bigamy, binoculars, bicycle, bipolar, biplane. We know what it means, and we use it correctly everywhere except for the word billion in the American system.

A trillion is a number with three lots of six zeros, the prefix "tri" meaning three. A quadrillion has four lots of six zeros, and so on for quintillion, sextillion, septillion, and larger numbers. It is all easy to remember and makes sense.

In the American system the nonsense starts with billion and continues with larger numbers. The American billion is a number with nine zeros (1,000,000,000), which has nothing whatever to do with the prefix "bi". The same applies to trillion, quadrillion, and all the rest of them.

However, we can still try to force some logic into the American

---

[7] See, for instance, Watson (1968) or Weisstein (2003)

system by doing some convoluted mathematical manipulations. For instance, subtract three zeros from nine zeros in a billion and you get six zeros. Divide six by three and you will have two, so you have managed to justify the use of the prefix "bi" in the American billion. As long as you can do quick subtractions and divisions you have a way of working out the names of large numbers in the American system.

The differences between the two systems are not trivial. The British billion is 1000 times the American billion; the British trillion is 1,000,000 times the American trillion, and so on. The higher the number the greater the difference between the two systems. It is surprising that we can still live with this confusion.

Apart from reasons of pedantry and logic, it would not matter which system we used as long as it were uniform throughout the world.

Table C-1 How to read large numbers

Name	British System		Number of zeros in the American System	
Billion	12	(2x6)	9	(2x3+3)
Trillion	18	(3x6)	12	(3x3+3)
Quadrillion	24	(4x6)	15	(4x3+3)
Quintillion	30	(5x6)	18	(5x3+3)
Sextillion	36	(6x6)	21	(6x3+3)
Septillion	42	(7x6)	24	(7x3+3)
Octillion	48	(8x6)	27	(8x3+3)
Nonillion	54	(9x6)	30	(9x3+3)
Decillion	60	(10x6)	33	(10x3+3)
Undecillion	66	(11x6)	36	(11x3+3)

Numbers in the parentheses show how the names are related to the number of zeros.

Example: Billion has 12 zeros in the British system (1,000,000,000,000), but only 9 zeros in the American system (1,000,000,000). The prefix *bi* means *two*. The number of zeros in the British system is 2x6 and in the American system 2x3+3 zeros.

Appendix D
A brief guide to energy units

The use of information about energy resources is hindered by the need to navigate between various energy units. For instance, British Petroleum (BP), the US Energy Information Administration (EIA), and the International Energy Agency (IEA) publish good updated information on the production and consumption of energy. However, these sources use different energy units, and comparing the numerical values is not always easy. For instance, BP uses barrels, tonnes, cubic metres, and tonnes oil equivalent. The IEA uses mainly tonnes oil equivalent and watt-hours. The EIA prefers British thermal units [Btu] and watt-hours, but it also uses other units.

The main source of energy is oil, and its primary units are barrels or tonnes. The conversion between such closely related units is easy: one tonne is 7.33 barrels.[8] The conversion between other units of energy poses also no problem,[9] until you start dealing with units of electricity. Here, in addition to the standard tabulated conversion factors, one has to use a conversion efficiency (production efficiency) factor. For instance, the tabulated conversion factor for converting tonnes oil equivalent [Mtoe] into terawatt-hours [TWh] is 11.63. However, if we assume that the conversion efficiency factor is 33 per cent we have to change the conversion factor to 3.84.

If you are lucky you will find a conversion efficiency factor somewhere in the small print. If it is not mentioned you will have to figure it out. Various sources use different values. For instance, Davis (1990) assumes 33 per cent for all sources of energy used in the generation of electricity; BP uses 33 per cent or 38 per cent depending on the year of publication. The IEA (2002) and the UNDP et al (2000) use 33 per cent for nuclear energy; 100 per cent for hydropower, wind, and solar energy; and 10 per cent for geothermal energy.

Conversion efficiency factors depend not only on the energy source, but also on where it is used. For instance, the world average conversion

---

[8] BP 2002a
[9] Conversion factors, which include tonnes oil equivalent, may be found for instance in IEA 2002 or UNDP et al 2000a, Part V

efficiency factor for coal in 1997 was 34 per cent, but in China it was 28 per cent, and in the Middle East and North Africa 39 per cent. The world average for natural gas was 37 per cent, but in China it was 30 per cent, and in the Middle East and North Africa 40 per cent.[10]

The moral to the story is: use the standard tabulated conversion factors. However, be aware that a conversion efficiency factor other than 100 per cent may have been used for the tabulated energy values expressed in watt-hours, and judiciously incorporate it in your recalculations of listed values.

With various energy units and conversion efficiency factors it is like reading documents written partly in Arabic, Chinese, or Hebrew, when all you want is plain English.

Table D-1 Conversion coefficients for energy units

To: From:	TJ	Gcal	Mtoe *Multiply by*	MBtu	GWh
TJ	1	238.8	0.00002388	947.8	0.2778
Gcal	0.0041868	1	0.0000001	3.968	0.001163
Mtoe	41 868	10 000 000	1	39 682 540	11 628
MBtu	0.0010551	0.252	0.0000000252	1	0.0002931
GWh	3.6	860	0.000086	3 412	1

TJ – terajoule (trillion joules)

Gcal – gigacalorie (billion calories)

Mtoe – mega (million) tonne oil equivalent

Mbtu – mega British thermal unit

GWh – giga watt-hour

Example: 1 TJ = 0.00002388 Mtoe

Source: IEA 2002

---

[10] UNDP et al 2000a, Part V

Appendix E
Biological warfare

A brief guide to some of the diseases caused by biological agents.

Anthrax An infectious disease in animals caused by spores of *Bacillus anthracis*, which can be transmitted to humans. It is often fatal even when treated. Anthrax infection affects the skin, causing small elevations containing pus (pustules) or skin oedema. However, there are other forms: cerebral anthrax (when spores invade the brain); pulmonary anthrax (when inhaled); and intestinal anthrax (when ingested).

Botulinum toxin Probably the most poisonous substance known. It is produced by the bacterium *Clostridium botulinum*. Initial symptoms of exposure may be in the form of weakness and dizziness. In severe cases it can lead to a paralysis of respiratory muscles and death. Botulinum toxin is used by the rich and famous (and others) for smoothing out wrinkles to make the face look young again. One person's poison is another's elixir of youth.

Brucellosis A disease caused by bacteria of the genus *Brucella* and characterised by rising and falling fever, sweating, aches; and pains.

Chikungunya fever A viral disease transmitted by mosquitoes; and resembling dengue fever.

Coccidioidomycosis Caused by the fungus *Coccidioides immitis*. Symptoms range from mild to fatal. They include flu-like symptoms, fever, cough, headache, muscle pain, and severe lung infection.

Dengue fever A viral disease usually transmitted by mosquitoes, and characterised by severe pain in the bones.

Encephalitis Inflammation of the brain. When the infection affects the whole nervous system it is called *encephalomyelitis*. Symptoms of the disease include fever, headache, lethargy, disorientation, seizures, focal paralysis, delirium, and coma.

Glanders   A disease in animals (horses, donkeys, mules, and cats), but it can be transmitted to humans. It is a debilitating condition caused by the bacterium *Pseudomonas mallei*. It attacks the mucous membranes of the nose and produces excessive mucous discharge. It also attacks the lymph glands of the lower jaw.

Histoplasmosis   A disease caused by the fungus *Histoplasma capsulatum*. It can be fatal for small children and the elderly. The disease mimics tuberculosis.

Melioidosis   A disease among rodents in India that is caused by the bacterium *Pseudomonas pseudomallei*. It can be transmitted to humans. It causes multiple abscesses and pneumonia.

Mooseri   Endemic typhus caused by *Rickettsia mooseri*.

Mycotoxins   These toxins are produced by fungi (mould). The effects of poisoning include sore throat, nausea, vomiting, diarrhoea, abdominal pain, and weakness. Exposure to mycotoxins can be fatal, but the fatality rate is considered moderate. In some cases, death may occur in minutes or hours.

Prowazeki   Epidemic typhus caused by *Rickettsia prowazeki*.

Psittacosis   Also known as parrot fever, the disease is caused by the bacterium *Chlamydia psittaci*, primarily from birds. Symptoms can range from mild flu-like reactions in younger people to serious bronchopneumonia in older people.

Q-fever   A disease caused by the bacterium *Coxiella burnetii*. Its symptoms include high fever, headache, malaise, sore throat, sweating, vomiting, diarrhoea, and abdominal and chest pain.

Ricin   A poison that comes from the castor bean plant, *Ricinus communis*. It is one of the most poisonous toxins. It causes abdominal pain, vomiting, and diarrhoea. Death may occur in three to five days.

Tsutsugamushi A disease caused by *Rickettsia tsutsugamushi*. Symptoms include fever, headache, swollen glands, and ulcers.

Tularemia A disease of wild and domestic animals that can be transmitted to humans. It is caused by the bacterium *Francisella tularensis*. It can cause ulcers, the enlargement of lymph nodes, throat infection, vomiting, diarrhoea, fever, chills, weakness, headache, and backache.

Appendix F
Acronyms and abbreviations

AAAS	American Association for the Advancement of Sciences
ABS	Australian Bureau of Statistics
ACUNU	American Council for the United Nations University
AGLW	Water Resources, Development and Management Service
AD	Anno Domini
AIDS	acquired immune deficiency syndrome
ARI	acute respiratory infections
ASA	American Society of Agronomy
BASC	Board on Atmospheric Sciences and Climate
BGS	British Geological Survey
BP	British Petroleum
BRWM	Board on Radioactive Waste Management
BSE	bovine spongiform encephalopathy (also known as mad cow disease)
CADRE	College of Aerospace Doctrine, Research, and Education
CCAP	Climate Change Action Plan
CBD	Convention of Biological Diversity
CDC	Centre for Disease Control
CDI	Centre for Defence Information
CDM	The Clean Development Mechanism
CERES	Coalition for Environmentally Responsible Economics
CFCs	chlorofluorocarbons
CIFPR	California Institute for Federal Policy Research
CII	Chartered Insurance Institute
CIP	competitive industrial performance
CIS	Commonwealth of Independent States
COP	Conference of the Parties (the annual Conferences of the Parties to the United Nations Framework Convention on Climate Change, identified as COP1, COP2, etc)
CSCC	Committee on the Science of Climate Change, NRC
CSD	Commission for Sustainable Development

CSIRO	Commonwealth Scientific and Industrial Research Organisation
CSIS	Centre for Strategic and International Studies
CSSA	Crop Science Society of America
CVD	cardiovascular disease
DAC	Development Assistance Committee
DALY	disability-adjusted life year
DDT	dichloro-diphenyl-trichloroethane
DEIA&EW	Division of Environmental Information, Assessment and Early Warning
DEWA	Division of Early Warning and Assessment
DFID	Department of International Development
DGDC	Belgian Development Co-operation
DOE	Department of Energy
DOTS	directly observed treatment, short course
DU	Dobson Unit
EDCs	endocrine disrupting chemicals
EIA	Energy Information Administration
ENS	Environmental News Service
EOPUS	Executive Office of the President of the United States
EPA	US Environmental Protection Agency
EUR	estimated ultimate recoverable
FAO	United Nations Food and Agriculture Organisation
FII	Finance Industries Initiative
FSU	Former Soviet Union
FY	Fiscal Year
G-8	Group Eight countries (the US, Britain, Germany, France, Italy, Canada, Japan, and Russia)
GCI	Green Cross International
GCM	general circulation model
GDP	gross domestic product
GDP (PPP)	GDP corrected for purchasing power parity
GEMS	Global Environmental Monitoring System
GHG	greenhouse gases
GNP	gross national product
GPI	genuine progress indicator
GPO	Government Printing Office
GWP	gross world product
HDI	Human Development Index
HIIK	Heidelberger Institut für International

	Konfliktforschung (Heidelberg Institute for International Conflict Research)
HIV	human immunodeficiency virus
HPI	Human Poverty Index
HPM	high-power microwave
IATP	Institute for Agriculture and Trade Policy
IBCT	interim brigade combat team
ICBL	International Campaign to Ban Landmines
ICLEI	International Council for Local Environmental Initiatives
ICS	Institute for Civil Society
IEA	International Energy Agency
IFA	International Fertiliser Industry Association
IFPRI	International Food Policy Research Institute
IIASA	International Institute for Applied Systems Analysis
ILO	International Labour Organisation
INPE	Instituto Nacional de Pesquisas Espaciais (National Space Research Institute)
INRA	Institute of Natural Resources in Africa
IPCC	Intergovernmental Panel on Climate Change
IPCS	International Program on Chemical Safety
IWMI	International Water Management Institute
IWRA	International Water Resources Association
JHSPH	The Johns Hopkins School of Public Health
LAC	Latin America-Caribbean
LDC	least-developed countries
MAD	mutually assured destruction
MDG	Millennium Development Goals
MDR-TB	multidrug-resistant tuberculosis
MENA	Middle East and North Africa
MRSA	methicillin-resistant *Staphylococcus aureus*
NAS	National Academy of Sciences
NATO	North Atlantic Treaty Organisation
NCDC	National Climate Data Centre
NCEH	National Centre for Environmental Health
NEA	Nuclear Energy Agency of OECD
NGO	nongovernmental organisation
NIC	National Intelligence Council
NLWRA	National Land and Water Resources Audit
NRC	National Research Council

NRDC	Natural Resources Defence Council
NRE	Natural Resources and Environment
NS	New Scientist
NYAS	New York Academy of Sciences
ODA	Official Development Assistance
ODG	Overseas Development Group
ODS	ozone depleting substances
OECD	Organisation for Economic Co-operation and Development
OPEC	Organisation of Petroleum Exporting Countries
OSB	Ocean Studies Board
OTA	Office of Technology Assessment
PAHs	polycyclic aromatic hydrocarbons
PAI	Population Action International
PCBs	polychlorinated biphenyls
PCDDs	polychlorinated dibenzo-p-dioxins
PCDFs	polychlorinated dibenzofurans
PEC	Pew Ocean Commission
PETEX	Petroleum Extension Service
PISDES	Pacific Institute for Studies in Development, Environment, and Security
PLEC	People, Land Management and Environmental Change Project
PM	particulate matter
POGO	Project on Government Oversight
POPs	persistent organic pollutants
POST	Parliamentary Office of Science and Technology
PPP	purchasing power parity
PRB	Population Reference Bureau
PRB	Polar Research Board
PRC	People's Republic of China
PSEG	Public Service Enterprise Group
R/P	reserves to production ratio
R/C	reserves to consumption ratio
RAFI	Rural Advancement Foundation International
RIIAEEP	Royal Institute for International Affairs, Energy and Environmental Program
RIVM	Rijksinstituut voor Volksgezonedheid en Milieu (National Institute of Public Health and the Environment)
SAPMP	Scientific Assessment Panel of the Montreal Protocol

SCA	South and Central America
SI	Système International
SIPRI	Stockholm International Peace Research Institute
SPM	suspended particulate matter
STBP	Stop TB Partnership
TB	tuberculosis
TEV	total economic value
UAE	United Arab Emirates
UBC	University of British Colombia
UCS	Union of Concerned Scientists
UEA	University of East Anglia
UN	United Nations
UNCED	United Nations Conference on Environment and Development
UNDP	United Nations Development Program
UNDESA	United Nations Department of Economic and Social Affairs
UNHCR	United Nations High Commissioner for Refugees
UNED	United Nations Environment and Development
UNEDP	United Nations Environment and Development Program
UNEP	United Nations Environmental Program
UNESCO	United Nations Educational, Scientific and Cultural Organisation
UNFCCC	United Nations Framework Convention on Climate Change
UNFPA	United Nations Population Fund
UNICEF	United Nations Children's Fund
UNIDO	United Nations Industrial Development Organisation
UNU	United Nations University
USAF	United States Air Force
USCB	United States Census Bureau
USDA	United States Department of Agriculture
USDD	United States Department of Defence
USFEC	Ukrainian State Foreign Economic Company
USGS	United States Geological Survey
USSR	Union of Soviet Socialist Republics
UWB	ultra-wide band
VISA	vancomycin-intermediate *Staphylococcus aureus*
VOCs	volatile organic compounds

VRE	vancomycin-resistant *Enterococci*
VRSA	vancomycin-resistant *Staphylococcus aureus*
WB	World Bank
WCD	World Commission on Dams
WCED	World Commission on Environment and Development
WEC	World Energy Council
WLI	Weapons Lethality Index
WMO	World Meteorological Organisation
WRC	Water and Rivers Commission
WRI	World Resources Institute
WSAA	Water Services Association Australia
WSI	Water scarcity index
WSSI	Water stress/scarcity index
WWF	World Wildlife Fund
WWI	World Watch Institute

Units of measurements

Btu	British thermal unit
EJ	exajoule (quintillion joules)
Gb	gigabarrel (billion barrels)
Gb/y	gigabarrel per year
GJ	gigajoule (billion joules)
GJ/p/y	gigajoule per person per year
Gm^3	giga cubic metre (billion cubic metres)
Gt	gigatonne (billion tonnes)
GW	gigawatt (billion watts)
ha/p	hectare per person
kg/ha	kilogram per hectare
kg/p/y	kilogram per person per year
l/p/d	litre per person per day
km^2	square kilometre
km^3	cubic kilometre
km^3/p/y	cubic kilometre per person per year
km^3/y	cubic kilometre per year
kWh	kilowatt-hour (thousand watt-hours)
kWh/p	kilowatt-hour per person
kWh/p/y	kilowatt-hour per person per year
m^2	square metre
m^3/ha	cubic metre per hectare
Mha	megahectare (million hectares)
Mm^2	mega square metre (million square metres)
m^3	cubic metre
m^3/p/y	cubic metre per person per year
MBtu	mega British thermal unit (million British thermal units)
mg/l	milligram per litre
$\mu g/m^3$	microgram per cubic metre
MJ	megajoule (million joules)
Mt/p/y	megatonne (million tonnes) per person per year
Mtoe	megatonne oil equivalent (million tonnes oil equivalent)
PBtu	peta British thermal unit (quadrillion British thermal units, also known as *quad*)

p/km^2 people per square kilometre
ppb   part per billion by volume
ppm   part per million
TJ   terajoule (trillion joules)
t/ha   tonne per hectare
t/p/y   tonne per person per year
toe   tonne oil equivalent
TWh   terawatt-hour (trillion watt-hours)
TWh/y   terawatt-hour per year (trillion watt-hours per year)
W/m^2 watt per square metre
Wh   watt-hour

SI Prefixes

M   mega   (million, 1,000,000)
G   giga   (billion, 1,000,000,000)
T   tera   (trillion, 1,000,000,000,000)
P   peta   (quadrillion, 1,000,000,000,000,000)
E   exa   (quintillion, 1,000,000,000,000,000,000)
m   milli   (1/1,000)
μ   micro   (1/1,000,000)

Notes

Introduction
[1] Meadows et al 1972; Pirages 1989.
[2] Malthus 1798; Marsh 1864, 1874.
[3] See, for instance, Brown 1954, Ehrlich 1968, Harrar 1970, Meadows et al 1992, Osborn 1948, and Sears 1935.
[4] NAS 1993.
[5] UCS 1992.
[6] "Prophecy is a good line of business, but it is full of risks." – Mark Twain, *Following the Equator*.
[7] See, for instance, EIA 2002a, Engelman et al 2000, NIC 2000, UNEP 2002, USDA 1997, and Wilson 2002.
[8] Raskin et al 2002.
[9] UNEP 2002.
[10] Crutzen and Stoermer 2000.
[11] The term *business as usual* is used widely in the literature and means the continuation of the current *status quo*—that is, the continuation of current technological, economic, political, and social trends.

Chapter 1 Environmental Degradation
[1] See for instance French 2000a, 2000b and UNDP et al 2003.
[2] Caldwell 1999
[3] Bazerman and Hoffman 1999, Caldwell 1999, Gilman 1992, and Watt 1974.
[4] ENS 2002a; Hinrichsen and Robey 2000; WHO 2002.
[5] Holt 2000; Lyons 1999; O'Sullivan and Alcock 1999; and Sabli_, A. 1989.
[6] UNEP 1999.
[7] WWF 2000.
[8] McGinn 2000; POST 1998; UNFPA 2001.
[9] Lewis and Tumlinson 1988; Takabayashi and Dicke 1996.
[10] DDT: dichloro-diphenyl-trichloroethane.
[11] PCBs: polychlorinated biphenyls.
[12] In the biological classification, heavy metals are defined as elements with an atomic mass greater than 40. Not all heavy metals are toxic (Kharikov and Smetana 2000).
[13] Ross and Birnbaum 2001.
[14] NCEH 2003.
[15] UNIDO 2002.
[16] I have calculated the percentage values using the average values of industrial outputs between 1985 and 1995. My calculated values are slightly different from the values listed in UNIDO 2002.
[17] We should remember that a large percentage of a small number can be smaller than a small percentage of a large number. Measured in dollars, the values of industrial outputs in developing countries are still small.
[18] French 2000a; Harrison and Pearce 2001; Pearce 2000.

Chapter 2 The Population Explosion

1   Cohen 1995.
2   PRB 2002.
3   Infant mortality rate—the number of deaths of children under one year of age per 1000 live births. Child mortality rate—the number of deaths of children under five years of age per 1000 live births.
4   PRB 2002.
5   PRB 2002.
6   UN 1999a.
7   Bongaarts and Bulatao 2000; Lutz 1996; Lutz et al 2001; O'Neill and Balk 2001; UN 1997, 1999b; USCB 1999
8   Bongaarts and Bulatao 2000.
9   Organization for Economic Cooperation and Development – see Chapter 6.
10   UN 1999a; UNDP 2002.
11   UN 1999a.
12   WB 2003.
13   In some countries the problem will be experienced earlier.
14   See, for instance Lubchenco 1998 or Wilson 1998.
15   Bongaarts and Bulatao 2000.
16   McGranahan et al 1999; UN 1999a; UNEP 2002;WB 2000.

Chapter 3 Diminishing Land Resources

1   Giampietro and Pimentel 1994; Lal 1989.
2   Engelman et al 2000.
3   Smil 1993.
4   Engelman et al 2000.
5   These estimates are based on the current footprints of various countries.
6   Acosta et al 1999; Buringh 1989; Fischer and Heilig 1998; Wackernagel et al 2002; WRI et al 1998.
7   Barbier 1999; UNEP 2002.
8   Harrison and Pearce 2001; UNEP 2000, 2002.
9   Wackernagel et al 1997, 2002.
10   WCED 1987.
11   WB 2003.
12   Barbier 1999; Bouwman and van Vuuren 1999; Stocking and Murnagham 2000; UNEP 2002.
13   Barbier 1999. See also Anderson et al 2001, Oldeman et al 1990, and WRI 1992.
14   UNEP 2002. See also Hinrichsen 1997.
15   Stocking and Murnaghan 2000; UNEP 2002.
16   Pimentel and Pimentel 1996.
17   Elwell 1985; OTA 1982; Pimentel et al 1995; Troeh et al 1991.
18   NLWRA 2001.
19   WRC 2000.
20   Wackernagel 2001; Wackernagel and Rees 1996; Wackernagel et al 1997, 2002.
21   Wackernagel et al 2002.
22   See, for instance, Schaeffer 1987 and UNEP 2001.

[23] Cincotta and Engelman 2000; Cracraft and Grifo 1999; Engelman et al 2000; UNEP 2002.

[24] UNDP et al 2000b.

[25] Freckman 1994.

[26] ENS 2002c.

[27] Erwin 1982.

[28] The number of plant species quoted as 270,000 in UNEP 2002 appears to be incorrect – see Cracraft and Grifo 1999, Graham 2002, and Pitman and Jorgensen 2002.

[29] UNDP et al 2000b.

[30] Pimm et al 1995.

[31] Pitman and Jørgensen 2002.

[32] Myers 1990; Myers et al 2000.

[33] PRB 2002.

[34] UNEP 2002.

[35] FAO 2000b; Shand 1997; UNEP 2002.

[36] In the FAO 2000b report, Europe includes also Belarus, Republic of Moldova, Russian Federation, and Ukraine. Oceania is made of Australasia (which includes Australia and New Zealand), Melanesia (which includes New Guinea and Fiji), Micronesia, and Polynesia.

[37] UNEP 2002. Shand 1997 estimates the loss of tropical forests at around 18 million hectares per year.

[38] FAO 2000b.

[39] FAO 2000b; UNEP 2002.

[40] Singh et al 2001.

[41] Hinrichsen, D. and Robey 2000.

[42] McCarthy et al 2001.

[43] WB 2003.

[44] Revenga et al 1998.

[45] Bryant et al 1997; UNEP 1999.

[46] Shand 1997.

[47] French 2000a.

[48] ENS 2003b.

[49] ENS 2002d.

[50] Costanza et al 1997.

[51] In 2000 US dollars.

[52] See references in Balmford et al 2002.

[53] Balmford et al 2002.

[54] Gardner 2000a.

[55] Rosegrant et al 2002.

[56] Rosegrant et al 2002.

[57] Dyson 1999.

[58] Tilman 1999. See also Rosegrant et al 2002.

[59] Harrison and Pearce 2001. See also Ghassemi et al 1995 and UNDP et al 2000b.

[60] Pinstap-Anderson 1999.

[61] Dyson 1999.

[62] Dyson 1999.

63 Postel 2000b.
64 Tilman 1999
65 Halweil 2001
66 Buringh 1989.
67 UNEP 1999.
68 UNFPA 2001.
69 McHarry et al 2002.
70 USDA 1997.
71 Schwarz 1995; Wada 1993.
72 UNEP 2002. See also FAO 2002a and McHarry et al 2002.
73 Alexandratos 1999; Dyson 1999. WB 2003. See also Chapter 7.
74 Halweil 2001; Mellon and Fondriest. 2001; Wallinga et al 2002; WHO 2000.
75 Mellon and Fondriest. 2001.
76 WHO 2000.
77 Wallinga et al 2002.
78 Harrison and Pears 2001; Porter 2000.
79 French 2000a.
80 Halweil 1999.
81 Al-Saleh 1994; French 2000a; Richter and Chlamtac 2002; Spiwak 2001;
   UNEP 2002.
82 Buranatrevedh and Roy 2001; Daniels et al 1997; Hopenhayn-Rich et al 2002;
   Ji et al 2001; Morrison et al 1992; Wesseling et al 1999.
83 See, for instance, Tamura et al 2001.
84 Betarbet et al 2000.
85 Brown 2001b.
86 Harrison and Pearce 2001.
87 Pinstrap-Anderson 1999.
88 Harrison and Pearce 2001; UNDP et al 2000b.
89 Bouwman and van Vuuren 1999; UNDP et al 2000b.
90 McHarry et al 2002.
91 Tilman 1999.
92 Halweil 2001.
93 Follet and Stewart 1985; OTA 1982; Pimentel 1993; Pimentel et al 1995;
   Tilman 1999.
94 McCarthy et al 2001.
95 McHarry et al 2002.
96 UNEP 2002.
97 UNDP 2002.
98 UNEP 2002.

Chapter 4 Diminishing Water Resources

1 Engelman and LeRoy 1993; UNEP 2002.
2 Shiklomanov and Rodda 2003. See also Clarke et al 1996, Clement et al 1997,
  and Shiklomanov 1998.
3 Shiklomanov and Rodda 2003. An estimate used in UNEP-DEWA 2003 is
  1,000 cubic kilometres per year.
4 Shiklomanov 2000; Shiklomanov and Rodda 2003.
5 Based on the data of Postel et al 1996.

[6] Engelman and LeRoy 1993 and references therein.
[7] Niemczynowicz 2000.
[8] Shiklomanov 2000
[9] Shiklomanov 2000. See also WMO and UNESCO 1997.
[10] UNDP 1999.
[11] Shiklomanov 2000. See also WHO and UNESCO 1997.
[12] WB 1999a.
[13] ABS 2000; NLWRA 2000.
[14] Gleick 1996. See also Gleick et al 2001.
[15] Rosegrant et al 2002.
[16] Dziegielewski 2000. See also Gleick 1996.
[17] Jobson 1999.
[18] Engelman et al 2000.
[19] Jobson 1999.
[20] Allan 1997. See also Turton 1999.
[21] WCD 2000; Gleick et al 2001.
[22] Shah et al 2000; Keller et al 2000.
[23] Gleick et al 2001; UNDP et al 2000b; WCD 2000.
[24] Wolf et al 1999.
[25] UNEP 2002.
[26] Shah et al 2000.
[27] Seckler and Amarasinghe 2000; UNEP-DEWA 2003.
[28] Shah et al 2000.
[29] Sampat 2000.
[30] Postel 2000a.
[31] Shah et al 2000.
[32] Engelman and LeRoy 1993; Postel 2000a.
[33] Shah et al 2000.
[34] Shah et al 2000.
[35] Postel 2000a.
[36] Postel 2000a.
[37] Shah et al 2000; Rosegrant et al 2002.
[38] Engelman and LeRoy 1993.
[39] UNEP 2002.
[40] UNEP-DEWA 2003.
[41] Sampat 2000; Sampat and Peterson 2000; WRC 1998.
[42] NRE 1999.
[43] Sampat 2000; Lazaroff 2000.
[44] UNEP-DEWA 2003.
[45] Engelman and LeRoy 1993.
[46] Shah et al 2000.
[47] Shah et al 2000; Sampat 2000.
[48] Economy 1997.
[49] Sampat and Peterson 2000.
[50] Two per cent of seawater added to fresh water would increase its salinity to around 1,200 mg/l, which would make it marginally saline. To make it brackish, one would need to add between around 3–13 per cent of seawater. For the salinity scale, see Box A4-1 in Appendix A.

[51] Engelman and LeRoy 1993; Shiklomanov 2000.
[52] Gardiner 2002a.
[53] For a list of water-related diseases see Box A4-2 in Appendix A.
[54] Calculated using statistical tables of UNDP 2002 and the global population of 6 billion in 2000.
[55] UNDP 2002. See also Gardiner 2002a.
[56] The statistical information is for the year 2000. I have calculated the number of people using the percentage values listed by Gardiner 2002a and the regional populations in 2000.
[57] Larsen 2001a.
[58] Engelman and LeRoy 1993.
[59] Björklund et al 2000; GCI 1998.
[60] Wolf et al 1999. See also UNEP 2002.
[61] Keller et al 2000; Shah et al 2000.
[62] Paton and Davis 1996.
[63] Ellis 2003.
[64] Björklund et al 2000; Hinrichsen and Robey 2000; Shiklomanov 2000.
[65] Shiklomanov 2000.
[66] UNEP 2002.
[67] Seckler et al 2000.
[68] Seckler et al 2000 and Figure 1 in Seckler 2000.
[69] Björklund et al 2000.
[70] ACUNU 2000.

Chapter 5 The Destruction of the Atmosphere

[1] Gribbin and Gribbin 1996; Kiehl and Trenberth 1997; Lugens and Tarbuck 2000.
[2] McCarthy et al 2001; OSB et al 2002.
[3] Broecker 1997; Ganopolski and Rahmstorf 2001; McCarthy et al 2001; OSB et al 2002; Rahmstorf 1997,1999; Stommel 1961.
[4] Ozone produced at ground level by man-made pollution such as car exhaust fumes.
[5] CSCC 2001; Hiblin 2002; Houghton et al 2001.
[6] Grubb et al 1999; Oberthur et al 1999; UNFCCC 1997.
[7] CSCC 2001.
[8] CSCC 2001. Houghton et al 2001.
[9] CSCC 2001.
[10] Houghton et al 2001; Wigley 1999.
[11] Sturges et al 2000.
[12] Dunn 2001b; EIA 2000.
[13] Mitchell and Hulme 2000.
[14] IPCC 2001. IPCC was created in 1988 and is made of around 2,500 of the world's leading scientists.
[15] Marland et al 2002. The data include carbon emissions from fossil fuel burning, cement production, and gas flaring for 1751–1999.
[16] EIA 2000.
[17] Houghton et al 1995; Houghton et al 2001.
[18] Watson et al 2000.

[19] Houghton et al 2001; Kump 2002.
[20] Houghton et al 2001; Kump 2002.
[21] Dunn 2001 b.
[22] Houghton et al 2002.
[23] UNEP 2002.
[24] Hansen et al 2001; Houghton et al 2001.
[25] The average temperature between 1961 and 1990.
[26] Hansen et al 2001; Houghton et al 2001.
[27] ABS 2001.
[28] Levitus et al 2000.
[29] Jones et al 2001.
[30] Watson et al 1995.
[31] McCarthy et al 2001.
[32] Mitchell and Hulme 2000.
[33] Jones et al 2003.
[34] Baumert and Kete 2002.
[35] Mitchell and Hulme 2000.
[36] Depledge 2002; Gardiner 2002 b
[37] Raskin et al 2002; UNEP 2002.
[38] WCED 1987.
[39] Houghton et al 1990; IPCC 1990; Tegart et al 1990.
[40] Recommendations for sustainable development (UNCED 1992).
[41] These are the annual Conferences of the Parties to the United Nations Framework Convention on Climate Change identified as COP1, COP2, etc.
[42] Houghton et al 1995; IPCC 1995; Watson et al 1995.
[43] Haughton et al 2001; McCarthy et al 2001; Watson et al 2001.
[44] Ott 2001.
[45] Nordhaus et al 2000; Rosenzweig et al 2002.
[46] According to Rosenzweig et al 2002, the first trial of ideas that would eventually lead to various schemes of emission reduction commitments was carried out as early as in 1993 under the so-called Climate Change Action Plan (CCAP).
[47] UNFCCC 2002.
[48] POST 2002.
[49] Annex I countries are: Australia, Austria, Belgium, Bulgaria, Canada, Croatia, Czech Republic, Denmark, Estonia, European Community, Finland, France, Germany, Greece, Hungary, Iceland, Ireland, Italy, Japan, Latvia, Liechtenstein, Lithuania, Luxembourg, Monaco, the Netherlands, New Zealand, Norway, Poland, Portugal, Romania, Russia, Slovakia, Slovenia, Spain, Sweden, Switzerland, Ukraine, the UK, and the United States.
[50] Alcamo and Kreileman 1996.
[51] Blanchard 2002.
[52] Houghton et al 2001.
[53] Houghton et al 2001.
[54] ENS 2001, 2002b; Munich Re 2002.
[55] ABS 2001.
[56] Swiss Re 2000; Walker 2001.
[57] Munich Re data as listed by Dunn 2000a, but converted to 2001 US dollars.

58 Dlugolecki 2001.
59 Dunn 2000a, in 2001 US dollars.
60 Data from the Munich Re database as listed by Dunn 2000a, but converted to 2001 US dollars.
61 Data listed by Brown 2001a, but converted to 2001 US dollars.
62 Munich Re 2001, 2002.
63 Munich Re 2002.
64 Munich Re 2003
65 Salt 2001.
66 Houghton 2001; Levitus et al 2000.
67 Hinrichsen 2000; McCarthy et al 2001.
68 Mastny 2001; McCarthy et al 2001.
69 Mastny 2001.
70 Davis and Saldiva 1999; Moore 1997; UNEP 2002; Walker 2002.
71 The diameter of a human hair is about 100 micrometres.
72 WB 2003.
73 Piver et al 1999.
74 Bouwman and van Vuuren 1999.
75 Houghton 2001.
76 NRC 1983.
77 Bouwman and van Vuuren 1999.
78 Juberg 1997.
79 WHO 2002. See the definition of DALY's in Chapter 7 and Table A7-4.
80 Houghton et al 2001; McCarthy et al 2001; IPCC 2001.
81 WMO and UNEP 2003.
82 If the ozone layer is 300 Dobson Units thick in the stratosphere it would be 3 millimetres thick if it were brought to ground level because the atmospheric pressure at ground level is higher than the pressure in the stratosphere.
83 Fraser 1999.
84 Shindell et al 1998.
85 Nagashima 2002.
86 SAPMP 2002.
87 EIA 2000.
88 UNEP 2002.
89 Houghton et al 2001.
90 UNEP 2002.
91 Blanchard 2002
92 Houghton et al 2001; McCarthy et al 2001.
93 Houghton et al 2001.

Chapter 6 The Approaching Energy Crisis

1 For a brief guide to the units of energy, see Appendix D.
2 Dunn 2001a; EIA 2002a; Meadows et al 1992. See also EIA 2000, 2001 and BP 2001, 2002a.
3 One quintillion—that is, one billion billions joules.
4 EIA 2002a. See also BP 2002a.
5 Davis 1990; EIA 2002a; Harris 2001; IEA 2000, 2002. A range of values indicates differences in estimates by various sources.

[6] UNDP et al 2000a, Part V.

[7] Berger and Anderson 1992; Van Dyke 1997.

[8] Australia, Austria, Belgium, Canada, the Czech Republic, Denmark, Finland, France, Germany, Greece, Hungary, Iceland, Ireland, Italy, Japan, Luxembourg, Mexico, the Netherlands, New Zealand, Norway, Poland, Portugal, Republic of Korea, the Slovak Republic, Spain, Sweden, Switzerland, Turkey, the UK, and the US. The leading country in this group is the US.

[9] Campbell and Laherrere 1998.

[10] MacKenzie 2000; Rogner 2000.

[11] BP 2002a.

[12] OPEC (Oil Producing and Exporting Countries): Algeria, Indonesia, Iran, Iraq, Kuwait, Libya, Nigeria, Qatar, Saudi Arabia, the United Arab Emirates, and Venezuela. The leading country in this group is Saudi Arabia.

[13] Davis and Diegel 2002.

[14] Odell 1997.

[15] BP 2002a.

[16] Larsen 2001b.

[17] Sablani et al 2001.

[18] Paton and Davis 1996.

[19] BP 2002a.

[20] BP 2002a.

[21] Based mainly on the data of EIA 2002a, 2002c. See also UNDP et al 2000a, Part V.

[22] The production capacity for the generation of electricity is expressed in terms of power of electricity-generating plants, and is measured in watts. The consumption of electricity is measured in watt-hours.

[23] The US, the UK, Germany, France, Italy, Canada, Japan, and Russia.

[24] ENS 2002e.

[25] BRWM 2001.

[26] NEA 1999; Williams 2000.

[27] DOE 2002.

[28] O'Hanion 2000.

[29] EIA 2002a. See also BP 2002a, 2003, IEA 2002, and Lenssen 2001.

[30] EIA 2002a; Rogner 2000. See also BP 2002a and IEA 2002.

[31] IEA 2002.

[32] Rogner 2000.

[33] Rogner 2000.

[34] IEA 2002.

[35] BP 2002a.

[36] Unless otherwise indicated, the discussion of these resources of energy is based on Turkenburg 2000.

[37] Flavin 2001.

[38] The capacity factor is the fraction of the installed capacity converted into electricity.

[39] IEA 1995, 1996.

[40] Flavin 2001.

[41] EIA 2002a.

[42] North, South, and Central America, including the West Indies.

[43]   Björnsson et al 1998.
[44]   Rogner 2000.
[45]   Kemp-Benedict et al 2002.
[46]   Dunn 2000b; Williams 2000.
[47]   EIA 2002a.
[48]   EIA 2002a.
[49]   WB 1999b.
[50]   Renner 2000a.
[51]   Davis and Diegel 2002; EIA 2002a.
[52]   Renner 2000a.
[53]   EIA 2002a; UNEP 1999; WB 1999b.
[54]   EIA 2002a.
[55]   UNEP 2002.
[56]   EIA 2002a.
[57]   A self-sustaining fusion process.
[58]   Nakicenovic 2000; Nakicenovic et al 1998.

Chapter 7 Social Decline

[1]   Meadows et al 1992.
[2]   UNEP 2002.
[3]   UNDP 1999.
[4]   UNICEF 2000b.
[5]   UNDP 2002.
[6]   UNDP 2002
[7]   UNDP 2002, UNFPA 1999
[8]   UNDP 2002
[9]   This indicator measures the number of people living in absolute poverty (see the *Absolute poverty* section).
[10]   UNDP 2002.
[11]   Dowrick and Quiggin 1998; Halstead 1998.
[12]   Cobb et al 1999, 2001; Hamilton and Denniss 2000.
[13]   Cobb et al 2001.
[14]   Hamilton and Denniss 2000; Newton et al 2001.
[15]   Feldstein, M. 1998.
[16]   Decornez 1998, Harrison and Bluestone 1988; Kreml 1997; Kutner 1983; Milanovic 2002; Milanovic, and Yitzhaki 2002; Peterson 1997; Thurow 1984.
[17]   Adelman and Taft Morris 1967; Landes 1998: Tilkidjiev 1998; Wallace and Haerpfer 1998.
[18]   Aristotle, 306 BC.
[19]   Incorrectly attributed to George Pope Morris (1802–1864), the origin of this well-known statement can be traced to 1796. It appears to belong to Patrick Henry (1736–1799).
[20]   Milanovic and Yitzhaki 2002.
[21]   Bandourian et al 2002; Kakwani 1980.
[22]   McDonald and Xu 1995.
[23]   Bandourian et al 2002. The three simple models identified by these authors as giving the best fits to income distribution date are: the Weibull two-parameter distribution, the Dagum three-parameter distribution, and the second-kind

generalised four-parameter beta distribution.

24 Gini 1912; Kakwani 1980
25 A Lorenz curve gives a cumulative fraction of the population as a function of income below a given value (see Kakwani 1980).
26 Milanovic 1997.
27 The Commonwealth of Independent States was created in 1991, and is made of Azerbaijan, Armenia, Belarus, Georgia, Kazakhstan, Kyrgyzstan, Moldova, Russia, Tajikistan, Turkmenistan, Uzbekistan, and Ukraine.
28 UNDP 2001.
29 UNDP 2003.
30 UNDP 2003
31 Bandourian et al 2002.
32 Milanovic 2002.
33 Milanovic 1999.
34 Pareto 1896.
35 Bouchaud and Mézard 2000. See also Buchanan 2000 and Hayes 2002.
36 McDonald and Xu 1995.
37 Levy and Levy 2003.
38 FII 2002; UNDP 1999.
39 French 2000a.
40 Renner 2000b.
41 UNDP 1999.
42 UNDP 1999.
43 Chapman 1999; Chapman et al 1999; Halweil 1999.
44 UNDP 1999.
45 The UNIDO 2002 report does not specify the reference years for the currency values. However, even if we assume that the values are in current dollars and if we apply the relevant corrections for the conversion of the currencies, we shall find that between 1985 and 1998 the gap in industrial outputs by industrialised and developing countries increased by around $1,000 billion, or by around $340 per person.
46 FII 2002.
47 UNDP 1999.
48 UNDP et al 2000b; WB 1999b.
49 Defined as the consumption of goods and services, but excluding the costs of buying a house.
50 All the GDP values have been corrected for purchasing power parity.
51 UNDP 1999.
52 Abbreviated as PPP, purchasing power parity is a correction that represents the actual purchasing power of the US dollar in various countries of the world. See WB 1999b, Table 4.11.
53 UNDP 2002. See also UNDP 2003; UNEP 2002; WHO 2002.
54 UNDP 2003. See also UNDP 2002.
55 Two other poverty lines are $4 and $11 a day. There is also a national poverty line, which depends on how the authorities of various countries decide to define it (UNDP 2000).
56 Hinrichsen and Robey 2000.
57 UNDP 2003. See also UNDP 2002.

58 WHO 2002.
59 WHO 1999b, 2002.
60 UNDP et al 2000b; WHO 2002.
61 UNDP 2002, 2003.
62 Central and Eastern Europe and the Commonwealth of Independent States.
63 In 2001 US dollars, and corrected for purchasing power parity.
64 UNDP 2003.
65 UNDP 2002, 2003.
66 UNDP 2003. GDP values have been corrected for purchasing power parity and are expressed in terms of 2001 US dollars.
67 UNDP 2000. In 2000 US dollars. Corrected for purchasing power parity.
68 UNDP 1999.
69 UN 2000; UNDP 2002.
70 Sheehan 2000.
71 UNFPA 1999, 2001, 2002; Sheehan 2000.
72 UNFPA 1999 expects it in 2005.
73 This projection agrees well with the projection by UNFPA 1999.
74 Harrison and Pearce 2001.
75 Harrison and Pearce 2001; Munich Re 2001; UNFPA 1999, 2001.
76 Harrison and Pearce 2001; UNFPA 2001; WB 2003.
77 Munich Re 2001.
78 Otto et al 2002.
79 Cincotta and Engelman 2000;
80 ACUNU 2000; UNDP 1999.
81 The discussion of infectious diseases in this and the following sections are based mainly on Heymann 2000 and WHO 1999a and 2000.
82 UNICEF 2000a.
83 Halweil 2000; Hwang 2001.
84 Vallanjon 2002.
85 The calculated figures are for 2004.
86 Potts and Walsh 2003.
87 PRB 2002 estimates the number of people living with HIV/AIDS at between 30 and 50 million, and STBP 2001 at over 36 million. UNDP 2002 estimates the number of women (aged 15-49 years) and children (aged 0-14 years) living with HIV/AIDS at 21.5 million.
88 UNFPA 2001.
89 UNDP 2002.
90 UNICEF 1999b.
91 UNICEF 2000a.
92 UNICEF 2000b.
93 UNFPA 2000.
94 Potts and Walsh 2003.
95 Hotchkiss Mehta 2002; Mastny 2000b; Reichman and Hopkins Tanne 2001; UNDP 2002; WHO 2002.
96 The Western Pacific region is made of 37 countries: American Samoa, Australia, Brunei Darussalam, Cambodia, China, Cook Islands, Fiji, French Polynesia, Guam, Hong Kong (China), Japan, Kiribati, Lao People's Democratic Republic, Macao (China), Malaysia, Marshall Islands, Federated

States of Micronesia, Mongolia, Nauru, New Caledonia, New Zealand, Niue, Northern Mariana Islands, Palau, Papua New Guinea, Philippines, Pitcairn Islands, Republic of Korea, Samoa, Singapore, Solomon Islands, Tokelau, Tonga, Tuvalu, Vanuatu, Viet Nam, and Wallis and Futuna.

[97] By definition, the MDR-TB bacterium has to be resistant to at least two most powerful anti-TB drugs, isoniazid and rifampicin.

[98] Lee 2002.

[99] Vallanjon 2002.

[100] STBP 2002.

[101] UNDP 2002; Vallanjon 2002; WHO 1999a, 1999b.

[102] Four species of *Plasmodium* are responsible for inflicting malaria fever: *Plasmodium falciparum, P. vivax, P. ovale* and *P. malariae.*

[103] I list here only countries with 15,000 or more cases per 100,000 people.

[104] UNDP 2002; WHO 1999b.

[105] Pimentel et al 1998.

[106] WHO 1999a.

[107] Pimentel et al 1998; WHO 1999a.

[108] DALYs (disability-adjusted life years) are used to measure the loss of productive life years due to injury or disease associated with various risk factors (Gold et al 1996; Murray and Lopez 1996). The burden of a disease or a harmful environmental factor can be expressed either as the number of years or as the percentage of total DALYs.

[109] WHO 1999b.

[110] WHO 2002.

[111] Gardner 2000b; Harrison and Pearce 2001.

[112] French 2000a; Mastny 2000a.

[113] CDC 2002a, 2002b.

[114] Tsiodras et al 2001.

[115] Potoski et al 2002.

[116] Heymann 2000.

[117] WHO 2000, Chapter 3.

Chapter 8 Conflicts and Increasing Killing Power

[1] Hughes 1999.

[2] Hughes 1999.

[3] Shinseki 1999

[4] Bush 1991.

[5] Dubik 2000.

[6] Filiberti 2001.

[7] Arthur T. Hadley, writer, journalist, and press secretary to three democratic presidential candidates, member of the Advisory Board of the International Security, University of Yale. Quoted by Hughes 1999.

[8] President Gerald Ford.

[9] President George Washington.

[10] CDI 1999a; Cordesman 1999; SIPRI 2003a and 2003b.

[11] I have converted all military expenditures to the constant currency of 2004 US dollars.

[12] Eisenhower 1953.

13 Berry et al 2003; EOPUS 2002, 2003a and 2003b.
14 NS 2001a.
15 Stix 2002.
16 As mentioned earlier, all military expenditures in this chapter are in the constant currency of 2004 US dollars.
17 Nordhous 2002
18 Cordesman 1999; SIPRI 2003a and 2003b.
19 Berry et al 2003.
20 KOSIMO database.
21 HIIK nd; 2004.
22 HIIK nd.
23 UNDP 2002.
24 CDI 1999c; UNDP 2002; UNICEF 1999a
25 Renner 2001a.
26 HIIK 2004.
27 UNEP 2002.
28 ACUNU 2000.
29 UNDP 2002.
30 In God We Trust.
31 UNDP 2002.
32 UNDP 2002. All the GDP values discussed in this section are corrected for purchasing power parity (PPP) and arc expressed in 2000 US dollars.
33 WB 2001.
34 Zuckerman 1996.
35 Norris and Arkin 2000.
36 Norris and Arkin 2000.
37 Carreño 2001.
38 Berry et al 2003.
39 Renner 2001b.
40 Lawler and Owens 1991.
41 CIFPR 2001; EOPUS 2002.
42 Vallanjon 2002. It is estimated that by 2007 about $15 billion per year will be needed to combat AIDS (Pegg 2003).
43 Cordesman 1997.
44 Cole 2002; Cordesman 1997.
45 A genus of fungi.
46 A genus of bacteria.
47 Boutwell and Klare 1999.
48 CDI 1999b.
49 UNDP 2002.
50 CDI 1999b.
51 ICBL 2002; Strada 2002.
52 Kopp 1996.
53 NS 2001b.
54 Edwards 2001.
55 MacKenzie 2001a, 2001b.
56 POGO 2001.
57 POGO 2002.

[58] UNEP 2002.
[59] Raskin et al 2002.
[60] Oberg 2001; Stix 2002.
[61] Albert Einstein.

Chapter 9 In a Nutshell
[1] For the definitions of water availability, see Appendix B.
[2] All values are in 2001 US dollars.
[3] In 2004 US dollars.

Chapter 10 Landmarks of Progress
[1] See Chapter 2.
[2] See Chapter 3.
[3] See Chapter 3.
[4] See Chapter 4.
[5] Ivanhoe 1997.
[6] See Chapter 3.
[7] Campbell and Laherrere 1998; Masters et al 1994.
[8] Masters et al 1994.
[9] Grainger and Garcia 1966.
[10] MacKenzie 2000.
[11] BP 2002b.
[12] EIA 2002b.
[13] BP 2002b.
[14] Wackernagel et al 2002. See also Chapter 3.
[15] Gardner 2000c.
[16] Gardner 2000a.
[17] WB 1999b.
[18] Dyson 1999, p. 5933.
[19] Brown 2000.
[20] Dyson 1999.
[21] Rosegrant et al 2002.
[22] Brown 2000.
[23] EIA 2001.
[24] EIA 2001.
[25] EIA 2001, 2003.
[26] BP 2002b.
[27] BP 2002b.
[28] Watson and Pauly 2001. See also Dayton et al 2002.
[29] Allen 2000.
[30] ENS 2003a.
[31] Pearce 2002.
[32] Grainger and Garcia 1996.
[33] See Chapter 6.
[34] BP 2002a
[35] BP 2002b.
[36] BP 2002b.
[37] Ivanhoe 1997.

[38] Burtlett 2000; Hubbert 1974, 1982.
[39] Campbell and Laherrere 1998; Ivanhoe 1996; Laherrere 1999.
[40] EIA 2002a.
[41] Grainger and Garcia 1996.
[42] Duncan 2001.
[43] See Chapter 3.
[44] See Chapter 7.
[45] Heymann 2000.
[46] Walsh 2000.
[47] BP 2002b.
[48] See Chapter 5.
[49] Albert Einstein, *What I Believe*, 1930.

References

ABS 2000, *Water Account for Australia 1993–94 to 1996–97*, ABS Catalogue No. 4610.0, ABS, Canberra, Australia.

ABS 2001, *2001 Year Book Australia*, ABS, Canberra, ACT, Australia.

Acosta, R., Allen, M., Cherian, A., Granich, S., Mintzer, I., Suarez, A. and von Hippel, D. 1999, *Climate Change*, Information Sheets, No. 30, UNEP, http://www.unep.ch/iuc/

ACUNU 2000, *Global Challenges for Humanity: UN Millennium Summit and Forum—Special Edition*, ACUNU, Washington, DC.

Adelman, I. and Taft Morris, C. 1967, *Society, Politics, and Economic Development: A Quantitative Approach*, Johns Hopkins Press, Baltimore.

Alcamo, J. and Kreileman, E. 1996, 'Emission Scenarios and Global Climate Protection', *Global Environmental Change*, **6**(4): 305–334.

Alexandratos, N. 1999, 'World Food and Agriculture: Outlook for the Medium and Longer Term', in *Colloquium on Plants and Population: Is There Time? Proceedings of the National Academy of Sciences of the United States of America (NYAS)*, Vol. 96, NYAS, Washington DC, pp. 5908–5914.

Allan, J. A. 1997, 'Virtual Water: A Long-Term Solution for Water-Short Middle Eastern Economies', paper presented at the 1997 British Association Festival of Science, University of Leeds, Water and Development Session, School of Oriental and African Studies, University of London.

Allen, D. 2000, *World Fisheries: Declines, Potential, and Human Reliance*, http://www.earthscape.org/t1/ald02/ald02.html

Al-Saleh, I. A. 1994, 'Pesticides: A Review Article', *J Environ. Pathol. Toxicol. Oncol.*, **13**(3): 151–161.

Anderson, A., Douglas, K., Holmes, B., Lawton, G., and Webb, J. (eds) 2001, 'Judgment Day: There Are Only Angels and Devils', *Global Environment Supplement, New Scientist*, 28 April, No. 2288.

Aslam, M. A. 2002, 'Equal per Capita Entailments: A Key to Global Participation on Climate Change?', in *Building on the Kyoto Protocol: Options for Protecting the Climate*, ed. K. A. Baumert with O. Blanchard, S. Llosa, and J.F. Perkaus, WRI, US, pp. 175–202.

Balmford, A., Bruner, A., Cooper, P., Costanza, R., Farber, S., Green, R. E., Jenkins, M., Jefferiss, P., Jessamy, V., Madden, J., Munro, K., Myers, N., Naeem, S., Paavola, J., Rayment, M., Rosendo, S., Roughgarden, J., Trumper, K., and Turner, R. K. 2002, 'Economic Reasons for Conserving Wild Nature', *Science*, **297**: 950–953.

Bandourian, R., McDonald, J. B., and Turley, R. S. 2002, *A Comparison of Parametric Models of Income Distribution across Countries and over Time*, Luxemburg Income Study Working Paper No 305, Syracuse University, Maxwell School, Syracuse.

Barbier, E. B. 1999, *The Economics of Land Degradation and Rural Poverty Linkages in Africa*, (UNU/INRA Annual Lectures 1998, Accra, Ghana), The United Nations University, Tokyo, Japan.

Bartlett, A. A. 2000, 'An Analysis of US and World Oil Production Patterns Using Hubbert-style Curves,' *Mathematical Geology* **32**(1): 1–17.

Baumert K. A. and Kete, N. 2002, 'Introduction: An Architecture for Climate Protection', in *Building on the Kyoto Protocol: Options for Protecting the Climate*, ed. K. A. Baumert with O. Blanchard, S. Llosa, and J.F. Perkaus, WRI, US, pp. 1–30

Bazerman, M. H. and Hoffman, A. J. 1999, 'Sources of Environmental Destructive Behavior: Individual, Organizational, and Institutional Perspectives', *Research in Organizational Behavior*, **21**: 39–79.

Berger, B. D. and Anderson, K. E. 1992, *Modern Petroleum: A Basic Primer of the Industry*, 3rd edn, PennWell Corporation, Tulsa, OK, US.

Berghuis, A. 2003, *Antibiotic Resistance*, McGill University, Montreal, Canada, unpublished.

Berry, N., Corbin, M., Hellman, C., Smith, D., Stohl, R., and Valasek, T. 2003, *Military Almanac 2001–2002*, CDI, Washington, DC.

Betarbet, R., Sherer, T. B., MacKenzie, G., Garcia-Osuna, M., Panov, A. V., and Greenamyre, J. T. 2000, 'Chronic Systemic Exposure Reproduces Features of Parkinson's Disease', *Nature Neuroscience*, **3**(12): 1301–1306.

Björklund, G., Ehlin, U., Falkenmark, M. Lundqvist, J., Swain, A., Röckström, J., and Brismar, A. 2000, *Water Development in the Developing Countries*, European Parliament, Luxembourg.

Björnsson, J., Fridleifsson, I. B., Helgason, T., Jonatansson, H., Mariusson, J. M., Palmason, G., Stefansson, V., and Thorsteinsson, L. 1998, 'The Potential Role of Geothermal Energy and Hydro Power in the World Energy Scenario in Year 2020', in *Proceedings of the 17th WEC Congress*. Houston, Texas, Cambridge University Press, Cambridge, UK.

Blanchard, O. 2002, 'Scenarios for Differentiating Commitments: A Quantitative Analysis', in *Building on the Kyoto Protocol: Options for Protecting the Climate*, ed. K. A. Baumert with O. Blanchard, S. Llosa, and J.F. Perkaus, WRI, US, pp. 203–222.

Bongaarts, J. and Bulatao, R. A. (eds) 2000, *Beyond Six Billion: Forecasting the World's Population*, National Academy Press, Washington, DC.

Bouchaud, J-P. and Mézard, M. 2000, 'Wealth Condensation in a Simple Model Economy', *Physica A*, 282: 536–545.

Boutwell, J and Klare, M. T. (eds) 1999, *Light Weapons and Civil Conflict: Controlling the Tools of Violence*, Rowman and Littlefield Publishers, Inc., Lanham, MD.

Bouwman, A. F. and van Vuuren, D. P. 1999, *Global Assessment of Acidification and Eutrophication of Natural Ecosystems,* UNEDP, Nairobi, Kenya (UNEDP/DEIA&EW/TR.99–6) and RIVM, Bilthoven, The Netherelands (RIVM 402001012).

BP 2001, *BP Statistical Review of World Energy: June 2001*, Group Media and Publications, London, UK.

BP 2002a, *BP Statistical Review of World Energy: June 2002*, Group Media and Publications, London, UK.

BP 2002b, *BP Statistical Review of World Energy: June 2002,* (Workbook), http://www.bp.com/centres/energy2002/

Broecker, W.S., 1997, 'Thermohaline Circulation, the Achilles Heel of our Climate System: Will Man-made CO_2 Upset the Current Balance?' *Science*, **278**: 1582–1588.

Brown, H. 1954, *The Challenge of Man's Future,* Viking Press, New York.

Brown, L. R. 2000, 'Grain Harvest Falls', in *Vital Signs 2000: The Environmental Trends that are Shaping Our Future*, ed. L. Starke, WWI, W. W. Norton and Company, New York, pp. 34, 35.

Brown, L. R. 2001a, 'World Economy Expands', in *Vital Signs 2001: The Trends that are Shaping Our Future*, ed. L. Starke, WWI, in cooperation with UNEP, W. W. Norton and Company, New York, pp. 56, 57.

Brown, L. R. 2001b, 'Fertilizer Use Rises', in *Vital Signs 2001: The Trends that are Shaping Our Future*, ed. L. Starke, WWI, in cooperation with UNEP, W. W. Norton and Company, New York, pp. 32, 33.

BRWM 2001, *Disposition of High-level Waste and Spent Nuclear Fuel: The Continuing Societal and Technical Challenges*, National Academy Press, Washington, DC.

Bryant, D., Nielsen, D., and Tangley, L. 1997, *Last Frontier Forests: Ecosystems and Economics on the Edge*, WRI, Washington, DC.

Buchanan, M. 2000, 'That's the Way the Money Goes', *New Scientists*, 19 August, No2252: 22–26.

Buranatrevedh. S., and Roy, D. 2001, 'Occupational Exposure to Endocrine-disrupting Pesticides and the Potential for Developing Hormonal Cancers', *J Environ Health*, **64**(3): 17–29.

Buringh, P. 1989, 'Availability of Agricultural Land for Crop and Livestock Production', in *Food and Natural Resources*, eds D. Pimentel and C.W. Hall, Academic Press, San Diego, pp. 69–83.

Bush, G.1991, 'In Defense of Defense', President George Bush's Speech to the Aspen Institute Symposium, 2 August 1990, in *Fundamentals of Force Planning, Vol. II: Defense Planning Cases*, ed. The Force Planning Faculty Naval War College, R. M. Lloyd, H. C. Bartlett, G. P. Holman, Jr, J. M. Kirby, T. Lawler, W. J. Neville, M. T. Owens, Jr, A. L. Ross, and T. E. Somes, Naval War College Press, Newport, RI, pp. 9–14.

Caldwell, L. K. 1999, 'Is Humanity Destined to Self-destruct?', *Politics and Life Sciences*, 18: 3–14.

Campbell, C. J. and Laherrere, J.H. 1998, 'The End of Cheap Oil', *Scientific American*, **278**(3): 78–83.

Carreño, A. L. C. de 2001, 'A Visit to the Dark Side', *Academy Update: New York Academy of Sciences Member Newsletter*, September, p. 7.

CDC 2002a, 'Vancomycin-resistant Staphylococcus Aureus—Pennsylvania, 2002', *Morbidity and Mortality Weekly Report* , **51**: 902.

CDC 2002b, 'Staphylococcus Aureus Resistant to Vancomycin—United States, 2002', *Morbidity and Mortality Weekly Report,* **51**: 565–567.

CDI 1999a, 'The Fiscal Year 2000 Military Budget', *The Defense Monitor* **28**(1): 1–5.

CDI 1999b, 'Small Arms: A Post-Cold War Disarmament Challenge', *The Defense Monitor* **28**(4): 1–3.

CDI 1999c, 'The Smallest Warriors: Child Soldiers', *The Defense Monitor* **28**(6): 1, 4–6.

Chapman, A. R. (ed.) 1999, *Perspectives on Genetic Patenting: Religion, Science and Industry in Dialogue*, AAAS, Washington, DC.

Chapman, A. R., Frankel, M. S., and Garfinkel, M. S. 1999, *Stem Sell Research and Applications: Monitoring the Frontiers of Biomedical Research*, AAAS and ICS, Washington, DC.

CIFPR 2001, *Analysis of the President's FY 2002 Budget*, California Institute Special Report, 28 February, CIFPR, Washington, DC.

Cincotta, R. P. and Engelman, R. 2000, *Nature's Place: Human Population and the Future of Biological Diversity*, PAI, Washington, DC.

Clarke R., Lawrence A. R and Foster S. S. D. 1996, *Groundwater: A Threatened Resource*, UNEP/GEMS Environment Library no 15, UNEP, Nairobi, Kenya.

Clement, J., Sigford, A., Drummond, R., and Novy, N. 1997, *World of Fresh Water: A Resource for Studying Issues of Freshwater Research*, EPA, Duluth, Minnesota.

Clinton, W. J. 2000, 'Armed Forces Day 2000', President Bill Clinton's address to Armed Forces, The White House, Washington, DC *Defense Link*, USDD, http://www.defenselink.mil/specials/outreachpublic/clintonafd.html

Cobb, C., Glickman, M., and Cheslog, C. 2001, *The Genuine Progress Indicator: 2000 Update*, Redefining Progress, San Francisco, CA.

Cobb, C., Goodman, G. S., and Wackernagel, M. 1999, *Why Bigger Isn't Better: The Genuine Progress Indicator*, Redefining Progress, San Francisco, CA.

Cohen, J. E. 1995, *How Many People Can the Earth Support?* W.W. Norton and Co., New York.

Cole, L. A. 2002, 'The Specter of Biological Weapons', in *The Science of War: Weapons*, Scientific American, Special Issue No 1, February, pp. 17–21.

Cordesman, A. H. 1997, *Terrorism and the Threat from Weapons of Mass Destruction in the Middle East*, CSIS, Washington, DC.

Cordesman, A. H. 1999, *Trends in Western Defense on the Edge of the New Millennium: A Comprehensive Summary of Military Expenditure; Manpower; Land, Air, Naval, and Nuclear Forces, and Arms Sales*, CSIS, Washington DC.

Costanza, d'Arge, R., de Grott, R., Farber, S., Grasso, M., Hannon, B., Limburg, K., Naeem, S., O'Neill, R. V., Paruelo, J., Ruskin, R. G., Sutton, P., and van den Belt, M. 1997, 'The Value of the World's Ecosystem Services and Natural Capital', *Nature*, **387**: 253–260.

Cracraft, J. and Grifo, F. T. (eds) 1999, *The Living Planet in Crisis: Biodiversity Science and Policy*, Columbia University Press, New York.

Crutzen, P. J. and Stoermer, E. F. 2000: 'The "Anthropocene"', *IGBP Newsletter*, **41**, 17–18.

CSCC 2001, *Climate Change Science: An Analysis of Some Key Questions*, National Academy Press, Washington, DC.

Daniels, J. L., Olshan, A. F., and Savitz, D. A., 1997, 'Pesticides and Childhood Cancers', *Environ Health Perspect.*,**105**(10): 1068–1077.

Davis, D. L. and Saldiva, H. N. 1999, *Urban Air Pollution Risks to Children: A Global Environmental Health Indicator*, Environmental Health Notes, WRI, Washington, DC.

Davis, G. R. 1990, 'Energy for Planet Earth', *Scientific American*, **263**(3): 55–62.

Davis, S. C. and Diegel, S. W. 2002, *Transportation Energy Data Book: Edition 22*, ORNL–6967, Oak Ridge National Laboratory, Oak Ridge.

Dayton, P. K., Thrush, S., and Coleman, F. C. 2002, *Ecological Effects of Fishing in Marine Ecosystems of the United States*, POC, Arlington, Virginia.

Decornez, S. S. 1998, *An Empirical Analysis of the American Middle Class (1968–1992)*, Ph.D. Dissertation, Vanderbilt University.

Depledge, J. 2002, 'Continuing Kyoto: Extending Absolute Emission Caps to Developing Countries', in *Building on the Kyoto Protocol: Options for Protecting the Climate*, ed. K. A. Baumert with O. Blanchard, S. Llosa, and J.F. Perkaus, WRI, US, pp. 31–60.

Dlugolecki, A. 2001, 'General Strategies', in *Research Report: Climate Change and Insurance*, CII Research Division, London, pp. 88–96.

DOE 2002, *Yucca Mountain Science and Engineering Report: Technical Information Supporting Site Recommendation Consideration*, DOE/RW-0539–1.

Dowrick, S. D. and John Quiggin, J. 1998, 'Measures of Economic Activity and Welfare: The Uses and Abuses of GDP', in *Measuring Progress: Is Life Getting Better?*, ed. R. Eckersley, CSIRO Publishing, Collingwood, Australia

Dubik, J. 2000, 'IBCT at Fort Lewis', Major General James Dubik's address to students attending Fort Leavenworthís Combined Arms and Services Staff School and School for Advanced Military Studies, *Military Review*, September/October, pp. 17–23.

Duncan, R. C. 2001, 'World Energy Production, Population Growth, and the Road to the Olduvai Gorge', *Population and Environment*, 22(5): 503–522.

Dunn, S. 2000a, 'Weather Damages Drop', in *Vital Signs 2000: The Environmental Trends that are Shaping Our Future*, ed. L. Starke, WWI, W. W. Norton and Company, New York, pp. 76, 77.

Dunn, S. 2000b, *The Hydrogen Experiment*, WWI, Washington, DC.

Dunn, S. 2001a, 'Fossil Fuel Use Falls Again', in *Vital Signs 2001: The Trends that are Shaping Our Future*, ed. L. Starke, WWI, in cooperation with UNEP, W. W. Norton and Company, New York, pp. 40, 41.

Dunn, S. 2001b, 'Carbon Emissions Continue Decline', in *Vital Signs 2001: The Trends that are Shaping Our Future*, ed. L. Starke, WWI, in cooperation with UNEP, W. W. Norton and Company, New York, pp. 52, 53.

Durand, J. D. 1974, *Historical Estimates of World Population: An Evaluation*, Analytical and Technical Reports, Number 10, University of Pennsylvania, Population Center.

Dyson, T. 1999, 'World Food Trends and Prospects to 2025', in *Colloquium on Plants and Population: Is There Time? Proceedings of the National Academy of Sciences of the United States of America (NYAS)*, Vol. 96, NYAS, Washington DC, pp. 5929–5936.

Dziegielewski, B. 2000, 'Efficient and Inefficient Uses of Water in North American Households', in *Proceedings of IWRA's Xth World Water Congress Held in Melbourne, Australia 12–16 March 2000*, 12–16 March, IWRA, Carbondale, IL. US.

Economy, E. 1997, *The Case Study of China—Reforms and Resources: The Implications for State Capacity in the PRC*, Occasional Paper, Project on Environmental Scarcities, State Capacity, and Civil Violence, American Academy of Arts and Sciences and the University of Toronto, Cambridge, MA, US.

Edwards, R. 2001, 'Plutonium for Sale: With Nuclear Smuggling on the Increase,

How Long before a Terrorist Builds a Bomb?, *New Scientist*, 26 May, No. 2292: 10 and 11.

Ehrlich, P. R. 1968, *The Population Bomb,* Ballantine Books, New York.

EIA 2000, *International Energy Outlook 2000*, EIA, Office of Integrated Analysis and Forecasting, US Department of Energy, Washington, DC.

EIA 2001, *Annual Energy Outlook 2000: With Projections to 2020*, EIA, Office of Integrated Analysis and Forecasting, US Department of Energy, Washington, DC.

EIA 2002a, *International Energy Outlook 2002*, EIA, Office of Integrated Analysis and Forecasting, US Department of Energy, Washington, DC.

EIA 2002b, U.S. *Crude Oil, Natural Gas, and Natural Gas Liquids Reserves: 2001 Annual Report*, EIA, Office of Oil and Gas, Washington, DC.

EIA 2002c, *Annual Energy Review 2001*, EIA, Office of Energy Markets and End Use, Department of Energy, Washington, DC.

EIA 2003, *Annual Energy Outlook 2003: With Projections to 2025*, EIA, Office of Integrated Analysis and Forecasting, US Department of Energy, Washington, DC.

Eisenhower, D. D. 1953, 'The Chance for Peace', President Dwight D. Eisenhower's address delivered before the American Society of Newspaper Editors, *The Dwight D. Eisenhower Library*, http://www.eisenhower.utexas.edu/chance.htm

Ellis, C. 2003, 'Hot Mist Strips Salt from the Sea', *New Scientist*, 12 July, No. 2403: 15.

Elwell, H.A. 1985, 'An Assessment of Soil Erosion in Zimbabwe', *Zimbabwe Science News* **19**: 27–31.

Engelman, R, Cincotta R. P., Dye, B. Gardner-Outlaw, T., and Wisnewski, J 2000, *People in the Balance: Population and the Natural Resources at the Turn of the Millennium*, Population Action International, Washington, DC.

Engelman, R. and LeRoy, P. 1993, *Sustaining Water*, Population Action International, Washington, DC.

ENS 2001, '2001 the Second Warmest Year on Record', *Environmental News Service*, 18 December, http://ens-news.com/ens/dec2001/ 2001L-12-18-04.html

ENS 2002a, 'Pollution Kills Thousand of Children', *Environmental News Service*, 10 May, http://ens-news.com/ens/may2002/ 2002L-05-10-06.html

ENS 2002b, '2002 Confirmed as 2nd Warmest Year on Record', *Environmental News Service*, 18 December, http://ens-news.com/ens/dec2002/2002-12-18-01.asp

ENS 2002c, 'Soil's Tiniest Organisms Could Solve Huge Problems', *Environmental News Service*, 29 November, http://ens-news.com/ens/nov2002/2002-11-29-02.asp

ENS 2002d, 'Myth of World Forest Cover Shattered', *Environmental News Service*, 4 April, http://ens-news.com/ens/apr2002/2002L-04-04-02.asp

ENS 2002e, 'Oil Supply, Nuclear Fusion Occupy G-8 Energy Ministers', *Environmental News Service*, 6 May, http://ens-news.com/ens/may2002/2002L-05-06-04.asp

ENS 2003a, 'Canada Lists Atlantic Cod as Endangered', *Environmental News Service*, 2 May, http://ens-news.com/ens/may2003/2003-05-02-03.asp

ENS 2003b, 'Satellite View Shows Amazon Rainforest Shrinking Fast', *Environmental News Service,* 27 June, http://ens-news.com/ens/june2003/2003-06-27-03.asp

EOPUS 2002, *Budget of the United States Government: Fiscal Year 2003,* EOPUS, Office of Management and Budget, The White House, US Government Printing Office, Washington, DC.

EOPUS 2003a, *Budget of the United States Government: Fiscal Year 2004,* EOPUS, Office of Management and Budget, The White House, US Government Printing Office, Washington, DC.

EOPUS 2003b, *Budget of the United States Government: Fiscal Year 2004: Historical Tables,* EOPUS, Office of Management and Budget, The White House, US Government Printing Office, Washington, DC.

Erwin, T. L. 1982, 'Tropical Forests: Their Richness in Coleoptera and Other Arthropod Species', *Coleoptera Bulletin,* **36**: 74–75

Falkenmark, M. 1989, 'The Massive Water Scarcity now Threatening Africa: Why Isn't it Being Addressed', *Ambio,* 18(2):112–118.

FAO 2000a, *The State of Food Insecurity in the World,* FAO, Rome, Italy.

FAO 2000b, *Global Forest Resources Assessment 2000: Main Report,* FAO Forestry Paper 140, FAO, Rome.

Feldstain, M. 1998, *Income and Poverty,* National Bureau of Economic Research Working Paper 6770, RePEc:nbr:nberwo:6770.

FII 2002, *Finance and Insurance: Industry as a Partner for Sustainable Development,* FII and UNEP, London, UK.

Filiberti, E. J. 2001, 'Chapter 1: Introduction', in *How the Army Runs: A Senior Leader Reference Handbook 2001–2002,* US Army War College, Carlisle, PA.

Fischer, G. and Heilig, G. K. 1998. *Population Momentum and the Demand on Land and Water Resources,* Report IIASA-RR-98-1, IIASA, Laxenburg, Austria.

Flavin, C. 2001, 'Wind Energy Growth Continues', in *Vital Signs 2001: The Trends that are Shaping Our Future,* ed. L. Starke, WWI, in cooperation with UNEP, W. W. Norton and Company, New York, pp. 44, 45.

Follett, R. F., and B. A. Stewart 1985, *Soil Erosion and Crop Productivity,* ASA, CSSA, Madison, WI.

Fraser, P. 1999, *Ozone Layer Damage to Continue until at Least 2050,* CSIRO, Atmospheric Research, http://www.dar.csiro.au/info/releases/1999/MR_9911.htm

Freckman, D. W. (ed.) 1994, *Life in the Soil—Soil Biodiversity: Its Importance to Ecosystem Process,* Report of a Workshop Held at the Natural History Museum, London, England, 30 Augus–1 September, College of Natural Resources, Colorado State University, Fort Collins, CO, US.

French, H. 2000a, *Vanishing Borders: Protecting the Planet in the Age of Globalization,* W. W. Norton and Company, New York.

French, H. 2000b, 'Coping with Ecological Globalisation', in *State of the World 2000,* ed. L. Stark, W. W. Norton and Company, New York, pp. 184–202.

Ganopolski, A. and Rahmstorf, S. 2001, 'Rapid Changes of Glacial Climate Simulated in a Coupled Climate Model', *Nature* **409**:153–158.

Gardiner, R. 2002a, *Freshwater: A Global Crisis of Water Security and Basic Water Provision,* freshwater briefing paper, Towards Earth Summit 2002,

Environment Briefing No 1, UNED Forum, London.

Gardiner, R. 2002b, *Earth Summit 2002 Explained*, Earth Summit 2002 Briefing Paper, Towards Earth Summit 2002, Stakeholder Forum for Our Common Future, London.

Gardner, G. 2000a, 'Grain Area Shrinks Again', in *Vital Signs 2000: The Environmental Trends that are Shaping Our Future*, ed. L. Starke, WWI, W. W. Norton and Company, New York, pp. 44, 45.

Gardner, G. 2000b, 'Prison Population Exploding', in *Vital Signs 2000: The Environmental Trends that are Shaping Our Future*, ed. L. Starke, WWI, W. W. Norton and Company, New York, pp. 150, 151.

Gardner, G. 2000c, 'Fish Harvest Down', in *Vital Signs 2000: The Environmental Trends that are Shaping Our Future*, ed. L. Starke, WWI, W. W. Norton and Company, New York, pp. 40, 41.

GCI 1998, *Averting a Water Crisis in the Middle East: Make Water a Medium of Cooperation Rather Than Conflict*, GCI, Geneva, Switzerland.

Ghassemi, F. Jakeman, A. J., and Nix, H. A. 1995, *Salinisation of Land and Water Resources*, University of New South Wales Press Ltd, Sydney.

Gilman, R. 1992, 'No Simple Answers', *In Context*, No 31: 10–13.

Gini, C. 1912, 'Variabilita e Mutabilita', *Studio Economicogiuridici*, Universita di Cagliari, Anno III, Parte 2a. Reprinted in Gini (1955).

Giampietro, M. and D. Pimentel 1994, 'Energy Utilization', in *Encyclopedia of Agricultural Science*, eds C.J. Arntzen and E.M. Ritter, Academic Press. San Diego, CA, pp. 73–76.

Gleick, P. H. 1993, *Water in Crisis: A Guide to the World's Freshwater Resources*, Oxford University Press, New York.

Gleick, P. H. 1996, 'Basik Water Requirements for Human Activities: Meeting Basic Needs', *Water International*, 21: 83–92.

Gleick, P. H., Singh, A., and Shi, H. 2001, *Threats to the World's Freshwater Resources*, PISDES in cooperation with UNEP, Oakland, California.

Gold, M. R., Siegel, J. E., Russel, L. B., Weinstein, M. (eds) 1996, *Cost-effectiveness in Health and Medicine*, Oxford University Press, New York.

Graham, S. 2002, 'Global Estimates of Endangered Plant Species Triples', *Scientific American, News*, http://www.sciam.com/article. cfm?articleID=000E29B2-967D-1DC1-94E2809EC5880108

Grainger, R. J. R. and Garcia, S. M. 1996, *Chronicles of Marine Fishery Landings (1950–1994)*, FAO Fisheries Technical Paper 359, FAO, Rome.

Gribbin, J. and Gribbin, M. 1996, *The Greenhouse Effect*, New Scientist Inside Science No. 92.

Grubb, M., Vrolijk, C. and Brack. D. 1999. *The Kyoto Protocol: A Guide and Assessment*. RIIAEEP, London.

Halstead, T. 1998, 'The Science and Politics of New Measures of Progress: A United States Perspective', in *Measuring Progress: Is Life Getting Better?*, ed. R. Eckersley, CSIRO Publishing, Collingwood, Australia

Halweil, B. 1999, *The Emperor's New Crops*, Worldwatch Institute, Washington, DC.

Halweil, B. 2000, 'HIV/AIDS Pandemic Hits Africa Hardest', in *Vital Signs 2000: The Environmental Trends that are Shaping Our Future*, ed. L. Starke, WWI, W. W. Norton and Company, New York, pp. 100, 101.

Halweil, B. 2001, 'Farm Animal Population Soar', in *Vital Signs 2001: The Trends that are Shaping Our Future*, ed. L. Starke, WWI, in cooperation with UNEP, W. W. Norton and Company, New York, pp. 100, 101.

Hamilton, C. and Denniss, R. 2000, *Tracking Well-being in Australia: The Genuine Progress Indicator 2000*, The Australia Institute, Canberra, ACT, Australia.

Hansen, J.E., Ruedy, R., Sato, Mki., Imhoff, M., Lawrence, W., Easterling, D., Peterson, T., and Karl, T. 2001. 'A Closer Look at United States and Global Surface Temperature Change', *J. Geophys. Res.* **106**: 23947–23963.

Harrar, J. G. 1970, 'Plant Pathology and World Food Problems', *Persp. Biol. Med.* **13**, 583–596.

Harris, J. M. 2001, *The Economics of Energy*, Tufts University, Medford, MA, United States.

Harrison, B. and Bluestone, B.1988, *The Great U-turn: Corporate Restructuring and the Polarizing of America*, Basic Books, New York.

Harrison, P. and Pearce, F. 2001, *AAAS Atlas of Population & Environment*, AAAS, Washington, DC.

Haub, C. 1995, 'How Many People Have Ever Lived on Earth?' *Population Today*, February, p. 5

Hayes, B. 2002, 'Follow the Money', *American Scientists*, **90**(5): 400–405.

Heymann, D. L. 2000, *The Urgency of a Massive Effort Against Infectious Diseases,* Statement by Dr. David L. Heymann Executive Director for Communicable Diseases, World Health Organization, before the Committee on International Relations, U.S. House of Representatives, 29 June, http://www.winterthurhealthforum.ch/PDF/testmo.pdf

Hiblin, B. 2002, *Climate Change and Energy: Can we Weather the Switch to Sustainable Energy?* Climate Change and Energy Briefing Paper, Towards Earth Summit 2002, Environment Briefing No. 2, UNED Forum, London.

HIIK nd, *Database KOSIMO 1945–1999*, http://www.hiik.de

HIIK 2004, *Konfliktbarometer 2003: 12. Järliche Konfliktanalyse*, Heidelberger Institut für Internationale Konfliktforschung e.V. am Institut für Politische Wissenschaft der Universität Heidelberg.

Hinrichsen, D 1997, *Winning the Food Race,* Population Reports **25**(4) December: M (13), Population Information Program, Center for Communication Programs, JHSPH, Baltimore, Maryland.

Hinrichsen, D. 2000, *The Oceans are Coming Ashore*, Worldwatch Institute, Washington, DC.

Hinrichsen, D. and Robey, B. 2000, *Population and the Environment: the Global Challenge* Population Reports **28**(3): M(15), Population Information Program, Center for Communication Programs, JHSPH, Baltimore, Maryland.

Holt, M. S. 2000, 'Sources of Chemical Contaminants and Routes into the Freshwater Environment', *Food and Chemical Toxicology*, **38**, Supplement 1: S21–S27.

Hopenhayn-Rich, C., Stump, M. L., Browning, S. R. 2002, 'Regional Assessment of Atrazine Exposure and Incidence of Breast and Ovarian Cancers in Kentucky', *Arch. Environ. Contam. Toxicol.*, **42**(1): 127–136.

Hotchkiss Mehta, L. 2002, 'Multi-drug-resistant TB: A Potential 21st Century Plague', *Update: NYAS Magazine for Members of NYAS*, January/February: 2–5.

Houghton, J. T., Ding, Y., Griggs, D. J., Noguer, M., Linden, P. J. van der, and

Xiaosu, D. (eds) 2001, *Climate Change 2001: The Scientific Basis*, IPCC, Cambridge University Press, Cambridge, UK.

Houghton, J. T., Jenkins, G. J., and Ephraums, J. J. (eds) 1990, *Scientific Assessment of Climate change—Report of Working Group I*, IPCC, Cambridge University Press, Cambridge, UK.

Houghton, J. T., Meira Filho, L. G., Callender, B. A., Harris, N., Kattenberg, A., and Maskell, K. (eds) 1995, *Climate Change 1995: The Science of Climate Change*, IPCC, Cambridge University Press, Cambridge, UK.

Hubbert, M.K. 1974, 'US Energy Resources, A Review as of 1972' A background paper prepared at the request of Henry M. Jackson, Chairman, Committee on Interior and Insular Affairs, United States Senate, Pursuant to Senate Resolution 45, A National Fuels and Energy Policy Study: Serial No. 93–40, (92–75) Part 1, GPO, Washington, DC.

Hubbert, M. K. 1982, 'Techniques of Prediction as Applied to the Production of Oil and Gas', in *Oil and Gas Supply Modelling*, ed. S. I. Gass, Proceedings of a symposium held at the US Department of Commerce, National Bureau of Standards, Washington, DC, 18–20 June; Report NBS Special Publication # 631, pp 16–141.

Hughes, P. M. 1999, 'Global Threats and Challenges: The Decades Ahead', Lieutenant General Patrick M. Hughes, U.S. Army, Director, Defense Intelligence Agency, Prepared Statement before the Senate Armed Services Committee, *Defense Link*, USDD, http://www.defenselink.mil/speeches/1999/s19990202-hughes.html

Hwang, A. 2001, 'AIDS Erodes Decades of Progress', in *Vital Signs 2001: The Trends that are Shaping Our Future*, ed. L. Starke, WWI, in cooperation with UNEP, W. W. Norton and Company, New York, pp. 78, 79.

ICBL 2002, *Landmine Monitor Report 2002: Toward a Mine-Free World*, Human Rights Watch, New York, NY,

IEA 1995, *World Energy Outlook,* IEA and OECD, Paris, France.

IEA 1996, *World Energy Outlook*. IEA and OECD, Paris, France.

IEA 2000, *World Energy Outlook*, IEA and OECD, Paris, France.

IEA 2002, *Key World Energy Statistics From the IEA*, IEA, Paris, France.

IPCC 1990, *The IPCC Response Strategies—Report of Working Group III*, Island Press, Covelo, CA, US.

IPCC 1995, *IPCC Second Assessment—Climate Change 1995*, IPCC, Geneva, Switzerland.

IPCC 2001, *Summary for Policy Makers: Climate Change 2001: Impacts, Adaptation, and Vulnerability*, IPCC, Geneva, Switzerland.

Ivanhoe, L. F. 1996, 'Updated Hubbert Curves Analyze World Oil Supply', *World Oil*, **217**(11): 91–94.

Ivanhoe, L. F. 1997, *King Hubert—Updated*, Hubert Center Newsletter # 97/1, M. King Hubert Center for Petroleum Supply Studies, Petroleum Engineering Department, Colorado School of Mines, Ojai, CA.

Jobson, S. 1999, *Water-Stressed Regions: The Middle East & Southern Africa—Global Solution*, Occasional Paper No. 16, Water Issues Study Group, School of Oriental and African Studies, University of London.

Ji, B. T., Silverman, D. T., Stewart, P. A., Blair, A., Swanson, G. M., Baris, D., Greenberg, R. S., Hayes, R. B., Brown, L. M., Lillemoe, K. D., Schoenberg,

J. B., Pottern, L. M., Schwartz, A. G., and Hoover, R. N. 2001, 'Occupational Exposure to Pesticides and Pancreatic Cancer', *Am. J. Ind. Med.*, **39**(1): 92–99.

Jones, P. D., Osborn, T. J., and Briffa, K. R. 2001, 'The Evolution of Climate over the Last Millennium', *Science*, **292**: 662–667.

Jones, C. D., Cox, P. M., Essery, R. L. H., Roberts, D. L., and Woodage, M. J. 2003, 'Strong Carbon Cycle Feedbacks in Climate Model with Interactive CO_2 ad Sulphate Aerosols', *Geophys. Res. Lett.*, **30**(9): 1479–1482.

Juberg, D. R. 1997, *Lead and Human Health,* American Council on Science and Health, New York.

Kakwani, N. 1980, *Income Inequality and Poverty*, Oxford University Press, Oxford.

Keller, A., Sakthivadivel, R., and Seckler, D. 2000, 'Water Scarcity and the Role of Storage in Development', in *World Water Supply and Demand*, IWMI, Colombo, Sri Lanka, pp. 72–83.

Kemp-Benedict, E., Heaps, C. Ruskin, P. 2002, *Global Scenario Group Futures: Technical Notes*, PoleStar Series Report no. 9, SEI, Boston.

Kharikov, A. M. and Smetana, V.V 2000, *Heavy Metals and Radioactivity in Phosphate Fertilizers: Short Term Detrimental Effects*, USFEC paper presented at the 2000 IFA Technical Conference, 1–4 October, preprint.

Kiehl, J. T. and Trenberth, K. E. 1997, 'Earth's Annual Global Mean Energy Budget', *Bulletin of the American Meteorological Society*, **78**: 197–206.

Kopp, C. 1996, 'The Electromagnetic Bomb—A Weapon of Electrical Mass Destruction', *USAF CADRE Air Chronicles*, October, http://www.airpower.maxwell.af.mil/airchronicles/kopp/apjemp.html

Kreml, W. P. 1997, *America's Middle Class: From Subsidy to Abandonment*. Carolina Academic Press, Durham NC.

Kump, L. R. 2002, 'Reducing Uncertainty about Carbon Dioxide as a Climate Driver', Nature, 419: 188–190.

Kuttner, B. 1983, 'The Declining Middle Class', *The Atlantic Po*st, July.

Laherrere, J. H. 1999, 'World Oil Supply—What Goes Up Must Come Down, but When Will It Peak?' *Oil and Gas Journal*, **97**(5): 57–62, 64.

Lal, R. 1989, 'Land Degradation and Its Impact on Food and Other Resources', in *Food and Natural Resources*, eds D. Pimentel and C.W. Hall, Academic Press, San Diego, CA. pp. 85–140.

Landes, D. 1998. *The Wealth and Poverty of Nations*, Norton, New York. NY.

Larsen, J. 2001a, 'Hydrological Poverty Worsening', in *Vital Signs 2001: The Trends that are Shaping Our Future*, ed. L. Starke, WWI, in cooperation with UNEP, W. W. Norton and Company, New York, pp. 94, 95.

Larsen, J. 2001b, 'Wheat/Oil Exchange Rate Skyrockets', in *Vital Signs 2001: The Trends that are Shaping Our Future*, ed. L. Starke, WWI, in cooperation with UNEP, W. W. Norton and Company, New York, pp. 120, 121.

Lawler, T. and Owens, M. T. 1991, 'Planning Strategic Nuclear Forces', in *Fundamentals of Force Planning, Vol. II: Defense Planning Cases*, ed. The Force Planning Faculty, Naval War College, Naval War College Press, Newport, RI.

Lazaroff, C. 2000, 'Hidden Groundwater Pollution Problem Runs Deep', *Environmental News Service,* 11 December, http://ens-news.com/ens/dec2000/2000L-12-11-06.html

Lee, J. W. 2002, 'Time Bomb: Multidrug-resistant Tuberculosis', *Stop TB News*, Summer: 1, 2.

Lenssen, N. 2001, 'Nuclear Power Inches Up', in *Vital Signs 2001: The Trends that are Shaping Our Future*, ed. L. Starke, WWI, in cooperation with UNEP, W. W. Norton and Company, New York, pp. 42, 43.

Levitus, S., Antono, J. I., Boyer, T. P., and Stephens, C. 2000, 'Warming of the World Ocean', *Science*, **287**: 2225–2229.

Levy, M. and Levy, H. 2003, 'Investment Talent and the Pareto Wealth Distribution: Theoretical and Experimental Analysis', *The Review of Economics and Statistics*, 85(3): 709–725.

Lewis, W. J., and Tumlinson, J. H. 1988, 'Host Detection by Chemically Mediated Associative Learning in a Parasitic Wasp', *Nature*, **331**: 257–259.

Lubchenco, J. 1998, 'Entering the Century of the Environment: A New Social Contract for Science', *Science* **279**: 491–497.

Lugens, F. K. and Tarbuck, E. J. 2000, *The Atmosphere: An Introduction to Meteorology*, 8th edn, Prentice Hall, Englewood Cliffs, N.J.

Lutz, W. (ed.) 1996, *The Future Population of The World: What Can We Assume Today?* earthscan Publications, London.

Lutz, W., Sanderson, W., and Scherbov, S. 2001, The End of World Population Growth', *Nature* **412**: 543–48

Lyons, G. 1999, *Chemical Trespass: A Toxic Legacy*, WWF-UK, Panda House, Weyside Park, Godalming, Surrey, UK.

MacKenzie, D. 2001a, 'Anatomy of Terror: How a Few Letters Sparked a Nationwide Panic', *New Scientist*, 20 October , No. 2313: 4 and 5.

MacKenzie, D. 2001b, 'Trail of Terror: Who Sent the Anthrax Letters and Are There More on the Way?' *New Scientist*, 27 October, No. 2314: 4 and 5.

MacKenzie, J. J. 2000, *Oil as a final resource: When is global production likely to peak*, WRI, http://www.wri.org/climate/jm_oil_000.html

Malthus, T. R. 1798, *An Essay on the Principle of Population*, Printed for J. Johnson, in St Paul's Church-Yard, London.

Marland, G., Boden, T. A., and Andres. R. J. 2002, 'Global, Regional, and National Fossil Fuel CO2 Emissions', in *Trends: A Compendium of Data on Global Change*, Carbon Dioxide Information Analysis Center, Oak Ridge National Laboratory, US Department of Energy, Oak Ridge, Tennessee, US.

Marsh, G. P. 1864, *Man and Nature*, Charles Scribner, New York.

Marsh, G. P. 1874, *The Earth as Modified by Human Action: Man and Nature, revised edn,* Charles Scribner's Sons, New York.

Masters, C. D., Root, D. H., Attanasi, E. D. 1994, 'World Petroleum Assessment and Analysis', in *Proceedings of the 14th World Petroleum Congress*, John Wiley and Sons, New York, NY, p. 529–541.

Mastny, L. 2000a, 'Tourism Growth Rebounds', in *Vital Signs 2000: The Environmental Trends that are Shaping Our Future*, ed. L. Starke, WWI, W. W. Norton and Company, New York, pp. 82, 83.

Mastny, L. 2000b, 'Tuberculosis Resurging Worldwide', in *Vital Signs 2000: The Environmental Trends that are Shaping Our Future*, ed. L. Starke, WWI, W. W. Norton and Company, New York, pp. 148, 149.

Mastny, L. 2001, 'World's Coral Reefs Dying Off', in *Vital Signs 2001: The Trends that are Shaping Our Future*, ed. L. Starke, WWI, in cooperation with

UNEP, W. W. Norton and Company, New York, pp. 92, 93.

McCarthy, J. J. Canziani, O. F. Leary, N. A. Dokken, D. J., and White, K. S. (eds) 2001, *Climate Change 2001: Impacts, Adaptation & Vulnerability*, IPCC, Cambridge University Press, Cambridge, UK.

McDonald J. B. and Xu, Y. J. 1995, 'A Generalization of the Beta Distribution with Applications', *Journal of Econometrics*, **66**: 133–152; Errata **69** (1995) 427–428.

McEvedy, C. and Jones, R. 1978, 'Atlas of World Population History', *Facts on File*, New York, pp. 342–351.

McGinn, A. P. 2000, 'Endocrine Disrupters Raise Concern', in *Vital Signs 2000: The Environmental Trends that are Shaping Our Future*, ed. L. Starke, WWI, W. W. Norton and Company, New York, pp. 130, 131.

McGranahan, G., Lewin, S., Fransen, T., Hunt, C., Kjellén, M., Pretty, J., Stephens, C., and Virgin, I. 1999, *Environmental Change and Human Health in Countries of Africa, the Caribbean and the Pacific*, Stockholm Environment Institute, Stockholm, Sweden.

McHarry, J., Scott, F., and Green, J. 2002, *Towards Global Food Security: Fighting Against Hunger*, Food Security Briefing Paper, Towards Earth Summit 2002, Social Briefing No. 2, UNEP and Stakeholder Forum's International Advisory Board, London.

Meadows, D. H., Meadows, D. L., Randers, J., and Behrens III, W. W. 1972, *The Limits to Growth*, Universe, New York, NY.

Meadows, D. H., Meadows, D. L., and Randers, J.1992, *Beyond the Limits: Confronting Global Collapse, Envisioning a Sustainable Future*, Chelsea Green, Boston.

Mellon, M. and Fondriest, S. 2001, 'Hogging It: Estimates of Animal Abuse in Livestock', *Nucleus,* **23**: 1–3.

Milanovic, B. 1997, 'A Simple Way to Calculate the Gini Coefficient, and some Implications', *Economic Letters,* **56**: 45–49.

Milanovic, B. 1999, 'Explaining the Increase in Inequality during Transition', *Economics of Transition*, 7(2): 299–341.

Milanovic, B. 2002, 'True World Income Distribution, 1988 and 1993: First Calculation Based on Household Surveys Alone', *Economic Journal*, **112**: 51–92.

Milanovic, B. and Yitzhaki, S. 2002, 'Decomposing World Income Distribution: Does the World Have a Middle Class?', *Review of Income and Wealth*, **48**(2): 155–178.

Mitchell, T.D. and Hulme, M. 2000, *A Country-by-Country Analysis of Past and Future Warming Rates,* Tyndall Centre Internal Report No.1, UEA, Norwich, UK.

Moore, C. 1997, *Dying Needlessly: Sickness and Death Due to Energy-Related Air Pollution*, http://www.earthscape.org/r1/moc02/moc02.html

Morrison, H. I., Wilkins, K., Semenciw, R., Mao, Y., Wigle, D. 1992, 'Herbicides and Cancer', *J. Natl Cancer Inst.*, **84**(24): 1866–1874.

Munich Re 2001, *Topics 2000: Natural Catastrophes—The Current Position*, Münchener Rückversicherungs Gesellschaft, München.

Munich Re 2002, *Topics: Annual Review: Natural Catastrophes 2001*, Münchener Rückversicherungs Gesellschaft, München.

Munich Re 2003, *Topics 2002/3*, Münchener Rückversicherungs Gesellschaft, München.

Murray, C.J.L. and Lopez, A.D. (eds) 1996, *The Global Burden of Disease: A Comprehensive Assessment of Mortality and Disability from Diseases, Injuries, and Risk Factors in 1990 and Projected to 2020*, Harvard University Press, Cambridge, MA.

Myers, N. 1990, 'The Biodiversity Challenge: Expanded Hot-spot Analysis', *The Environmentalist*, **10**(4): 243–256.

Myers, N., Mittermeier, R. A., Mittermeier, C. G., Da Fonseca, G. A. B., and Kent, J. 2000, 'Biodiversity Hotspots for Conservation Priorities', *Nature*, **403**: 853–858.

Nagashima, T. 2002, 'Future Development of the Ozone Layer Calculated by a General Circulation Model with Fully Interactive Chemistry', *Geophysical Research Letters*, **29**: 31–34.

Nakicenovic, N. (ed.) 2000, 'Energy Scenarios', in UNDP, UNDESA, and WEC, *World Energy Assessment: Energy and the Challenge of Sustainability*, UNDP, New York, 335–366.

Nakicenovic, N., Grübler, A., and McDonald, A. (eds) 1998, *Global Energy Perspectives*, Cambridge University Press, Cambridge, UK.

NAS 1993, *Report: Population Summit of The World's Scientific Academies*, National Academy of Sciences Press, Washington, DC.

NCDC 2003, *Worldwide Weather and Climate Events*, http://www.ncdc.noaa.gov/oa/reports/weather-events.html

NCEH 2003, *Second National Report on Human Exposure to Environmental Chemicals*, NCEH Pub. No. 02-0716, NCEH, Atlanta, Georgia.

NEA 1999, *Progress Towards Geologic Disposal of Radioactive Waste:Where Do We Stand: An International Assessment, NEA,* Issy-les-Moulineaux, France.

Newton, P. W., Baum, S., Bhatia, K., Brown, S. K., Cameron, A. S., Foran, B., Grant, T., Mak, S. L., Memmott, P. C., Mitchell, V. G., Neate, K. L., Pears, A., Smith, N., Stimson, R. J., Tucker, S. N. and Yencken, D. 2001, *Human Settlements*, Australia State of the Environment Report 2001 (Theme Report), CSIRO Publishing on behalf of the Department of the Environment and Heritage, Canberra, Australia.

NIC 2000, *Global Trends 2015: A Dialog about the Future with Nongovernment Experts*, US Government Printing Office, Washington, DC.

Nielsen, R. 2000, *The 21st Century: Reaching to the Sky,* 2nd edn, (first published in Australia 1999), RN Publishing, Gold Coast, Australia.

Nielsen, R. 2002, *The 21st Century: The Seven Thunders: A Summary of the Current Critical Global Events,* 2nd edn, (first published in Australia 2001, reprinted 2001), RN Publishing, Gold Coast, Australia.

Niemczynowicz, J. 2000, 'Present Challenges in Water Management: A Need to See Connections and Interactions', *Water International*, **25**(1): 139–147.

NLWRA 2000, *Water in a Dry Land: Issues and Challenges for Australia's Key Resources*, Commonwealth of Australia, Canberra, ACT, Australia.

NLWRA 2001, *Australian Dryland Salinity Assessment 2000: Extent, Impacts, Processes, Monitoring and Management Options*, Commonwealth of Australia, Canberra, ACT, Australia.

Nordhaus, R. R., Danish, K. W., Rosenzweig, R. H., and Fleming, B. S. 2000,

'International Emissions Trading Rules as a Compliance Tool: What
Is Necessary, Effective, and Workable?' Environmental Law Institute,
Environmental Law Reporter, **30** (October): 10837–10855.

Nordhaus, W. D. 2002, "The Economic Consequences of a War with Iraq", *War
with Iraq: Costs, Consequences, and Alternatives*, Committee on International
Security Studies, American Academy of Arts and Sciences, Cambridge, MA,
pp 51–86.

Norris, R. S. and Arkin, W. M. 2000, 'Global Nuclear Stockpiles, 1945–2000',
The Bulletin of Atomic Scientists, **56**(2): 75.

NRC 1983, *Polycyclic Aromatic Hydrocarbons: Evaluation of Sources and
Effects*, National Academy Press, Washington, DC.

NRE 1999, *What Can You Do to Prevent Groundwater Pollution*, Groundwater
Notes, Note Number 15, State of Victoria Department of Natural Resources
and Environment, Melbourne, Vic., Australia.

NS 2001a, 'Not Just Star Wars: New Ways to Militarise Space are Going
Unnoticed', Editorial, *New Scientist*, 2 June, No. 2293: 3.

NS 2001b, 'Roots of Terror: Smart Bombs and Satellites Cannot Beat This
Enemy', Editorial, *New Scientist*, No. 2309, 22 September, p. 3.

Oberg, J. 2001, 'The Heavens at War', *New Scientist*, 2 June, No. 2293: 26–28.

Oberthur, S., Hermann E., Ott, H. E., and Tarasofsky, R.G. (eds) 1999, *The Kyoto
Protocol: International Climate Policy for the 21st Century*, Springer Verlag,
New York.

Odell, P. 1997, *A Guide to Oil Reserves and Resources*, Energy Advice Ltd,
London.

O'Hanion, L. 2000, 'The Time Travelling Mountain', *New Scientist*, 1 July, No.
2245: 30–33.

Ohlsson, L. 1998, *Water and Social Resource Scarcity,* an issue paper
commissioned by the FAO AGLW (Rome).

Oldeman, L. R., van Engelen, V. W. P., and Pulles, J. H. M. 1990, 'The Extent
of Human-induced Soil Degradation', in Annex 5, *World Map of the Status
of Human-Induced Soil Erosion: An Explanatory Note*, 2nd edn, eds L.
R. Oldeman, R. T. A. Hakkeling, and W. G. Sombroek, International Soil
Reference and Information Centre, Wageningen, The Netherlands

O'Neill, B. and Balk, D. 2001, 'World Population Futures', *Population Bulletin*,
56(3) September 2001

OSB, PRB, and BASC 2002, *Abrupt Climate Change: Inevitable Surprises*,
National Academy Press, Washington, DC.

Osborn, F. 1948, *Our Plundered Planet*, Little, Brown and Co., Boston. Reprinted
1988, Island Press, Washington, DC.

O'Sullivan, C. and Alcock, S. J. 1999, 'BIOSET: Biosensors for Environmental
Technology EU Workshop "Biosensors for Environmental Monitoring:
Technology Evaluation" Held in Kinsale, Ireland, 12–15 May 1998,
Biosensors and Bioelectronics, **14**(6): 541–544.

OTA 1982, *Impacts of Technology on U.S. Cropland and Rangeland Productivity*,
OTA, US Congress, Washington, DC.

Ott, H. E. 2001, 'The Bonn Agreement to the Kyoto Protocol—Paving the Way
for Ratification', *International Environmental Agreements: Politics, Law and
Economics*, **1**(4): 469–476.

Otto, B., Ransel, K., Todd, J., Lovaas, D., Stutzman, H., and Bailey, J. 2002, Paving Our Way to Water Shortages: *How Sprawl Aggravates Drought*, Washington, DC., Natural Resources Defense Council, New York, and Smart Growth America, Washington, DC.

Pareto, V. 1896, *Cours d'Economie Politique*, Droz, Genva, Switzerland.

Paton, A. C. and Davis, P. 1996, 'The Seawater Greenhouse for Arid Lands', *Proc. Mediterranean Conference on Renewable Energy Sources for Water Production*, Santorini, Greece, 10–12 June.

Pearce, F. 2000, 'Tails of Woe', *New Scientist*, 11 November, pp. 46–49.

Pearce, F. 2002, 'Thanks for All the Fish', *New Scientist*, 12 January, p. 15.

Pegg, J. R. 2003, 'Exploring the Link between Health and Environment', *Environmental News Service*, 28 May, http://ens-news.com/ens/may2003/2003-05-28-10.asp

Peterson, W. C. 1997, 'Class Warfare and Middle Class Decline in America', *Journal of Income Distribution*, 7(2): 175–210.

Pimentel, D. 1993, *World Soil Erosion and Conservation*, Cambridge University Press, Cambridge, UK.

Pimentel, D. and Pimentel, M. 1996, *Food, Energy and Society*, revised edn, University Press of Colorado, Niwot, Colorado.

Pimentel, D., Harvey, C., Resosudarmo, P., Sinclair, K., Kurz, D., McNair, M., Crist, S., Shpritz, L., Fitton, L., Saffouri, R. and Blair, R. 1995, 'Environmental and Economic Costs of Soil Erosion and Conservation Benefits', *Science* 267, 1117–1123.

Pimentel, D., Tort, M., D'Anna, L., Krawic, A., Berger, J., Rossman, J., Mugo, F., Doon, N., Shriberg, M., Howard, E., Lee, S., and Talbot, J. 1998, 'Ecology of Increasing Disease: Population Growth and Environmental Degradation', *BioScience*, 48(10): 817–826.

Pimm, S. L. Russell, G. J., Gittleman, J. L., and Brooks, T. M. 1995, 'The Future of Biodiversity', *Science*, 269: 347–350.

Pinstup-Anderson, P. 1999, Selected Aspects of the Future Food Situation, 25th Enlarged Council Meeting, Rome, Italy, 30 November–3 December, IFA, Paris, France.

Pirages, D. 1989, *Global Technopolitics*, Brooks/Cole Publishing Company, Pacific Grove, CA.

Pitman, N. A. and Jørgensen, P. M. 2002, 'Estimating the Size of the World's Threatened Flora', *Science*, 298: 989.

Piver, W.T., Ando, M. Ye, F., and Portier, C.J. 1999, 'Temperature and Air Pollution as Risk Factors for Heat Stroke in Tokyo, July and August 1980–1995', *Environmental Health Perspectives*, 107: 911–916.

POGO 2001, *US Nuclear Weapons Complex: Security at Risk*, http://www.pogo.org/p/environment/eo-011003-nuclear.htm

POGO 2002, *Nuclear Power Plant Security: Voices from Inside the Fences*, http://www.pogo.org/p/environment/eo-020901-nukepower.html

Porter, S. 2000, 'Pesticide Trade Nears New High', in *Vital Signs 2000: The Environmental Trends that are Shaping Our Future*, ed. L. Starke, WWI, W. W. Norton and Company, New York, pp. 48, 49.

POST 1998, *Hormone-Mimicking Chemicals*, POST Technical Report 108, January, POST, House of Commons, London.

POST 2000, *Water Efficiency in the Home*, POST Technical Report 135, March, POST, House of Commons, London.

POST 2002, *Ratifying Kyoto*, POST Technical Report 176, April, POST, House of Commons, London.

Postel, S. 2000a, 'Groundwater Depletion Widespread', in *Vital Signs 2000: The Environmental Trends that are Shaping Our Future*, ed. L. Starke, WWI, W. W. Norton and Company, New York, pp. 122, 123.

Postel, S. 2000b, 'Troubled Waters', *The Sciences* (NYAS members' edn), **40**(2): 19–24.

Postel, S. L., Daily, G. C., and. Ehrlich, P. R. 1996, 'Human Appropriation of Renewable Fresh Water', *Science*, **9**: 785–788.

Potoski, B. A., Mangino, J. E., and Goff, D. A. 2002, 'Clinical Failures of Linezolid and Implications for the Clinical Microbiology Laboratory', *Emerging Infectious Diseases*, **8**(12): 1519, 1520.

Potts, M. and Walsh, J. 2003, 'HIV Epidemic: Lessons from Africa', *British Medical Journal*, **326**: 1389–1392.

PRB 2002, *2002 World Population Data Sheet*, PRB, http://www.prb.org.

Rahmstorf, S. 1997, 'Ice-cold in Paris', *New Scientist,* 8 February, No. 2068: 26–30.

Rahmstorf, S. 1999, 'Shifting Seas in the Greenhouse?' *Nature*, **399**: 523–524.

Raskin, P., Banuri, T., Gallopín, G., Gutman, P., Hammond, A., Kates, R., and Swart, R. 2002, *Great Transition: The Promise and Lure of the Times Ahead*, PoleStar Series Report No. 10, Stockholm Environment Institute-Boston, Boston,

Reichman, L. B. and Hopkins Tanne, J. 2001, *Timebomb: The Global Epidemic of Multi-drug Resistant Tuberculosis*, McGraw-Hill.

Renner, M. 2000a, 'Vehicle Production Increases', in *Vital Signs 2000: The Environmental Trends that are Shaping Our Future*, ed. L. Starke, WWI, W. W. Norton and Company, New York, pp. 86, 87.

Renner, M. 2000b, 'Corporate Mergers Skyrocket', in *Vital Signs 2000: The Environmental Trends that are Shaping Our Future*, ed. L. Starke, WWI, W. W. Norton and Company, New York, pp. 142, 143.

Renner, M. 2001a, 'War Trends Mixed', in *Vital Signs 2001: The Trends that are Shaping Our Future*, ed. L. Starke, WWI, in cooperation with UNEP, W. W. Norton and Company, New York, pp. 82, 83.

Renner, M. 2001b, 'Limited Progress on Nuclear Arsenals' in *Vital Signs 2001: The Trends that are Shaping Our Future*, ed. L. Starke, WWI, in cooperation with UNEP, W. W. Norton and Company, New York, pp. 86, 87.

Revenga, C., Murray, S., Abramovitz, J., and Hammond. A. 1998, *Watersheds of the World: Ecological Value and Vulnerability*. WRI, Washington, DC.

Richter, E. D., and Chlamtac. N. 2002, 'Ames, Pesticides, and Cancer Revisited', *Int. J. Occup. Environ. Health*, **8**(1): 63–72.

Rogner, H-H. (ed.) 2000, 'Energy Resources', in UNDP, UNDESA, and WEC, *World Energy Assessment: Energy and the Challenge of Sustainability*, UNDP, New York, 137–172.

Rosegrant, M. W., Cai, X., and Cline, S. A. 2002, *World Water and Food to 2025: Dealing with Scarcity*, IFPRI, Washington, DC.

Rosenzweig, R., Varilek, M., Feldman, B., Kuppalli, R., and Janssen, J. 2002, *The

Emerging International Greenhouse Market, Pew Center on Global Climate Change, Arlington, VA, US.

Ross, P. S. and Birnbaum, L. S. 2001, 'Persistent Organic Pollutants (POPs) in Humans and Wildlife', *Integrated Risk Assessment*, report prepared for the WHO/UNEP/ILO by IPCS, http://www.earthscape.org/r3/who01

Sablani, S.S., M.F.A. Goosen, W.H. Shayya, C. Paton, and H. Al-Hinai. 2001. 'Thermodynamic and Economic Considerations in Solar Desalination: Seawater Greenhouse Development', *Proceedings of Sharjah Solar Energy Conference*, Sharjah, UAE, 19–22 February.

Sabli, A. 1989, 'Quantitative Modelling of Soil Sorption for Xenobiotics Chemicals', *Environ. Health Perspect.*, **83**: 179–190.

Salt, J. 2001, 'Investment Policies of Insurance Companies', *Research Report: Climate Change and Insurance*, CII Research Division, London, pp. 71–80

Sampat, P. 2000, *Groundwater Shock*, Worldwatch Institute, Washington, DC.

Sampat, P. and Peterson, J. A. 2000, *Deep Trouble: The Hidden Threat of Groundwater Pollution*, Worldwatch Paper 154, Worldwatch Institute, Washington, DC.

SAPMP 2002, *The 2002 UNEP/WMO Scientific Assessment of Ozone Depletion*, UNEP/WMO, Geneva, Switzerland

Schaeffer, P. V. 1987, 'A Dynamical Model of Labor-market Change in International Labor Migrations when Demand for Labor is Exogenous', *Eviron Plan A*, 19(1): 51–57.

Schwarz, M. 1995, *Soilless Culture Management*, Springer-Verlag Publisher, New York.

Sears. P. 1935, *Deserts on the March*, University of Oklahoma Press, Norman.

Seckler, D. 2000, 'Introduction', *Word Water Supply and Demand*, IWMI, Colombo, Sri Lanka, pp. 1–3.

Seckler, D. and Amarasinghe, U. 2000, 'Water Supply and Demand, 1995 to 2025: Water Scarcity and Major Issues', *Word Water Supply and Demand*, IWMI, Colombo, Sri Lanka, pp. 35–52.

Seckler, D., Molden, *D.,* Amarasinghe, U., and de Fraiture, Ch. 2000, 'Overview of the Data and Analysis', *Word Water Supply and Demand*, IWMI, Colombo, Sri Lanka, pp. 5–34.

Shah, T., Molden, D. Sakthivadivel, R., and Seckler, D. 2000, *The Global Groundwater Situation: Overview of Opportunities and Challenges*, IWMI, Colombo, Sri Lanka.

Shand, H. 1997, *Human Nature: Agricultural Biodiversity and Farm-based Food Security*, RAFI, Ottawa, Ontario, Canada.

Sheehan, M. O. 2000, 'Urban Population Continues to Rise', in *Vital Signs 2000: The Environmental Trends that are Shaping Our Future*, ed. L. Starke, WWI, W. W. Norton and Company, New York, pp. 104, 105.

Shiklomanov. I. A, 1998, *World Water Resources: A New Appraisal and Assessment for the 21st Century*, International Hydrological Programme, UNESCO, Paris.

Shiklomanov, I. A. 2000, 'Appraisal and Assessment of World Water Resources', *Water International*, **25** (1): 11–32.

Shiklomanov, I. A. and Rodda, J. C. 2003, *World Water Resources at the Beginning of the Twenty-first Century*, Cambridge University Press.

Shindell, D. T., Rind, D., and Lonergan, P. 1998, 'Increased Polar Stratospheric Ozone Losses and Delayed Eventual Recovery Owing to Increasing Greenhouse-gas Concentrations', *Nature*, **392**: 589–592.

Shinseki, E. K. 1999, *The Army Vision: Soldiers on Point for the Nation: Persuasive in Peace, Invincible in War*, General Eric K. Shinseki's statement, Chief of Staff of the Army, Department of the Army, US, Office of the Chief of Staff and Office of the Secretary of the Army, Washington, DC.

Singh, A., Shi, H., Zhu, Z., and Foresman, T. 2001, *An Assessment of the Status of the World's Remaining Closed Forests*, UNEP-DEWA, Nairobi, Kenya.

SIPRI 2003a. *SIPRI Yearbook: Armaments, Disarmaments, and International Security*, Oxford University Press.

SIPRI 2003b, *The SIPRI Military Expenditure Database*, http://www.sipri.org/

Smil, V. 1993, *Global Ecology: Environmental Change and Social Flexibility*, Routledge, London

Spiewak, R. 2001, 'Pesticides as a Cause of Occupational Skin Diseases in Farmers', *Ann. Agric. Environ. Med.*, **8**(1): 1–5.

STBP 2001, 'The Urgent Need for New TB Drugs', *Stop TB News*, Issue 4: 5–7.

STBP 2002, 'DOTS-Plus: Defusing the Threat: New Enemy—New Tactics', *Stop TB News*, Issue 7: 3.

Stix, G. 2002, 'Fighting Future Wars', *The Science of War: Weapons*, Scientific American, Special Issue No 1, February, pp. 22–27.

Stocking, M. and Murnagham, N. 2000, *Land Degradation: Guidelines for Field Assessment*; advised by A. Tengberg, UNEP, Nairobi, Kenya, and Humphreys, G., Macquarie University, Sydney, Australia; in cooperation with UNEP, UNU, PLEC, DFID, ODG/UEA, and Ministry of Foreign Affairs of the Royal Government of Norway, earthscan Publications, London.

Stommel, H. 1961, 'Thermohaline Convection with Two Stable Regimes of Flow', *Tellus*, **13**: 224–230.

Stork, N. E. 1999, 'The Magnitude of Global Diversity and its Decline', in *The Living Planet in Crisis: Biodiversity Science and Policy*, eds J. Cracraft and F. T. Grifo, Columbia University Press, New York.

Strada, G. 2002, 'The Horror of Land Mines', *The Science of War: Weapons*, Scientific American, Special Issue No 1, February, pp. 12–15.

Sturges, W.T., Wallington, T. J., Hurley, M. D., Shine, K. P., Sihra, K., Engel, A., Oram, D. E., Penkett, S. A., Mulvaney, R., and Brenninkmeijer, C. A. M. 2000, 'A Potent Greenhouse Gas Identified in the Atmosphere: SF_5CF_3', *Science*, **289**: 611–613.

Swiss Re 2000, *Natural Catastrophes and Manmade Disasters in 1999*, Sigma Report 2, Swiss Re, Zürich.

Takabayashi, J., and Dicke, M. 1996, 'Plant-carnivore Mutualism through Herbivore-induced Carnivore Attractants', *Trends Plant Science,* **1**: 109–113.

Tamura, H., Maness, S. C., Reischmann, K., Dorman, D. C., Gray, L. E., and Gaido, K. W. 2001, 'Androgen Receptor Antagonism by the Organophosphate Insecticide Fenitrothion, *Toxicological Sciences*, **60**: 56–62.

Tegart, W.J. McG., Sheldon, G. W., and Griffiths, D. C. (eds) 1990, *Impacts Assessment of Climate Change—Report of Working Group II*, Australian Government Publishing Service, Canberra, Australia.

Thurow, L. 1984, 'The Disappearance of the Middle Class', *New York Times*, 5

February, Section 3, p. 2.

Tilkidjiev, N. (ed.) 1998, *The Middle Class as a Precondition of a Sustainable Society*, Association for Middle Class Development, Sofia.

Tilman, D. 1999, 'Global Environmental Impacts of Agricultural Expansion: The Need for Sustainable and Efficient Practices', *Colloquium on Plants and Population: Is There Time? Proceedings of the National Academy of Sciences of the United States of America (NYAS)*, Vol. 96, NYAS, Washington DC, pp. 5995–6000.

Troeh, F.R, Hobbs, J.A., and Donahue, R.L. 1991, *Soil and Water Conservation*, 2nd edn, Prentice Hall, Englewood Cliffs, NJ.

Tsiodras, S., Gold, H. S., Sakoulas, G., Eliopoulos, G. M., Wennersten, C., Venkataraman, L., Moellering, R. C., Ferraro, M. J. 2001, 'Linezolid Resistance in a Clinical Isolate of Staphylococcus Aureus', *Lancet*, **358**(9277): 207–208.

Turkenburg, W. C. (ed.) 2000, 'Renewable Energy Technologies', in UNDP, UNDESA, and WEC, *World Energy Assessment: Energy and the Challenge of Sustainability*, UNDP, New York, 219–272.

Turton, A. R. 1999, *Water Scarcity and Social Adaptive Capacity: Towards an Understanding of the Social Dynamics of Water Demand Management in Developing Countries*, Occasional Paper No. 9, Water Issues Study Group, School of Oriental and African Studies, University of London.

UCS 1992, *World Scientists' Warning to Humanity*, UCS, Cambridge, MA.

UN 1997, *World Population Prospects, 1950–2050: The 1996 Revision*, Annex I and II, UN Population Division, New York.

UN 1999a, *World Population Prospects: The 1998 Revision*, Vol. 1, Comprehensive Tables, UN, New York.

UN 1999b, *Long-Range World Population Projections: Based on the 1998 Revision*, UN, New York.

UN 2000, *Report of the Open-ended Working Group on the Question of Equitable Representation and on Increase in the Membership of the Security Council and Other Matters Related to the Security Council*, Document A/54/57, General Assembly Official Records, fifty-fourth session, New York.

UNCED 1992, *Agenda 21: Programme of Action for Sustainable Development*, UN, Rio de Janeiro, Brazil.

UNDP 1999, *Human Development Report 1999*, Oxford University Press, New York.

UNDP 2001, *Human Development Report 2001*, Oxford University Press, New York.

UNDP 2002, *Human Development Report 2002*, Oxford University Press, New York.

UNDP 2003, *Human Development Report 2003*, Oxford University Press, New York.

UNDP, UNDESA, and WEC 2000a, *World Energy Assessment: Energy and the Challenge of Sustainability*, UNDP, New York.

UNDP, UNEP, WB, and WRI 2000b, *World Resources 2000–2001: People and the Fraying Web of Life*, Oxford University Press, New York, US.

UNDP, UNEP, WB, and WRI 2003, *World Resources 2002–2004: Decisions for the Earth: Balance, Voice, and Power*, WRI, Washington, DC.

UNEP 1999, *Global Environment Outlook 2000*, Earthscan Publications, London, UK.

UNEP 2000, 'The Urban Environment Facts and Figures', *Industry and Environment*, **23**(2): 4–11.

UNEP 2001, *Globalisation and Sustainable Development: Opportunities and Challenges for Financial Services Sector*, International Roundtable Meeting on Finance and the Environment, Deutsche Bank, Frankfurt am Main, Germany, 16–17 November, UNEP Finance Initiatives, Geneva.

UNEP 2002, *Global Environment Outlook 3: Past, Present, and Future Perspectives*, UNEP, Nairobi, Kenya.

UNEP-DEWA 2003, *Groundwater and its Susceptibility to Degradation: A Global Assessment of the Problem and Options for Management*, UNEP-DEWA, in partnership with DFID, DGDC, and BGS, Nairobi, Kenya.

UNFCCC 1997, *Kyoto Protocol,* UNFCCC Secretariat, Bonn, Germany.

UNFCCC 2002, *Report of the Conference of the Parties on its Seventh Session, Held at Marrakech from 29 October to 10 November 2001*, Decision 15/CP.7. Document numbers: FCCC/CP/2001/13/Add.1 to Add.4.

UNFPA 1999, *The State of World Population 1999: 6 Billion: A Time for Choice*, UNFPA, New York.

UNFPA 2000, *The State of World Population 2000: Lives Together, Worlds Apart: Men and Women in a Time of Change,* UNFPA, New York.

UNFPA 2001, *The State of World Population 2001: Footprints and Milestones: Population and Environmental Change*, UNFPA, New York.

UNFPA 2002, *State of World Population 2002: People, Poverty and Possibilities*, UNFPA, New York.

UNICEF 1999a, *The Progress of Nations*, UNICEF, New York.

UNICEF 1999b, *The State of the World's Children 1999*, UNICEF, New York.

UNICEF 2000a, *The Progress of Nations*, UNICEF, New York.

UNICEF 2000b, *The State of the World's Children 2000*, UNICEF, New York.

UNIDO 2002, *Industrial Development Report 2002/2003: Competing through Innovation and Learning*, UNIDO, Vienna, Austria.

USCB 1999, *World Population Profile: 1998*, Department of Commerce, Washington, DC.

USDA 1997, *Food Security Assessment*, International Agriculture and Trade Reports, USDA, Washington DC

Vallanjon, M. (ed.) 2002, *Scaling up the Response to Infectious Diseases: A Way out of Poverty*, WHO, Geneva.

Van Dyke, K. 1997, *Fundamentals of Petroleum*, 4th edn, PETEX, Division of Continuing Education, University of Texas at Austin, Austin, Texas.

Wackernagel, M. 2001, *What We Use and What We Have: Ecological Footprint and Ecological Capacity*, Redefining Progress, San Francisco, CA.

Wackernagel, M. and Rees, W. E. 1996, *Our Ecological Footprint: Reducing Human Impact on the Earth*, New Society Publishers, Gabriola Island, BC.

Wackernagel, M., Monfreda, C., and Deumling, D. 2002, *Ecological Footprint of Nations: November 2002 Update: How Much Nature Do They Use? How Much Nature Do They Have?* Redefining Progress, San Francisco, CA.

Wackernagel, M., Onisto, L., Linares, A. C., Falfán, I. S.L., García, J. M., Guerrero, A. I. S., and Guerrero, M. G. S. 1997, *Ecological Footprints of*

Nations: How Much Nature Do They Use?—How Much Nature Do They Have? ICLEI, Toronto.

Wada, Y. 1993, *The Appreciated Carrying Capacity of Tomato Production: The Ecological Footprint of Hydroponic Greenhouse versus Mechanical Open Field Operations*, MA thesis, UBC School of Community and Regional Planning, Vancouver

Walker, G. 2000, 'Wild Weather', *New Scientist*, 16 September, No. 2256: 26–31.

Walker, M. 2002, *Benchmarking Air Emissions of the 100 Largest Electric Generation Owners in the US—2000*, CERES, in collaboration with NRDC and PSEG, Boston, US.

Walker, T. 2001, 'Developments Since 1994', in *Research Report: Climate Change and Insurance*, CII Research Division, London, pp. 4–14.

Wallace, C. and Haerpfer, C. 1998, *Some Characteristics of the New Middle Class in Central and Eastern Europe: A 10 Nation Study*, Sociological Series No. 30, Institute for Advanced Studies, Vienna.

Wallinga, D., Bermudez, N., and Hopkins, E. 2002, *Poultry on Antibiotics: Hazard to Human Health*, 2n edn, IATP, Minneapolis, MN and Sierra Club, Washington, DC.

Walsh, J. H. 2000, *Parabolic Projection of World Conventional Natural Gas Production Based on Year 2000 Resource Assessment of the US Geological Survey*, http://pages.ca.inter.net/~jhwalsh/wusgs2.html, and private communication.

Watson, O. (ed) 1968, *Longman Modern English Dictionary*, Longman Group Limited, London.

Watson, R. I. N., Bolin, B., Ravindranath, N. Verardo, D., and Dokken, D. (eds) 2000, *Land Use, Land-Use Change, and Forestry*, IPCC, Cambridge University Press, Cambridge, UK.

Watson, R. T., Zinyowera, M. C., and Moss, R. H. (eds) 1995, *Climate Change 1995: Impacts, Adaptations and Mitigation of Climate Change: Scientific-Technical Analyses*, IPCC, Cambridge University Press, Cambridge, UK.

Watson, R., and Pauly. D. 2001, 'Systematic Distortions in World Fisheries Catch Trends', *Nature,* **414**:534–536.

Watson, R. T. and the Core Writing Team (eds) 2001, *Climate Change 2001: Synthesis Report*, IPCC, Geneva, Switzerland.

Watt, K. E. F. 1974, *The Titanic Effect: Planning for the Unthinkable*, Dutton, New York.

WB 1999a, *World Development Report 1998/99: Knowledge for Development*, Oxford University Press, New York.

WB 1999b, *World Development Indicators 1999*, WB, Washington, DC.

WB 2000, *World Development Indicators 2000*, WB, Washington, DC.

WB 2001, *World Bank Governance Indicators Dataset*, http://www.worldbank.org/wbi/governance/govdata2001.htm

WB 2002, *Sustainable Development in a Dynamic World: Transforming Institutions, Growth*, and Quality of Life, Oxford University Press, New York.

WB 2003, *World Development Report 2003: Sustainable Development in a Dynamic World: Transforming Institutions, Growth, and Quality of Life*, Oxford University Press, New York, NY.

WCD 2000, *Dams and Development: A New Framework for Decision-Making:*

The Report of the World Commission on Dams, Earthscan Publications Ltd, London.

WCED 1987, *Our Common Future*, Oxford University Press, Oxford.

Weisstein, E. W. 2003, *Eric Weisstein's World of Mathematics*, Wolfram Web Resources, Wolfram Research Inc., Champaign, IL, http://mathworld.wolfram.com/LargeNumber.html

Wesseling, C., Antich. D., Hogstedt, C., Rodriguez. A. C., and Ahlbom. A. 1999, 'Geographical Differences of Cancer Incidence in Costa Rica in Relation to Environmental and Occupational Pesticide Exposure', *Int. J. Epidemiol.*, **28**(3): 365–374.

WHO 1999a, *Removing Obstacles to Healthy Development: World Health Report on Infectious Diseases*, WHO, Geneva.

WHO 1999b, *The World Health Report 1999: Making a Difference*, WHO, Geneva.

WHO 2000, *Overcoming Antimicrobial Resistance*, WHO, http://www.who.int/infectious-disease-report/2000/

WHO 2002, *The World Health Report 2002: Reducing Risks, Promoting Healthy Life*, WHO, Geneva.

Wigley, T. M. L. 1999, *The Science of Climate Change*, Pew Center on Global Climate Change, Arlington, VA, US.

Williams, R. H. (ed.) 2000, 'Advanced Energy Supply Technologies', in UNDP, UNDESA, and WEC, *World Energy Assessment: Energy and the Challenge of Sustainability*, UNDP, New York, 273–332.

Wilson, E. O. 2002, *The Future of Life*, Random House, UK.

Wilson, E. O. 1998, 'Integrated Science and the Coming Century of the Environment', *Science* **279**: 2048–2049.

WMO and UNEP 2003, *Scientific Assessment of Ozone Depletion: 2002*, Global Ozone Research and Monitoring Project—Report No. 47, WMO, Geneva.

WMO and UNESCO 1997, *The World's Water: Is There Enough?* WMO-No. 857, WMO/UNESCO, Geneva.

Wolf, A. T., Natharius, J. A., Danielson, J. J., Ward, B. S., and Pender, J. K. 1999, 'International River Basins of the World', *International Journal of Water Resources Development*, **15**(4): 387–427.

WRC 1998, *Water Facts 10*, WRC, Perth, WA, Australia.

WRC 2000, *Water Facts 15*, WRC, Perth, WA, Australia.

WRI 1992, *World Resources 1992–93*. Oxford University Press, New York, US.

WRI, UNDP, UNEP, and WB 1998, *World Resources 1998–99*. Oxford University Press, Oxford.

WSAA 2001, *The Australian Urban Water Industry: WSAA Facts 2001*, WSAA, Melbourne, Australia.

WWF 2000, *Toxic Hot Spots: Polluting Planet Earth*, WWF, Washington, DC.

Zuckerman, P. A. 1996, *Beyond the Holocaust: Survival or Extinction? A Survival Manual for Humanity*, Human Progress Network, Washington DC, http://www.hpn.org/beyond/

Index

Page numbers in italics refer to Tables.

See also following index by country and region.

Index by Country and Region